ALTERNATIVE PATHWAYS TO COMPLEXITY

ALTERNATIVE PATHWAYS TO COMPLEXITY

A Collection of Essays on Architecture, Economics, Power, and Cross-Cultural Analysis in Honor of Richard E. Blanton

EDITED BY
Lane F. Fargher and Verenice Y. Heredia Espinoza

UNIVERSITY PRESS OF COLORADO
Boulder

Published by University Press of Colorado
5589 Arapahoe Avenue, Suite 206C
Boulder, Colorado 80303

Printed in the United States of America

The University Press of Colorado is a proud member of
Association of American University Presses.

The University Press of Colorado is a cooperative publishing enterprise supported, in part, by Adams State University, Colorado State University, Fort Lewis College, Metropolitan State University of Denver, Regis University, University of Colorado, University of Northern Colorado, Utah State University, and Western State Colorado University.

∞ This paper meets the requirements of the ANSI/NISO Z39.48-1992 (Permanence of Paper).

ISBN: 978-1-60732-532-1 (cloth)
ISBN: 978-1-60732-533-8 (ebook)

Library of Congress Cataloging-in-Publication Data

Names: Blanton, Richard E., honouree. | Fargher, Lane, editor. | Heredia Espinoza, Verenice Y., editor.
Title: Alternative pathways to complexity : a collection of essays on architecture, economics, power, and cross-cultural analysis / edited by Lane F. Fargher and Verenice Y. Heredia Espinoza.
Description: Boulder : University Press of Colorado, [2016] | Includes bibliographical references and index. | Includes bibliographical references.
Identifiers: LCCN 2016020071 | ISBN 9781607325321 (cloth) | ISBN 9781607325338 (ebook)
Subjects: LCSH: Social archaeology. | Economic anthropology. | Architecture and archaeology. | Archaeology—Cross-cultural studies. | Technological complexity.
Classification: LCC CC72.4 .A47 2016 | DDC 930.1—dc23
LC record available at https://lccn.loc.gov/2016020071

Cover illustration: Detail of *Prosperous Suzhou* by Xu Yang. Public domain image.

Contents

Figures

Tables

ALTERNATIVE PATHWAYS TO COMPLEXITY

Introduction

Lane F. Fargher, Verenice Y. Heredia Espinoza, and Cynthia L. Bedell

Although archaeology experienced a major reorientation with the rise of New Archaeology and the incorporation of neoevolutionist theory beginning in the 1960s and early 1970s (e.g., Binford 1962; Flannery 1972, 1973; Sanders and Price 1968; Watson et al. 1971), it was only somewhat later that a mature, theoretically and epistemologically complex, processual archaeology began to take shape. This mature processual archaeology, also called "alternative pathways to complexity," moved beyond neoevolutionism's obsession with explaining centralization, power, and exploitation based on environmental conditions, to recognize that other factors including agency, negotiation, and cooperation are important factors shaping complex societies. The development of "alternative pathways to complexity" can be attributed to a number of key scholars, but we think Richard Blanton merits special recognition for both his contribution and leadership. His work is especially important because he looked to economics, sociology, political science, and geography in order to broaden his thinking on complexity. Inspired by research in these disciplines, he worked to develop a holistic approach that applied his wide theoretical and methodological purview to understanding the role of households, urbanism, regions, markets, world-systems, and political economics in cultural evolution. In the process, he has developed an impressive, robust, and flexible toolkit for understanding the evolution of social complexity that has inspired scholars working

DOI: 10.5876/9781607325338.c000

in diverse world areas, including Mesoamerica and the Old World, as well as scholars engaged in cross-cultural comparative research, to look at social complexity in new ways.

Therefore, in order to highlight the contribution that he has made to anthropological and archaeological thinking on the evolution of complexity, we gathered a group of distinguished scholars and asked them to prepare a series of chapters that apply Rich's ideas to the study of architecture, economics, and power in Mesoamerica, the Old World, and cross-cultural analysis. Here, we document how Rich became interested in archaeology, as well as the individuals, publications, and field research that coalesced to shape his research and theoretical paradigm.

A LITTLE HISTORY

Rich's interest in archaeology is rooted in his family history and experiences growing up in Colorado. Rich's paternal grandfather was a miner who had moved the family there to work in the gold mines. During the Great Depression, Ed, Rich's father, and Helen Maxine, his mother, graduated from high school and got married. Ed and Maxine were too poor to attend college, and the only work Ed could find was in a gold mine even though it was dangerous work, as evidenced by the fact that his father had been trapped for several days in a mine collapse. In spite of the challenging work he did in the mines and his own father's traumatic experience, Ed developed a life-long interest in Colorado mines and mining. As a boy, Rich and Ed, as well as other family members, would hike up into abandoned mining towns in the Rockies west of Denver (figure 0.1). While Ed explored historical mines, Rich explored the "ghost towns." As he sifted through what people had left behind in their houses, Rich developed a fascination for understanding how people lived from the study of material remains. Rich's interest in archaeology received another boost from a trip to Mexico when he was 15 years old. Ed and Maxine loved to travel, taking the four children on long adventures. One of these, a lengthy driving trip through Mexico, exposed Rich to Central Mexican archaeological sites, such as Cuicuilco and Teotihuacan. In his own words, Rich discovered on those trips that he liked, "old stuff that is trashed out."

After graduating from Denver's Abraham Lincoln High School in 1962, Rich accepted a gymnastic scholarship from the University of Michigan with the intention of studying anthropology. At the University of Michigan, Leslie White, Elman Service, Marshall Sahlins, Roy Rappaport, and Eric Wolf encouraged him to think about anthropology holistically, to look at large-scale

Figure 0.1. *Richard Blanton (on the left) at age 18 on a climbing expedition in the Rockies with a friend.*

processes, and to immerse himself in theory. James B. Griffin encouraged him to think in new ways and to go beyond the faculty. In the Museum of Anthropology's graduate program he was exposed to probability statistics in the quantitative methods course taught by Bob Whallon. Jeffrey Parsons (his major professor) introduced him to a new survey methodology that was beginning to provide important new insights on regional systems and sociocultural processes in prehispanic Central Mexico. Henry Wright, who was becoming interested in regional systems in the Near East as well, suggested that he look into cultural geography, spatial analysis, and regional market systems as potentially useful material for understanding the regional systems he was studying in archaeology.

After finishing his doctorate in 1970 under the guidance of Jeff Parsons, Rich moved first to Rice University and then to Hunter College at the City University of New York. In New York, he joined a distinguished group scholars who converged there in the early 1970s, including, among others, fellow Michigan graduates John Speth and Greg Johnson; a Penn State student,

Chris Hamlin; as well as Melvin and Carol Ember, Susan Lees, Daniel Bates, Robert Sussman, Jane Schneider, and Eric Wolf. Chris Hamlin, a computer expert, showed Rich how to use computers and introduced him to statistical analysis software. Melvin and Carol Ember introduced him to systematic cross-cultural analysis and the potential it held for addressing questions related to scale and complexity. From Hunter College, Rich moved to Purdue in 1976, where he benefited from participating in a joint Sociology-Anthropology department, which included a number of scholars with strong backgrounds in statistical analysis as well as resources for software and statistical support. He also had the opportunity to interact with Tenzing Takla, who introduced him to classical social theory, especially that of Max Weber.

Rich also became involved in the Society of Economic Anthropology in the early 1980s and eventually served on the board and as its president. At the society's meetings, he interacted with a number of stimulating scholars, including Sutti Ortiz, Frank Cancian, Robert McC. Netting, Frances Berdan, Stuart Plattner, Carol Smith, and Harold Schneider, among others. These scholars stimulated and contributed to his thinking and research regarding markets and commercialization in ancient states and civilizations.

At each turn in his career, Rich has shown a singular capacity not only to learn from both his professors and his colleagues but to bring together disparate research and thinking from across the social sciences to provide a deeper and more holistic understanding of premodern complex societies within a scientific epistemology that is rigorous, empirical, and oriented toward testing and falsification. An approach he often encouraged his students and colleagues to adopt, including the editors of this volume.

This approach has also contributed to his ever-dynamic theoretical and research paradigm and increasingly complex empirical projects. Rich initially worked on Jeff Parson's Texcoco settlement pattern project (Parsons 1971) and then directed a regional survey of the Ixtapalapa peninsula (Blanton 1972), both in the Basin of Mexico. From these projects, he was invited by Kent Flannery to bring the regional survey methodology developed in the Basin of Mexico to the Valley of Oaxaca. He first directed an intensive survey and mapping project at Monte Albán (Blanton 1978) and then a regional survey of the southern arm of the Valley of Oaxaca (Blanton et al. 1982). Incorporating former students and colleagues as codirectors, Rich encouraged the expansion of the Valley of Oaxaca Settlement Pattern Project to the entire Valley of Oaxaca (Kowalewki et al. 1989). From Oaxaca, Rich turned his research attention to systematic cross-cultural research on the built environment (households) and, most recently, rational choice and collective action theory (Blanton 1994; Blanton and Fargher 2008; Blanton and Taylor 1995). He also returned to the field to carry out a regional survey in Turkey and an intensive site survey at Tlaxcallan (Blanton 2000; Fargher, Blanton, et al. 2011).

Through this research and more synthetic works, Rich continued to refine and expand his—and in the process, scholarly—understanding of social complexity. Specifically, his academic production has brought to bear ideas concerning markets and commercialization, world-systems, political economic and egalitarian behavior (especially cooperation), households, demography and settlement patterns, urbanism, scale issues, boundedness, social integration, architectural analysis, public goods, bureaucratization, and rational decision-making on theories concerning the evolution of social complexity and states (e.g., Blanton 1975, 1976, 1978, 1983a, 1983b, 1985, 1989, 1994, 1995, 1996, 1998a; Blanton and Fargher 2008; Blanton and Feinman 1984; Blanton et al. 1982, 1993, 1996, 2005; Fargher and Blanton 2007; Fargher et al. 2010; Fargher, Heredia Espinoza, and Blanton 2011; Feinman et al. 1984, 1985; Kowalewski et al. 1983, 1989). All of this using a comparative and systematic methodology geared toward robust (statistical-based) testing and falsification. The impact of Rich's scholarly endeavors on archaeology and other social sciences, especially scholars interested in ancient or premodern states, is amply evidenced by the more than 4,200 citations that his publications have received at the time of this writing.

THE CONTENTS OF THIS VOLUME

Thus, to honor his contribution to and leadership in the study of the evolution of social complexity and ancient states, we invited a group of highly distinguished scholars to prepare a series of theoretically and empirically robust chapters. These chapters focus on at least one of the research themes that have interested Rich (e.g., architecture, economics, power, and cross-cultural analysis) and employ methodologies involving variously regional studies, testing, falsification, and/or comparison. We asked these scholars to address issues in novel ways and to experiment with new explanations. We think that "pushing the envelope" in terms of explanation is the best way to honor Rich's contribution because he has been a constant innovator across his career.

Given the diversity of areas and themes in this volume, organizing the chapters thematically proved overly complex. Thus, in order to avoid a confusing array of sections and subsections, we opted for a simple ordering based on world areas. The first section is dedicated to Mesoamerica, the second section to the Old World (Europe, Africa, and South Asia), and the final section to cross-cultural comparison. Given Rich's primary focus on Mesoamerica, the number of chapters dealing with Mesoamerica is slightly more than the Old World; yet the Old World contribution is substantial and illustrates the broad appeal and extensive impact of his scholarship.

Mesoamerica

The first two chapters in this section are dedicated to Oaxaca. In chapter 1, Stephen Kowalewski argues that markets significantly impacted Mesoamerica before the conquest and identifies six implications of a market-dominated economy. He then looks for material evidence of market economies in the prehispanic Mixteca Alta of Oaxaca. He concludes that markets affected the spatial distribution of cities, regional specialization, economic integration, wealth stratification, consumption, and economic cycles. In chapter 2, Arthur Joyce and Sarah Barber compare Monte Albán and Río Viejo during the later Formative (350 BC–AD 250). They argue that during the later Formative political architects at both sites initially built complex political structures around corporate and collective strategies. But by the end of the Terminal Formative, these structures came under attack by exclusionary strategies resulting in major reorganization at the outset of the Classic period. Río Viejo collapsed, while the political elite at Monte Albán built a hierarchical structure that persisted for another 400 years. They conclude that the differences in the ability of the ruling elite to transform local corporate structures into regionwide integrative institutions resulted in the different pathways followed by each polity.

The second pair of chapters is dedicated to West Mexico. In chapter 3, Christopher Beekman works to link corporate and exclusionary strategies with regional data in the southern Tequila valleys. He uses Pierre Bourdieu's concept of fields to tie specific strategies to particular physical spaces. Within this framework, Beekman concludes that individualizing rituals, marked by elaborate shaft tombs, declined as more and more communities in Jalisco adopted circular architecture and corporate rituals after AD 200. In chapter 4, Verenice Heredia Espinoza addresses the role of corporate political strategies in the northern Tequila valleys during the Postclassic. This region was under threat from the Tarascans and the Caxcanes, yet it maintained its independence by building small collective polities that could be mobilized for defense but that could not be dominated by a single individual or lineage. Conversely, the more exclusionary polities located in neighboring regions were easily conquered and incorporated by the Tarascans.

The remaining chapters in this section focus on Veracruz, the Basin of Mexico, and the Maya area. In chapter 5, Barbara Stark uses data on "Standard Plan" (SP) architectural arrangements from the Classic-period Gulf Lowlands to develop a more nuanced application of the corporate-network continuum. She concludes that the SP provided a physical arena where a tug-of-war between divergent corporate/collective and network strategies played out. Hence, a middle-ground political strategy where corporate groups were important but not dominant best explains the Classic period in south-central Veracruz. In chapter 6, Frances Berdan looks at the ways that a single commodity complex, feathered ornaments, transformed an array of secondary production activities that fed into the manufacture of these ornaments. In chapter 7, Lisa LeCount examines the development of markets in the Mopan River valley, Belize. She concludes that while marketplaces were present throughout the Maya Lowlands during the Late and Terminal Classic, commercialization was more limited. The political elite manipulated the flow of highly elaborate ceramics and obsidian by controlling the markets in which they could be sold and by price fixing. As a result, rural households were more poorly supplied with these goods despite being well supply with plain ware ceramics.

Old World

In the second part of the book we grouped chapters on Europe (Sweden), sub-Saharan Africa (Nyoro), the Near East (Mari), and South Asia (Indus Valley and Sri Lanka). In chapter 8, T. L. Thurston argues that both collective ideologies and political strategies have deep roots in the Swedish state. Focusing on the emergence of the first state in Sweden, Thurston analyzes the conflict between the Svear crown and the strongly collective organization of Småland pastoralists, who occupied a region that housed natural resources (e.g., iron ore) coveted by the Crown. As the Crown moved in, the Smålanders faced increasingly unfair and brutal tax oppression. At first, they responded by moving into higher and more remote valleys to escape voracious tax-famers and thugs employed by the Crown. Then, when they had exhausted their exit options, they violently pushed back against the Crown and successfully maintained the ambitions of absolutist rulers in check over many centuries. Thus, Thurston argues, the Smålanders were instrumental in laying the foundation for modern democracy in Sweden. In chapter 9, Peter Robertshaw examines the history of the Nyoro state from the perspective of collective action theory and corporate-power strategies. He argues that the Nyoro state originally developed with a more corporate political economy and shifted toward a high degree of "despotism" in the nineteenth century. Yet, the Nyoro state expressed a "tension" between the people and the ruler and, thus, between more collective and more exclusionary strategies throughout its history. In chapter 10, Rita Wright compares the ways in which objective bases of power and corporate cognitive codes limited individual power in Mari, in the Near East, and the Indus civilization, in South Asia. She concludes in both cases that corporate cognitive codes, built around the collective ideologies of pastoralists in the case of Mari and craft-producer and merchant communities in the case of the Indus, were important for limiting the power of individual rulers. In chapter 11, Deborah Winslow examines changes in Sinhalese houses in the village of Walangama, Sri Lanka, from the perspective of canonical and indexical communication. She notes that although economic changes over the last 30 years have brought much more wealth to the community, wealth display on household façades has remained muted, a pattern consistent with the maintenance of a strong collective ethic in the village.

Cross-Cultural Comparison

In chapter 12, Peter Peregrine and Carol Ember evaluate the degree to which the corporate-network continuum is related to socialization for mistrust, unpredictable natural disasters, and external warfare. Their cross-cultural analysis finds support for the hypothesis that network strategies are associated with unpredictable natural disasters as well as more frequent external warfare, but they did not find support for the hypothesis that xenophobia is more strongly associated with network than corporate strategies. In chapter 13, Gary Feinman and Linda Nicholas argue that archaeological thought on the origins of hierarchical societies should focus on patterns of diversity as opposed to uniform types, and on historical sequences instead of individual stages. In chapter 14, Tim Earle explores how resource mobilization, especially productive "bottlenecks," are related to diverse political-economic strategies (e.g., the corporate-network continuum). He concludes that property rights over productive bottlenecks are key aspects of political economy. The degree to which the state or ruling elite monopolizes bottlenecks affects the degree to which resources may be mobilized for exclusionary political economies. In the final chapter, Lane Fargher examines the relationship between corporate political strategies and collective action. Building on Blanton's work on corporate strategies and statistical assays, he determines that corporate strategies are strongly correlated with several aspects of collectivity (e.g., internal revenues, public goods, and control of principal agency). Accordingly, he concludes that corporate strategies are an important tool that can be deployed in building collective states, especially for controlling rulers or a powerful nobility.

CONCLUSION

Each one of the chapters included in this volume investigates the myriad pathways to complexity followed by human societies across the globe and throughout history in new and provocative ways. Following Rich's leadership, they show that multiscalar analysis, recognition of human agency, and a robust and diverse theoretical toolkit are necessary for understanding cultural evolution and complexity. Especially important is the accumulation of knowledge in this volume that demonstrates that cooperation and market development are as much a part of the development of complex societies and states as coercion and exploitation, regardless of geographical area. The chapters in this volume collectively show, in accordance with Rich's theoretical arguments, that collective action and competitive market systems played a decisive role in the cultural evolution of social complexity and civilization, regardless of world area.

SECTION 1

Mesoamerican Cases

1

It Was the Economy, Stupid

Stephen A. Kowalewski

Were the major cycles of growth and decline in Mesoamerican civilization (and other urban societies) caused by uncommon (in alphabetical order) aliens, droughts, eruptions, exhaustions, invaders, or raptures? This chapter reviews a theory of the ancient Mesoamerican economy, tests some of its expectations, and proposes that common economic forces would be a reasonable and sufficient cause for episodes of growth and decline. Despite problems of archaeological specification, there is sufficient reason to develop this line of research, in which preindustrial urban societies are treated as subject not to exotic forces but to things familiar to our own experience. Our field is weak in general theory concerning the long-term dynamics of urban societies. Explanations tend to be ad hoc, particularistic, long-since discredited, or reliant on exogenous causes; social science needs to identify regularities and processes in its domain, society itself—and central to social life is economics. Hence the title of this chapter: in 1992 the political advisor James Carville kept admonishing the Bill Clinton campaign to stick to the main issue, insisting it's "the economy, stupid" (Kelly 1992).

In a recent essay (Kowalewski 2012), I explored in conceptual or theoretical terms how the ancient Mesoamerican economy worked. Here I develop observable implications of the theory, using data from five decades of regional archaeological survey in highland Oaxaca.

DOI: 10.5876/9781607325338.c001

THEORETICAL BASE

What follows is a general model of the ancient Mesoamerican economy. As a model it is not an empirical description. To distinguish the two, I use present tense for the model and past tense for the past observable world. The theoretical model is general, not designed for one place or time in Mesoamerican history. It is formal in the sense of using general concepts or principles not tied to specific cultural contexts. And in its economic anthropology style it is formalist rather than substantivist, as explained by Cook (1966).

Formal models such as what follows are not made from sensory data, but from principles; they are not data but they tell you what data are. The purpose of models is to run them up against real situations so one can see whether something was acting as if it were doing so according to one model's principles or those of another. The model is an ideal type designed for comparison—data against model, and by means of data, model against model (e.g., Apostel 1961; Braudel 1970; Clarke 1972; Weber 1947).

As I present parts of the theory here, I include a few key references to the relevant Mesoamerican archaeology and history. In themselves these sources do not "prove" the principle or premise, but only show that applying the premise here is reasonable and that it can have connection to Mesoamerica.

The following paragraphs describe this theoretical model of the ancient Mesoamerican economy. This economy works by market principles of supply and demand. Many scholars would deny that it is a market economy. Perhaps calling it a commercial economy (commerce means goods exchange) would be more agreeable. The real task is to explain how the economy functions and how economy and society shape each other.

I begin with the actors, which are households, mostly but not entirely smallholders. Households do not produce all the goods and services they consume. They desire to consume and they produce for exchange. Goods and services (including labor) circulate. The household is the firm, the marketer, and the consumer all in one. These premises are realistic given the abundant archaeological studies showing that in Mesoamerica the household or house was the unit of production and consumption (e.g., Hendon 1996; Hirth 2009, 2013; Robin 2003a; Santley and Hirth 1993). (That the household is the firm and the locus of specialization, and that there are so many household-firms, suggests that the ancient economy is a better case of the economist's "perfect competition" than is oligopoly capitalism.)

Products in ancient Mesoamerica are elaborated, specialized, differentiated, subject to fashion, and consumed in great quantities. Notice that the properties just listed are not exclusive to industrial manufacturing. This blurs the distinction in Western economics between agricultural and industrial sectors and in anthropology between industrial and preindustrial economies. Again, these premises about products are realistic (see Berdan, chapter 6, this volume), and for examples from the large literature on Mesoamerican technologies, Sahagún (1950–1982) for sixteenth-century historical descriptions, Feinman and Nicholas (2000) on shell, Healan (2011) on obsidian, and Tarkanian and Hosler (2011) on rubber.

Households produce for the market and consume from the market. Urban concentration increases exchange and market dependence. Given Mesoamerica's large urban and rural populations, the demand for goods and services requires widespread and daily participation in exchange. There is a high volume of transactions. In fact, urbanization rates in Mesoamerica were comparable in many respects to those in preindustrial Mesopotamia, China, and Europe (Kowalewski 1990). On theoretical grounds, Kohler et al. (2000) argued for anonymous market exchange rather than personal reciprocal exchange for late prehistoric pueblos in New Mexico, where populations and population density were much lower than in Mesoamerica.

Exchange takes place by means of goods markets (in and outside of marketplaces) in which prices are determined by supply and demand (a key characteristic of market economy, e.g., Pryor 1977). Mesoamerica's systems of periodic markets are well known. Comparative research by Richard Blanton (1985) showed that the density of market places was equal to or higher than that in preindustrial Europe and China. Recent archaeological studies of markets and marketing in Mesoamerica include for example LeCount (chapter 7, this volume) and Stark and Garraty (2010). Some prices may be set by tradition or local law but these too must vary in response to supply and demand in the longer run. Exchange also takes place through nonmarket mechanisms. The relative importance and relationship between market and nonmarket mechanisms is an important factor in the dynamics of the economy, as discussed below.

Given the high volumes of exchange, it is improbable that transactions take place without credit and culturally defined media of exchange. This is not barter. (Graeber 2011:21–41 argues that the barter economy is a myth.) There are multiple commodity monies, used as means of exchange, standards of value, stored value, and means of credit or account. For Mesoamerican monies, Millon (1955) compiled historic descriptions of cacao as money (see also Smith 2012).

The degree to which Mesoamerica diverged from the Old World in the matter of money has not been seriously investigated. Mesoamerica did not have state-minted coins nor carefully measured bullion. Nor is there evidence of the strong silver-staple grain nexus seen in the Near East (e.g., Davies 2002;

Powell 1996; Wray 2004). Whether Mesoamerican commodity monies were a more popular, less statist method of exchange is an interesting speculative problem best left aside for now.

The preponderance of transactions takes place outside the tax or tribute system, that is, among the masses of household producer-marketer-consumer firms. The economy is largely structured by the principles governing the household consumption-exchange-production circuit, not tribute. On tribute in Mesoamerica, see for example Berdan and Anawalt (1992) for the Aztec Matrícula de Tributos/ Codex Mendoza and Landa (1941) for Yucatán. The impressive total volume of the Aztec tribute needs to be tempered with per capita measurements (e.g., Kowalewski et al. 2010). The weight of the tribute burden would have varied with state power, core versus periphery position, local resources, and other factors.

The degree of regional specialization and division of labor is a function of transaction costs. This premise follows models developed by Krugman and colleagues (Fujita et al. 1999; Krugman 1980). Although these models explicitly privilege industry over agriculture, I see no reason to assume that the process only works with industry, since Mesoamerican agriculture and forestry could themselves be quite differentiated and dynamic.

The production, consumption, and exchange activities of firms—households, that is—drives the economy. In turn the aggregate household behavior sets the conditions that actors have to deal with in their affairs. The economy works by the invisible hand, or the aggregate effect of households consuming, exchanging, and producing. Lords and the wealthy, operating as large houses, manipulate and take advantage of exchange and accumulated labor, but this is not a state or command economy.

Consumption, production, savings, and investment are variable, not constant. Likewise, exchange through market versus nonmarket institutions is variable, not constant. A key factor influencing household behavior in these things is access to efficient markets. If access is poor or markets cannot deliver goods at acceptable prices, households can withdraw from participation; if access and efficiency are better, that encourages participation. High levels of demand encourage more market participation, all other things being equal, and withdrawal from the market, if sufficiently prevalent, makes the market less efficient. Participation depends on prices, expectations, demand, trust, and confidence. Here we have the ingredients for volatility, for good times and bad, and cycles of boom and bust.

These ideas were more developed in the longer article (Kowalewski 2012). In this essay, I ground aspects of the larger theory to archaeological data from highland Oaxaca, especially several features that should be manifest at the regional scale. Testing a broad theory of economy has to be done piece by piece, some of the pieces are more amenable than others; features that are better addressed at macro, local, or household scales are beyond the scope of this chapter.

If the ancient economy operated by market principles, then certain expectations follow. The distribution of cities should conform more to commercial than political needs. Regional specializations should take hold in response to market mechanisms. Economic cycles of growth and decline should affect rates of production and consumption; that is, output and consumption should be variable, not fixed, and they should be related to market integration. The distribution of material wealth among households should be strongly influenced by market participation. Some of these expectations can be tested fairly easily but others are more difficult because of the magnitude of the data requirements.

CITY SYSTEMS

To what degree was the distribution of cities determined by commercial factors—the invisible hand? Alternatively, did the landscape reflect the vision of the kingdom—the visible hand of power? We can assess whether the invisible hand or the naked hand was the stronger, because the two processes lead to distinctive settlement patterns.

In preview, the settlement pattern difference is this: if the hand of power is stronger, exchange is oriented toward a single center in an exclusive territory. If the invisible hand is stronger, exchange is distributed among multiple nodes in a network, boundaries are permeable, and commerce draws participants close together regardless of political affiliation.

Here I develop and use a simple model to measure where regional city systems fit on a continuum between these two polar positions, the political and the commercial. In the former, rulers place their capitals at a maximum distance from one another in order to have exclusive sovereign control over as much territory and as many subjects as possible, with a buffer zone between themselves and their counterparts. Christaller thought of this as a sociopolitical, noneconomic principle, which he called separation: "The ideal . . . has the nucleus as the capital (a central place of a higher rank), around it, a wreath of satellites places of lesser importance, and toward the edge of the region a thinning population density—and even uninhabited areas" (Christaller 1966:77).

Christaller's separation principle resembles closely an idealized, isolated Mesoamerican city-state, in Aztec terms, the *altepetl*. Mesoamerica had many cities, states tended to be small although they could be combined into larger alliances or empires, the state had a capital that was the largest city in its domain, and the city-state had a longstanding territorial nucleus. In Mesoamerica, the city did not have the legal autonomy that many cities in Europe did, and the state was not always defined as a contiguous territory, since hereditary rulers sometimes had subjects or holdings in scattered places (Hirth 2003; Smith 2008). These particularities aside, the altepetl was a political, autonomous entity, defined by itself without relation to its neighbors, and it resided in a persistent core territory. The separation principle is thus a good representation of the Mesoamerican political vision.

I needed another ideal model for the other, commercial end of the continuum. I considered Christaller's supply or marketing principle: "The system of central places has been developed, on the basis of the range of the central goods, from the point of view that all parts of the region are supplied with all conceivable central goods from the minimum possible number of functioning central places" (Christaller 1966:72). This principle entails four assumptions that were difficult for me to make: that I knew the central goods (estimated population size of centers is what we have), goods were supplied everywhere, they were supplied by a minimal number of centers (Christaller wanted efficiency), and land was an isotropic plain (highland Oaxaca is not). I needed a simpler model.

I found the basic idea for a simpler model in Christaller's results and conclusions. He had observed that the wealthier, more populous regions in southern Germany had more numerous high-order central places, which tended to cluster and be close to one another, but the poorer regions had fewer high-order places and other central places were more widely spaced. For Germany as a whole, cities also tended to cluster together in the wealthy regions (such as the Rhine-Ruhr Valleys), whereas poorer regions had fewer, more separated, but larger high-order centers (Berlin and Munich, for example) (Christaller 1966:193; Smith and Branom 1937). This contrast gave me the direction I needed.

When commercial activity dominates, cities tend to cluster, as buyers and sellers of central-place goods try to increase the number of exchange partners within their reach. Cities are contagious, they are attracted to one another. But they are not totally drawn into one megacenter, because neither economic nor political power is monopolized and because competing centers serve and draw from hinterland customers and producers. Unlike the conceptual model of the independent kingdom, cities created in the commercial world are expressly situated in relation to each other in a wider world. The result is a galaxy or cluster of roughly equal-size centers located near one another, with overlapping wedge- or pie-shaped hinterlands expanding outward. In this manner the entire developed region is served not by one monopolistic center but by competing centers that may offer varying ranges of specialized goods. The borders of political territories are permeable and population does not thin out toward borders. In essence, the commercial economy develops multiple high-order central places relatively close to one another, without regard to borders.

City distributions can be assessed against the contrasting expectations of the ideal separation (altepetl) model versus this ideal commercial model. Richard Blanton (1996) mapped the distribution of Aztec altepetl centers overlaid on a reconstructed understanding of the regional market system in the Basin of Mexico. He was able to show that the market system was already an influence on the location of centers in Early Aztec times, and that the market network filled in and became more integrated in the Late Aztec period.

In the next several pages, I evaluate the spatial distribution of central places using three slightly different approaches: at a macroregional scale using only the largest cities; a regional scale that adds middle-size central places; and regional-scale analysis of large and middle-size central places that examines the location of centers with regard to the borders of political territories. Factors such as land quality, topography, and transport differentials influence city distributions. I can control these other factors partially by comparing change over several periods of time in the same place.

Figure 1.1 shows the regional archaeological survey coverage, now over 7,500 km^2, from the Mixteca Alta to the Valley of Oaxaca. The data are from the following sources: Tamazulapan/Tejupan (Byland 1980); Coixtlahuaca (García Ayala 2011) and work by this author's project in progress; Cuicatlán (Spencer and Redmond 1997); Central Mixteca Alta (Kowalewski et al. 2009; Pérez Rodríguez, Anderson, and Neff 2011; Spores and Robles García 2007); Nochixtlán (Byland and Pohl 1994; Plunket 1983; Pohl and Byland 1990; Spores 1972); Peñoles (Finsten 1996; Smith 1993); Sosola/Tenango (Drennan 1989); Valley of Oaxaca (Blanton 1978; Blanton et al. 1982; Kowalewski et al. 1989); Sola (Balkansky 2002); Ejutla (Feinman and Nicholas 1990); Miahuatlán (Markman 1981); Guirún (Feinman and Nicholas 2004).

There are limitations on the comparability of data sets gathered over five decades by different projects, but if I restrict the analysis to those things that can be compared, especially the largest settlements, these survey data meet present purposes quite well. The older surveys have broader periodizations than the more recent ones. For example the Nochixtlán survey used a general Ramos period (Late and Terminal Formative) whereas the Central Mixteca Alta survey split Ramos into Early and Late. Because of this Nochixtlán may be somewhat underrepresented in the discussion that follows. I am not including

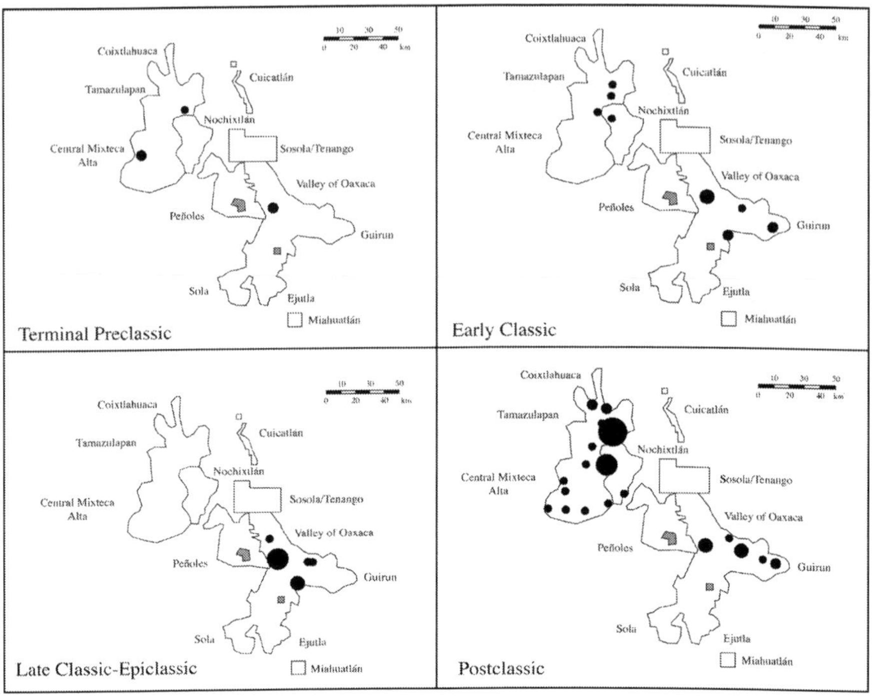

FIGURE 1.1. *Spatial distribution and sizes of cities. (Data sources are cited in the text.)*

the Mixteca Baja, west of the area under consideration (Rivera Guzmán 1999), because the surveys there are not contiguous to those in the Mixteca Alta. For discussion of comparability problems with highland Mesoamerican survey data see Smith (2002).

The first approach concentrates on just the largest cities—that is, all settlements with estimated populations greater than 5,000. I am using this limit of 5,000 inhabitants for all time periods to improve comparability, to exclude sites that were just overgrown villages without the full array of urban central-place functions, and to simplify the visual presentation. Figure 1.1 shows the distributions and relative sizes of the cities as I defined them, the top-ranked central places with populations over 5,000; the maps do not show the many smaller cities and towns. The time periods are the Terminal Formative (Monte Albán II in the Valley of Oaxaca and Late Ramos in the Mixteca Alta), Early Classic (IIIA and Early Las Flores), Late Classic/Epiclassic (IIIB-IV and Late Las Flores), and Postclassic, which is mostly but not exclusively Late Postclassic (V and Natividad).

Multiple urban centers emerged in the Terminal Formative. The principal ones were Monte Albán and Huamelulpan, and the third, Ñiaxugue-Naduza in Coixtlahuaca, has the minimum size but still might be suspect as an urban place. Monte Albán had a well-developed system of towns in its hinterland but the same cannot be said for the Mixteca Alta, where in this time period much of the rural area was abandoned and population concentrated into the immediate environs of the large center. Compared to the later periods, the two (or three) cities were quite distant from one another, as in the ideal type of the political landscape.

Both the Valley of Oaxaca and the Mixteca Alta had developed city systems in the Early Classic. There were clusters of four leading cities in both regions. These largest cities were quite close to one another (20 km in the Valley of Oaxaca and 8–18 km in the Mixteca Alta). The maximum distance between these centers—the size of the cluster along its long axis—was 47 km in the Valley and 23 km in the Mixteca Alta. Both regions had substantial development of smaller cities and towns not shown on the map (internally and also extending to the south in both cases). Using the political/commercial landscape model, I suggest two observations: (1) that within each region commercial forces attracted the largest centers toward each other; and (2) that there was still quite a bit of distancing between the major regions (the Mixteca Alta and the Valley of Oaxaca).

In the Late Classic/Epiclassic, the Mixteca Alta did have some settlement, but many areas seem to have been totally abandoned. The Valley of Oaxaca had Monte Albán, which grew to its maximum and then collapsed in this period, two places that seem to have grown early and then collapsed, and two that seem to have grown larger as the others fell. All five of these disappeared as cities by the end of the Epiclassic.

Finally, the Postclassic had a much larger urban system. This was the time of maximum population in all regions. There were two large clusters of top-ranked cities, and all regions had strong development of lesser cities and towns. The Valley of Oaxaca had a more dispersed settlement pattern, while in the Mixteca Alta a greater proportion resided in large cities. The second-largest city in the Mixteca Alta is Cerro Jazmín, which according to the most recent study had most of its occupation a bit earlier in the Postclassic (Pérez Rodríguez, Anderson, and Neff 2011). The other very large city was Inguiteria or Coixtlahuaca, known to be quite late.

In the Postclassic, cities clustered together. One clump in the Mixteca Alta measures 77 km in maximum length, the other in the Valley of Oaxaca extends over 49 km, suggesting attraction and the commercial side of the continuum.

Within each of these clusters, spacing was rather close—the average nearest-neighbor distance between these top-ranked cities was about 10 km, a two-way trip easily walked in a day. Perhaps the very large sizes of Jazmín and Inguiteria indicate another level emerging at the top of the urban hierarchy.

The Postclassic settlement patterns were not simply a result of population growth and filling-in of available space. Instead, people aggregated into cities that were distributed in a particular way. Within regional systems the largest cities were closely spaced. The mutual attraction of the top cities, that is, the persistent clustering into hierarchically organized, regional systems of cities, suggests commercial forces operating at the regional and intraregional scales.

Moving to the second approach, I expand the range of central places to include middle-size places. Three decades ago Jill Appel (1986) used Christaller's central-place models to study Valley of Oaxaca central-place patterns. She tested whether the Early Classic and Late Postclassic conformed more to a k = 3 pattern (each center equally spaced from three higher-level centers, ideally) or a k = 4 pattern (each center equally spaced between two higher ranking places). A k = 3 system is most efficient for rural consumers where transport is not developed; k = 4 optimizes the movement of bulk goods along straight roads between centers. Appel found that the settlement pattern fit the k = 3, rural-retail model, especially the Late Postclassic but also the Early Classic.

The Mixteca Alta has many small valleys surrounded by mountains, quite different from the open plain of the Valley of Oaxaca. I use our Central Mixteca Alta survey area to examine central place distributions. The size of this study area is 1622 km^2, roughly 44 × 37 km. The top-ranked population centers in the Classic and Postclassic had about 17,000 and 32,000 inhabitants, respectively, and considering the top 30 centers, the smallest had about 1,000 inhabitants (table 1.1).

Half of the total population of the area resided in the top 20 centers (in both time periods), which is a high rate of urbanization. The top population centers were physically close to one another—the average distances from each center to its nearest neighbor center were only 3.3–5.0 km (depending on time period), less than an hour's walking time. This aggregation into central places situated close to one another is consistent with the model of commercial exchange.

The spatial distribution of population centers is expected to be even or regular under the separation principle but more clustered under the commercial principle. To assess this expectation, I use the nearest-neighbor statistic *R* (the simplest uncorrected measure in Pinder et al. 1979). The conventional interpretation of *R* is that 0 is perfectly clustered, 1.0 is random (random

Table 1.1. Spacing and size range of centers in the central Mixteca Alta (NN = nearest neighbor)

	Classic	*Postclassic*	*1990*
Top Twenty Centers			
Pop. of Largest Center	17,180	31,996	9,555
Pop. of Smallest Center	1305	2145	437
Average NN Distance (km)	5.0	4.0	5.9
NN Statistic R	1.1	0.89	1.32
Top Thirty Centers			
Pop. of Largest Center	17,180	31,996	
Pop. of Smallest Center	963	1475	
Average NN Distance (km)	3.6	3.3	
NN Statistic R	0.80	0.74	

Source: Kowalewski et al. 2009.

distributions have some clustering), and 2.15 is perfectly even or regular. Since *R* is quite sensitive to the size of the study area, and since the Mixteca Alta's natural resources are clumped and not evenly distributed, a strict interpretation of the numeric values is not as informative as their relative tendencies. For the top 20 population centers, the nearest neighbor *R* values are 1.10 for the Early Classic, 0.89 for the Postclassic, and 1.32 for 1990; for the top 30 centers, *R* = 0.80 for the Classic and 0.74 for the Postclassic (table 1.1). (I did not calculate, *R* for the top 30 centers in 1990 because that list would extend to small places with no central functions). In the prehispanic periods, centers tended toward clustering, with *R* dipping below 1.0 (random) in three of the four samples. Postclassic cities were more clustered than Early Classic cities. The recent settlements are more evenly distributed, as expected since most are the administrative centers for their municipalities. My interpretation is that the nearest-neighbor statistic is discriminating between the dominant administrative principle of recent times versus the more commercial principle of prehispanic times.

My third analytical approach to the spatial distribution of central places considers political territories and their borders. Our work in the Central Mixteca Alta (Kowalewski et al. 2009) used both historical and archaeological information to identify and describe the development of the *ñuu*, the small kingdoms that could be independent states or could be combined with other ñuu into a larger state (*yuhuitayu*). The ñuu is like the altepetl. If human settlement were dictated by administrative needs, there would be civic-ceremonial buildings at a capital, the capital should be central to the territorial core of the polity, and the demographic center of gravity should also be in the territorial center. Capitals of different polities should be as far from one another as the degree of packing of polities permits (the separation principle). Thus the relevant variables are the specific locations of the civic-ceremonial centers (the palaces), the borders and geographical centers of these irregularly shaped territories, and the locations of the major populated places. Not counting the polities that we only partially surveyed because they extended beyond our study area to the east or to the west, we identified 17 ñuu in the Central Mixteca Alta.

In the Late Postclassic, the palaces were usually at the demographic center of the ñuu (e.g., Teposcolula), in the largest settlement (Magdalena/San Isidro Peñasco), or at another place more demographically central (Achiutla) (figure 1.2). This relationship conforms to the political model.

But there was also a strong tendency for the leading demographic center not to be situated in the geographical center of the polity. Instead the demographic center was attracted toward similar population concentrations in neighboring polities. For example the "inner basin" group of the 10 ñuu in the southern third of the study area all had populations strongly attracted to their neighbors. Apparently the political center was not in this cluster at all—it was in Tlaxiaco, 15 km to the west. A second cluster is made up of Yucuxaco, Huamelulpan, and Tayata. In the north all the polities have their populations skewed toward population centers in Teposcolula, which was also the seat of a yuhuitayu lord. (Yodobada and Lagunas also have a second attraction toward the Tejupan center, just north of the study area.)

Quantitative tests reinforce these observations. Cities and large towns tended to be pulled toward the borders of the ñuu instead of being located in the polity's geographic center. The mean distance from geographic center to nearest border is 2.7 km. This represents the maximum potential separation of cities from a neighboring territory. Since I measured from the centers of the settlements, the minimum possible distance to the border is a bit larger than zero, about 0.2 km for the smaller of these centers. How do the observed locations for the largest population centers in each polity fall along this scale? In the Early Classic the mean distance to border was 1.7 km; in the Postclassic it was 1.5 km, and in 1990 it was 2.0 km ($p = < 0.05$ for the three groups). The largest cities or towns were pulled from the center toward the edge of the territory in prehispanic times, and in the Postclassic they were actually closer to border than to the geographical center. The recent locations conform much more to the administrative expectation than the prehispanic.

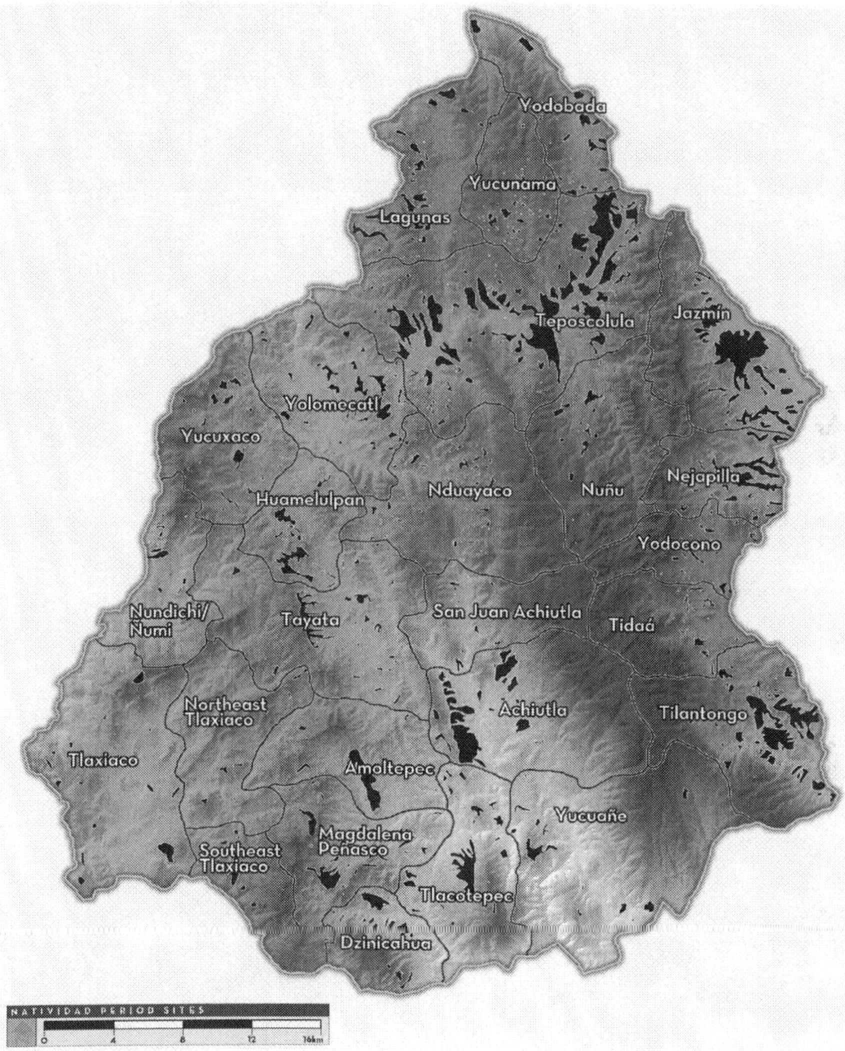

Figure 1.2. *Postclassic sites in the Central Mixteca Alta. (Map by John F. Chamblee and John C. Burns.)*

That populations would be attracted away from the geographical center of the polity and toward each other is not accounted for by the purely political model of the ñuu or altepetl. Economic—commercial/labor/consumption—forces were stronger than the political ideal. One might wryly propose that the people located as they wished and the kings came to perch on top of them, but lords were also involved in commerce and wanted to be where the action was.

To summarize, if the commercial economy determined settlement pattern more so than political forces, then cities should cluster. The three means of analysis all indicate the same result. At the broad macroregional scale the largest cities did tend to cluster, somewhat in the Early Classic and more so in the Postclassic. Within regions, the top- and middle-ranked central places also tended to cluster, noticeably in the Early Classic and more so in the Postclassic. Large and middle-ranked population centers were situated near each other, regardless of political borders. In the Late and Terminal Formative, political forces were relatively stronger, but by the Early Classic and especially in the Postclassic, settlement patterns were more heavily determined by commercial forces.

Mesoamerican stelae, codices, and lienzos depict the indigenous lord's concept of the kingdom. Apparently no one commissioned monuments to the invisible hand. But the archaeological evidence indicates that by Classic times commercial forces were powerful determinants of settlement pattern.

Regional Specialization

Nineteenth-century economists understood regional or zonal specialization (Ricardo 1821:ch. VII). In market economies, zonal specialization works by comparative advantage. It originates and is maintained by low transaction costs, which include transportation and other costs such as tariffs, bribes, arrangements, and so on. Fargher (2009) provides a cross-cultural analysis of five premodern or early modern states in which there developed strongly differentiated core zones of intensive agriculture and peripheral zones of mixed extensive agriculture along with other activities such as craft production. Since the marginal returns to labor in the more fertile areas are high with intensification, farmers in these zones specialize in farming, even though they could do other things; likewise, producers in less-fertile zones should turn sooner to other activities besides crops because the returns will be comparatively higher (Fargher 2009).

In the early periods in highland Oaxaca the hinterlands of the first urban centers did not show much evidence of this zonal specialization. Agriculture and other sectors were intensified, but this took place everywhere in the urban hinterland and without the clear emergence of zones of specialization.

Zonal specialization in the Valley of Oaxaca emerged in the Early Classic. Settlement patterns indicate intensive agriculture in the fertile northern arm of the valley. Extensive agriculture, cultivation of xerophytic plants, and intensive chipped-stone production and use took place in the drier southern and eastern valley (Kowalewski et al. 1989; Feinman et al. 2007).

There is less evidence of Early Classic zonal specialization in the Mixteca Alta. Differences in settlement types and locations strongly suggest extensive upland farming and intensive farming using terraces, but the zonal patterning does not appear to have been as pronounced as in the Valley of Oaxaca (Kowalewski et al. 2009).

The growth and decline during the Late Classic and Epiclassic created short-lived and fairly small zones of specialization. This was the first time the mountains west of the Valley of Oaxaca were intensively occupied (Finsten 1996; Garvin 1994).(There had been similar high-elevation settlement farther west, in the Central Mixteca Alta, earlier.) It is likely that charcoal, fuel-wood, construction timber, flowers, fruit, other mountain products, and perhaps labor were being drawn into valley markets. Some places in the Valley of Oaxaca, such as Lambityeco, Macuilxochitl, and a number of Etla towns, had remarkable spurts of growth and construction for a few generations, as did several new subdivisions at Monte Albán (Blanton 1978; Lind and Urcid 2010). Growth was fueled by intensive canal irrigation in Etla, floodwater farming in Tlacolula, extensive agriculture in uplands, and exploitation of mountain products. At times, hilltop-terrace towns and villages in the drier south and east may have provided surplus labor. These systems of zonal specialization were fairly small in spatial scale and they probably did not last very long.

In the Late Postclassic zonal specialization is evident in all regions. In the Valley of Oaxaca the northern Etla arm was a zone of intensive agriculture, with much less specialization in other crafts. The eastern (Tlacolula) and southeastern (Ocotlán) zones of the Valley have abundant evidence for craft specialization in chipped and ground stone, pottery, salt, lime, and xerophytic plants. Production in the mountains west of the Valley was at an all-time high.

Interestingly the Valley of Oaxaca regained a strong zonal specialization by the nineteenth century. The zonal pattern mirrors that of the Postclassic: Etla had intensive agriculture and dairy oriented to the urban market; Tlacolula and Ocotlán had more craft specialization (Kowalewski 1995).

In the Mixteca Alta, especially in the Postclassic, labor-and-land intensive agriculture was carried out in core areas, especially on the fertile, loose soils of the Yanhuitlán-Jaltepec geological formations. This specialization involved making and maintaining cross-drainage and contour terraces. These works were so extensive that they covered and transformed the whole landscape. The highest populations are found in these core agricultural areas.

In several outlying subregions of the Central Mixteca Alta, we found very extensive chipped-stone quarries and workshops. Some of the products were destined to population centers in core areas and some may have been used in the processing of upland products.

Archaeological surveys in the Mixteca Alta have not found much evidence of pottery-making places, compared to the Valley of Oaxaca. Although the surveys were conducted with the same methods, the lack of evidence for pottery-making places in the Mixteca Alta conceivably could be an unknown identification problem. But another possibility is that the Mixteca Alta surveys have concentrated in core areas and that the pottery making took place in marginal zones. The recent pottery-making villages are located (with only one exception) in upland places outside the core areas where intensive agriculture is practiced (figure 1.3). Whether this pattern in the historical era was true of the Postclassic is not known.

Alliances between petty kingdoms, or blocks of ñuu subject to one ruling house, could have promoted internal exchange. Perhaps a consequence of state or royal house expansion was to lower the costs of transactions within the territory of the state or among its subjects. One example is the yuhuitayu of Teposcolula, which was centered in the agricultural core zones in the Teposcolula and Tamazulapan valleys and had subject ñuu in all the surrounding upland valleys. (The holdings of a royal house were not always contiguous, nor were alliances always between contiguous polities, but on the archaeological time scale the patterns of core and periphery are detectable and persistent.)

Another institutional mechanism for lowering transactions costs and promoting specialization and exchange is ritual obligation (e.g., Wells and Davis-Salazar 2007). At the intravillage scale, Monaghan (1995) showed how much exchange was generated by calendrical and other ritual events. At the regional scale in Michoacán, Castilleja (2011) has been documenting ritual exchanges that link Purépecha towns and villages to one another. A few years ago I attended the major fiesta in San Juan Bautista Coixtlahuaca, Oaxaca, a town of a thousand inhabitants, for which the mayordomo spent the equivalent of $150,000 USD, a figure that does not include the offerings, contributions, and expenses of scores of other families. Zonal specialization is symbolized and reinforced in the ritual cycle of this fiesta by outlying villages supplying considerable labor, animals, food, and materials. The fiestas also entail pilgrimages, offerings, and commitments from participants in Veracruz, Puebla, and Mexico City. Participants say these practices are a continuation of precolumbian ties that Coixtlahuaca had as a major international marketplace.

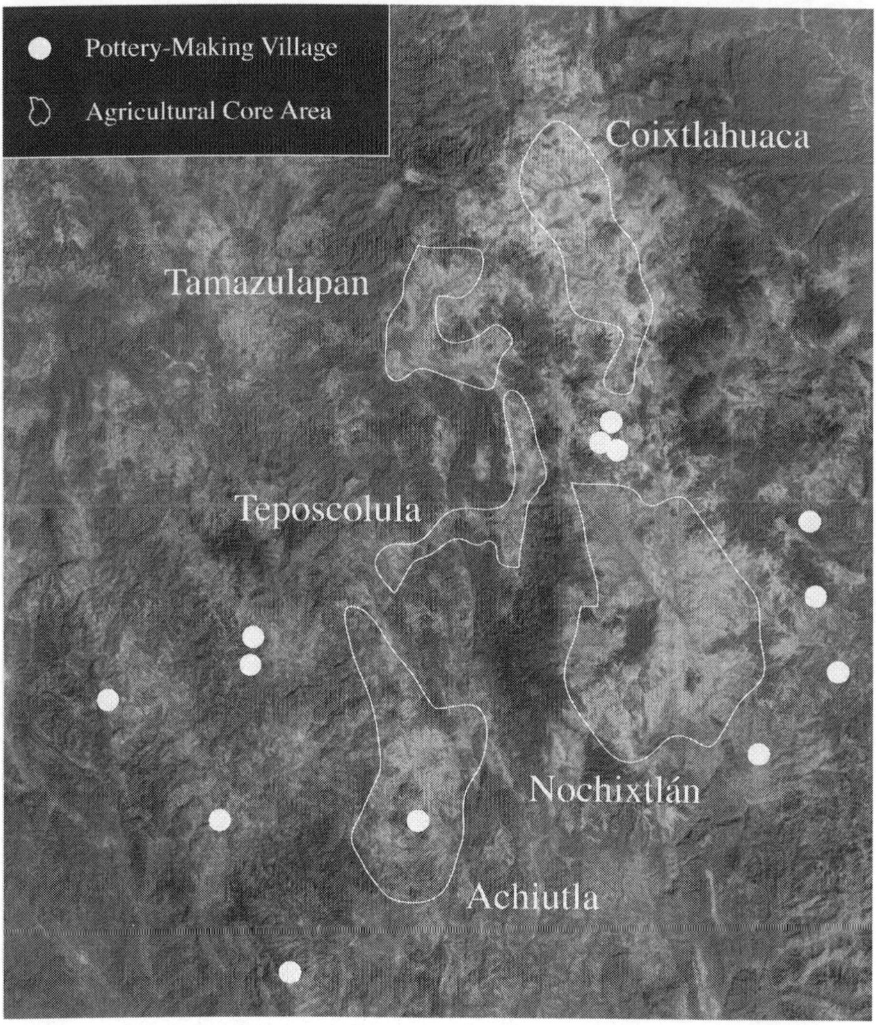

FIGURE 1.3. *Twentieth-century pottery-making villages and agricultural core areas in the Mixteca Alta. The villages are San Miguel Adequez, Santa Inés del Río, Buenavista Jaltepec, Santo Domingo Tonaltepec, Vista Hermosa Tonaltepec, Río Blanco Tonaltepec, Magdalena Peñasco, San Juan Mixtepec, Rancho Morelos Tinú, San Antonio Nduaxico, Santa María Cuquila, Atatlahuaca, and San Juan Numí (Spores 2007:108; Warner 1976).*

The role of ritual in the economy is somewhat contradictory because participation is very demanding of time and effort, and it is a drain on savings that could potentially go toward capital investment. Yet ritual obligation involves many nonmarket exchanges that stimulate production and consumption. Ritual obligations act as stimulus because they turn household savings into investment, and they pave the way for nonritual exchange. Plus, nonmarket offerings of goods and services contribute to the success of the fiesta, which as a whole has a major and widespread effect on economic activity.

Market Integration

Since at least Early Classic times, regional urban systems had well-developed systems of market places. Market places need accessible, appropriate spaces. We studied potential market places in the Central Mixteca Alta (Pluckhahn and Kowalewski 2003), where the terrain is quite mountainous. The lack of naturally occurring flat spots means that if people were to have market places in appropriate, accessible locations they had to build the space. Our survey found many of these public spaces or plazas. Of course we cannot be sure that each plaza was used as a market, but we can say that these were likely the best potential or possible market places. For the Classic and Postclassic, we identified potential market plazas at a rate of one for every 2,000–4,000 people, or one for every 40 km^2. This would be a high market density even if only half the places functioned as markets (cf. Blanton 1985). By comparison, in the 1960s the Nochixtlán cyclical market area had 20 markets (Warner 1976)—one market for every 5,400 people and one for every 280 km^2.

At times market systems in adjacent regions are poorly connected; at other times the connections may be so good that we can consider it a single pool of producers and consumers. Greater integration spreads risk; smaller systems are subject to higher probabilities of supply-demand mismatches. The degree of connectedness was a major variable in the cycles of development in civilizations (see Smith 2002 for Mesoamerica).

Market integration is understood as the degree to which prices are coordinated over space and among different commodities. For present purposes, I focus on the spatial aspect of market integration. Under preindustrial conditions for the integration of markets in two different regions, consumer demand should be continuous in the intervening space, transportation and transaction costs should not be prohibitive, and there should be nodes of exchange such that local demands might be satisfied by competitors from both regions.

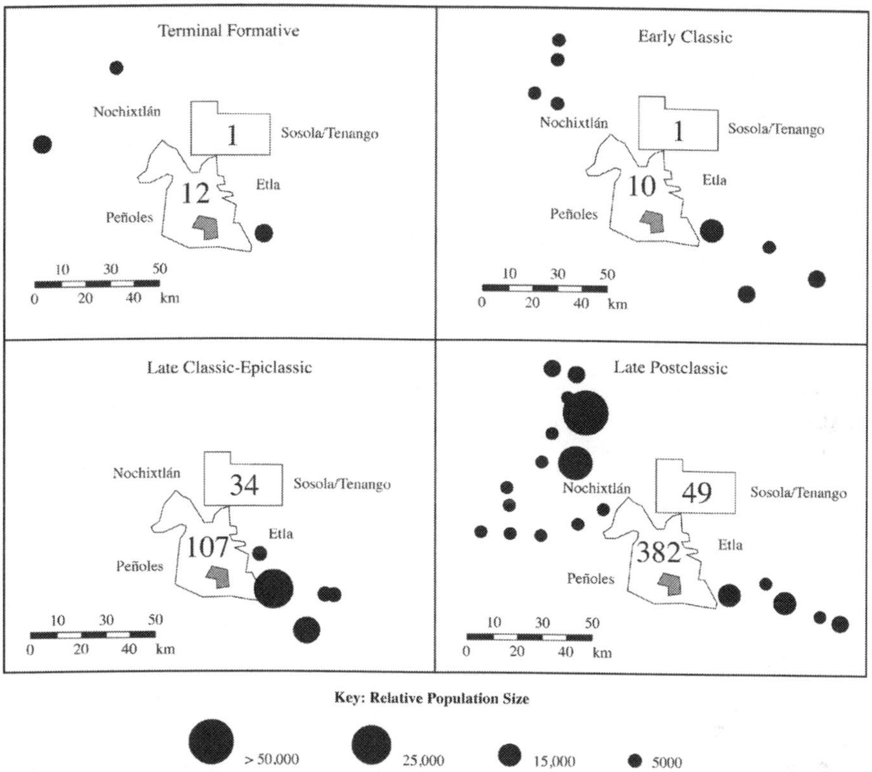

Figure 1.4. *Relative interconnectedness of Mixteca Alta and Valley of Oaxaca regions as gauged by the numbers of sites in the intervening mountains.*

Market integration would thus be unlikely if the two regions were separated by uninhabited area and if market places were too far apart.

We have appropriate data to assess whether some of the conditions for market integration were met in highland Oaxaca. Consider the Mixteca Alta and the Valley of Oaxaca as two regions. Did the intervening area have the population and market centers sufficient to allow the regular exchange of goods and services and were producers and consumers in the same market? The Valley of Oaxaca and the Mixteca Alta are separated by 30 km of mountains, some of which, the Sosola-Tenango area, was surveyed by Drennan and another area, Peñoles, by Finsten and myself (Drennan 1989; Finsten 1996). Again it is helpful to evaluate the relative potential for integration from one period to the next. Figure 1.4 shows the locations of the largest cities, just as in figure 1.1;

the numbers in the Sosola-Tenango and Peñoles regions are the numbers of settlements. The Nochixtlán and Etla Valleys are the sectors of the Mixteca Alta and the Valley of Oaxaca closest to the intervening area.

The Terminal Formative was the least-likely period for market integration, as the mountains between the Mixteca Alta and the Valley of Oaxaca were very sparsely populated.

The Etla arm of the Valley of Oaxaca had some settlement in the Early Classic, a string of towns down the middle of the valley, and this string may have continued with several villages in Peñoles. But overall the mountains were not much occupied and Etla was less important than the highly developed eastern and southern arms of the Valley of Oaxaca and parts of the Mixteca Alta. Substantial growth and development occurred in the two regions, but the intervening area did not play much of a role.

The Late Classic/Epiclassic panel in figure 1.4 shows that in the Mixteca Alta no large cities are identified; except for a few places in the Nochixtlán Valley much of the region appears to have been little occupied. In the Valley of Oaxaca there was strong but stuttering growth in the Late Classic before collapse at the end of the Epiclassic. Monte Albán grew larger than ever and then collapsed. Tlacochahuaya and towns in Etla thrived early and then collapsed; Jalieza and Lambityeco were at their largest a bit later, then declined. The mountains outside the Valley to the north and west had villages and a few towns—some earlier, some later, some a bit of both.

The Late Postclassic had strong urban and rural growth in both the Mixteca Alta and the Valley of Oaxaca, and the strongest development ever in the intervening mountains. Sosola-Tenango and especially Peñoles had multiple towns and villages with plazas. In Peñoles the market orientation over most of the area was toward the Valley of Oaxaca, except for the northwestern quarter, which was oriented west to the Nochixtlán Valley. That most of Peñoles traded more with the Valley of Oaxaca is interesting because historically it is Mixtec-speaking, like its neighbors to the west, and its prehispanic ritual artifacts are mostly in the Mixteca Alta style. Yet the Valley of Oaxaca was predominately Zapotec speaking. Economic ties thus criss-crossed linguistic and cultural boundaries.

In sum the Late Postclassic had the most intensive and most extensive market integration, in which the Valley of Oaxaca and the Mixteca Alta, more than any previous time, were connected together into a single system.

It is more difficult to assess market integration in the sense of equivalent prices for goods and thus equivalent cost for the factors of their production. Blanton and colleagues (Blanton et al. 2005) and Williams (2004), have proposed that certain strategic or world-system goods marked broadscale exchange of common commodities and shaped economic relations. These goods provide another line of evidence for market integration. Obsidian is one such commodity. It was much more common in the two periods of greater spatial integration, the Early Classic and (more so) the Late Postclassic; it is less common in the Terminal Formative and the fragmenting Late Classic/ Epiclassic, even in the regions experiencing some growth during that time.

And then there is pottery. I think that in market disintegration, standard bulk pottery and its styles are of poor quality and do not travel very far. This stands to reason and one need look no farther than the G-35 type for a bulk good that perhaps for very good reason did not travel very far, whereas the Postclassic G-3M and polychrome did. Likewise the predominant and tell-tale orange look of Early Classic pottery was followed by multiple producers and the style was widely known in common goods.

But pottery and styles traveling far is scarcely a late thing; otherwise, cross-dating would not work as well as it does in all phases from the Early Formative on. We need to be cautious here. Perhaps with urbanization market integration could have been a factor in pottery quality, but it is not the only relevant variable and there is not a very good understanding of why Formative pots and styles traveled as far and as regularly as they did.

Wealth Stratification

The economy shaped the distribution of wealth and livelihood. Mesoamerican social organization is often described in terms of its class system of nobles, a small class of merchants and luxury crafters with intermediate privileges, commoners landed and landless, and slaves. These are ideal or legal categories; these categories may not correspond to the distribution of wealth (Chase and Chase 1992). An indication of the importance of the market would be the degree of divergence between legal class categories and the distribution of household wealth. Feinman and colleagues studied this problem in detail with well-controlled, excavated houses at El Palmillo, a Classic-period town in the Valley of Oaxaca (Feinman et al. 2007). They observed the distribution of 21 different everyday, uncommon, and ritual objects and found that there were differences in household wealth, but that the differences were continuous and quantitative. Our review (Steere and Kowalewski 2012) of house and artifactual data from excavations and mapping projects in Central Mexico, Oaxaca, and the Maya area concludes that wealth distributions were indeed variable, they tended to be continuous, and they often had a stronger middle than predicted under the nobility/commoner model of social stratification.

Variable Consumption

My understanding of the ancient Mesoamerican economy implies that household consumption was not fixed but variable, and that it varied with market dependence and participation. In 1970 I worked for John Paddock at Lambityeco, a town in the Valley of Oaxaca dating to around AD 700. Paddock talked about Lambityeco as decadent, usually pronouncing the word "decay-dent" (Paddock 1970), because the ancient Lambityecans, to his mind, made almost nothing of artistic quality even though they consumed enormous quantities of stuff. For a relatively short occupation—most of it in just a few generations—Lambityeco certainly did leave a lot of débris. Ancient Mesoamericans did that at some times and places, but not always. The Lambityeco bubble was not only relatively brief, it was also localized.

Measuring household consumption rates is not easy because it requires good chronological control, a way of getting at consumption per unit time, and large, representative samples. It requires quantitative measurement. This is why early professional economists in Western countries campaigned for better statistics (i.e., reliable numbers on such things as manufacturing output, sales, employment, and the flow of money). Understanding a modern economy requires quantitative measurement with reliable data. Why should we think the ancient Mesoamerican economy was any different?

Economic Cycles

In our experience today we understand that poorly controlled markets fluctuate between exuberance, stability, and depression. This is not a new revelation. In the fourteenth century Ibn Khaldûn described economic cycles of growth and decline that lasted about 40 years. He suggested a sophisticated explanation based on movements in wages, prices, supply, demand, and expenditures in different sectors (Boulakia 1971; Khaldûn 1967:2:291ff.; Soofi 1995). Cycles of various periodicities are recognized by modern historians and there is considerable literature on their causes (reviewed by Berry 1991).

Did Mesoamerica have a business cycle? If so, was it faster or slower than in economies with metal coinage? Were there economic waves of longer periods? Does the cyclical notion of time characteristic of some civilizations have to do with people trying to understand and control the good and bad times of economic cycles?

Civilizations have longer-period cycles of growth and decline, unification and disintegration. Prolonged stability or growth is not known. All civilizations cycle. Older ideas about rise and fall (Spengler 1926–1928) were discredited long ago, but they keep returning (as in drought causing the Classic Maya collapse, e.g., Haug et al. 2003). Modern social science has not dealt with the problem very well (McAnany and Yoffee 2010). Anthropologists have typically opted to emphasize particular histories instead of general regularities, or they assert that preindustrial societies were a different species from modern societies and not subject to uniformitarian processes. Admittedly this is a large and difficult problem, and I for one do not have the cross-cultural expertise in this area, but our abdication has left the field open for all sorts of even lesser-qualified advocates for exogenous causes and externalities. The professionals leave the field and the amateurs take over.

DISCUSSION

In this chapter I have tried to ground in archaeological evidence six implications of a general theory of a market-dominated Mesoamerican economy: (1) the spatial distribution of cities, (2) regional specialization, (3) market integration, (4) wealth stratification, (5) variable consumption, and (6) economic cycles. These are observable at the regional scale but some are better studied with locality and household data. The present analysis is limited because it is confined to highland Oaxaca. Some of the implications are fairly well understood and supported; others are reasonable but still speculative.

If cities were located according to the political needs of rulers, they should have been spaced far from one another. This was true in the Terminal Formative. But by the Early Classic they were not far from one another, they clustered together. Cities were even more attracted to one another in the Late Postclassic. Urban attraction is a characteristic of commercial economies but it is not anticipated in the idealized political landscape of the Mesoamerican city-state.

Market systems can grow in spatial scale and regional market systems can have different degrees of articulation with each other. In highland Oaxaca, Terminal Formative market systems did not extend beyond their capital city's hinterland (small for Huamelulpan, larger for Monte Albán). Regional systems were quite integrated in the Early Classic but they did not extend farther than the region in terms of daily, high-volume exchange. In the Late Postclassic, regional systems were highly integrated internally and the two core regions—the Mixteca Alta and the Valley of Oaxaca—had the highest degree of interconnectedness. The Late Classic/Epiclassic was a time of spatially smaller systems, sometimes well integrated internally, but not so integrated at the interregional scale.

In market economies zones of specialized economic activity develop as transactions costs are sufficiently low and producers find comparative advantage in mutual exchange. This happened in Mesoamerica. In Oaxaca, archaeological evidence of regional specialization is strongest in the Late Postclassic and is also found for the Valley of Oaxaca in the Classic period. It was weakly developed in the Early Classic in the Mixteca Alta and there is little indication of it in the Terminal Formative. Because regional or zonal specialization has such an impact on local and household economies, linking variation in domestic production to role in regional system is a promising area for further testing.

Market economies create wealth disparities and continua in wealth distributions that deviate from expectations that wealth would be determined by legal status. Data on household wealth in numerous archaeological cases from different times and places in Mesoamerica do indicate both substantial variation and continuous (as opposed to categorical) variation. Variability in household wealth needs to be linked to other economic factors, such as market integration.

The general theory does not assume a static peasant with fixed need or wants; instead, consumption should be variable, depending in large part on market integration. In principle variable consumption rates are measurable, but practically this seems to me to be a knotty problem with archaeological data.

The commercial economy structured Mesoamerican society in fundamental things: where people lived, how they made a living, their standard of living, the rhythms of life, and so on. By the later periods the distribution of cities, towns, and villages and patterns of regional specializations were consistent with the expectations of market economics. I use the term "market" in a broad sense that does not imply capitalist mechanisms or special institutions derived in the experience of industrialism. Integration of commercial systems was associated with increased production, consumption, market dependence, and demographic growth; market disintegration was associated with decline. I argue that the Mesoamerican economy, like other urbanized economies, underwent such cycles. Still, the intensity and speed of these movements in this commodity-money economy are not yet understood.

Mesoamerica does seem to belong in that set of civilizations within which it is totally appropriate to explore how market economies behave and how they might be better understood than if we relied only on the narrower range of experience of select Western countries in the past couple of centuries. Such a "modernist" stance, in which the past is not a caste apart but a relevant kindred of our own experience, has enlivened studies of the economic history of the Near East. Silver's summary comment on Polanyi's view of markets in the Near East applies just as well to Mesoamerica: "It is incorrect to magnify the economic flows of temple or palace to Amazonian proportions while shrinking the market to a mere brook" (Silver 1983:829). Temin (2013) has used Roman data to challenge the claim that the ancient economy was incapable of real growth and had only Malthusian growth. Similarly for Mesoamerica, Stark and colleagues (1998) pointed out nearly 20 years ago that Mesoamerica's cotton sector exhibited growth and dynamism sufficiently powerful to affect important aspects of society and the regional and interregional economy. Stark has explicitly taken up the question of economic growth, describing difficulties in archaeological measures, but concluding nevertheless that in precolumbian Mesoamerica "economic growth is largely episodic or unstable, but nevertheless exhibits some cumulative effects" (Stark 2013).

Anthropologists have as one of their special duties to science, the critical examination and broadening of theories that were originally developed in more culture-bound, usually Western contexts. Classical economic geography clearly improved when it was broadened to non-Western cases (e.g., Smith 1976). Anthropological study of ancient urban economics might broaden and enrich general economic theory by reexamining basic concepts, including money, price and value, credit and debt, savings and investment, and cycles of boom and bust.

2

Alternative Pathways to Power in Formative Oaxaca

Arthur A. Joyce and
Sarah B. Barber

Richard Blanton has been at the forefront of research aimed at broadening archaeological approaches to the development and organization of early complex societies (Blanton 1998a; Blanton and Fargher 2008; Blanton et al. 1996). In particular, he has challenged archaeologists to expand views of political complexity beyond traditional conceptions of highly integrated polities with well-developed hierarchies led by powerful rulers. Instead, Blanton has sought to problematize the concept of social integration in complex societies and to consider collective or corporate forms of political organization by examining the ways that rulers and subordinates collaborate to develop and maintain polities. While collective political organization can entail assembly or council-based government without individual rulers, it also includes governments with powerful rulers who gain and retain authority by complying with collective moral codes.

Blanton and his colleagues (Blanton and Fargher 2008; Fargher et al. 2010) argue that a key process in complex polities involves bargaining between rulers and subjects. They assert that rulers will be more likely to bargain with subjects to the degree that the former are dependent on the surplus production, labor, and taxable commercial transactions of the latter. When rulers largely depend on resources mobilized from followers (internal revenue), rather than external sources like imperial tribute or imported valuables (external revenue), followers are in a stronger position to make

DOI: 10.5876/9781607325338.c002

demands from rulers in return for their compliance. Rulers, in turn, provide public goods and services like military defense, judicial services, and sponsorship of public rituals. Subaltern compliance can also be achieved through coercion, although the greater the force used to control subjects the greater the incentives for resistance and rebellion. Rulers and subjects therefore pursue a continuous process of negotiation through which a temporary and contingent form of social contract is constructed. We find Blanton's perspective especially effective because, unlike traditional cultural evolutionist perspectives, it leaves room for agency as well as social and historical contingency.

In this chapter, we draw on Blanton's work to compare the negotiation of political authority during the emergence of two complex, urban polities at the end of the Formative period in Oaxaca: Monte Albán in the Valley of Oaxaca and Río Viejo in the lower Río Verde valley in the Pacific coastal lowlands. Following from our previous research (Barber 2013; Barber and Joyce 2007; Joyce 2000, 2010), we examine evidence from both regions for tension and conflict between more traditional, corporate, and egalitarian (*sensu* Blanton 1998a) forms of authority and leaders who were trying to extend their political influence to the broader region. We argue that the outcomes of attempts to institutionalize more expansive forms of authority were dramatically different. The Río Viejo polity collapsed circa AD 250, perhaps due to internal conflict resulting ultimately from what Blanton and Fargher (2008:112) term "collective action problems." In contrast, the rulers of Monte Albán were successful in establishing hierarchical institutions that persisted for centuries, although the evidence suggests that their success may have come via the suppression of internal enemies and a greater reliance on external sources of revenue. Drawing on the collective action perspective of Blanton and Fargher (2008), we consider some of the factors that may account for the divergent histories of these two polities.

POLITICAL AUTHORITY AND INTEGRATION IN LATER FORMATIVE OAXACA

The Later Formative period (400 BC–AD 250) throughout much of Mesoamerica was a time of emerging political centralization and the expansion of political authority over broader regions and larger populations (e.g., Joyce 2010; Pool 2008; Sugiyama 1993). In Oaxaca, archaeologists have shown that the social changes that occurred with the emergence of early urban centers included increased inequality, warfare, the mobilization of labor for the construction of monumental architecture, changes in settlement patterns and social organization, and innovations in religion, ideology, and economy.

Monte Albán

The founding and early development of Monte Albán represents a dramatic transformation in social and political relations in the Oaxaca Valley. Monte Albán was founded circa 500 BC on a previously unoccupied series of hilltops in the center of the Oaxaca Valley (Blanton 1978; Joyce 2010; Marcus and Flannery 1996; Winter 2001) and quickly grew into an urban center. By 300 BC Monte Albán far exceeded any other site in the valley in size, population, and scale of monumental architecture. During the Late Formative (300–100 BC) the city grew to cover 442 ha with an estimated population of 10,200–20,400 (Blanton 1978:44). Data pertaining to the first several centuries after the founding of Monte Albán indicate that hierarchical rulership emerged in the context of more traditional, communal forms of authority (Joyce 2010). Indeed, hereditary status distinctions in the Valley of Oaxaca are clearly evident only a century or two prior to the founding of Monte Albán (Blanton et al. 1999:36–42; Joyce 2010:111–114; cf. Marcus and Flannery 1996:93–110).

Late Formative residential and mortuary data indicate increasing status differentiation although it appears that commoners could acquire significant wealth, a pattern consistent with more corporate forms of political organization. Excavations at Monte Albán suggest that high-status residences were concentrated in areas around the North Platform, a public ceremonial space (Joyce 2010:142–143, 156). The most completely excavated Late Formative high-status house in the region is the Area I residence at the site of El Palenque, which Spencer and Redmond (2004) argue was a ruler's residence. It covered an area of 16 m x 16 m including eight rooms arranged around a central patio measuring 8 m x 8 m. Ritual feasting may have occurred in a paved courtyard east of the residence. Presumably, rulers' houses at Monte Albán were at least as elaborate as the Area I residence at El Palenque. At Tomaltepec, Whalen (1988) excavated the residence of a local elite family. In contrast to the possible royal residence at El Palenque, the Tomaltepec residence is similar in size to low-status houses (Winter 1986), but included a stone masonry tomb with an elaborate offering. The Tomaltepec residence was located near the site's ceremonial center, a pattern also seen with high-status residences at El Palenque and Monte Albán.

Although royal residences seem to have been considerably more elaborate than typical houses, mortuary data suggest that commoners could acquire significant wealth (Whalen 1988:300–301; Winter 1995). Most interments were associated with residences and consisted of simple graves, *fossa* (graves lined with stones or adobes), cists, adobe tombs, and stone masonry tombs. The most elaborate interments were those in stone masonry tombs, which were likely interments of hereditary nobles such as Monte Albán Tomb 43 with 72 ceramic vessels and Tomb 111 with 51 vessels. There is considerable variability in offerings associated with other types of interments and even simple graves could include elaborate offerings such as Monte Albán Burial VI-12 with an offering of 29 vessels and onyx drill cores.

Another indicator of collective action and a corporate form of authority during the Late Formative involves resources mobilized by rulers to fund administrative institutions and to acquire wealth. The evidence suggests that the most important resource was probably labor provided by people from Monte Albán and nearby communities, which constituted a form of internal revenue consistent with collective action. The concentration of approximately three-quarters of the valley's population within 20 km of Monte Albán (Kowalewski et al. 1989) would have facilitated labor mobilization and tribute collection. The scale of monumental public buildings and spaces at Monte Albán during the early years of the site was considerable (Winter 2001). The initial version of the Main Plaza, dating to the late Middle Formative and Late Formative (500–100 BC), consisted of the plaza, along with the western row of buildings and much of the eastern half of the massive North Platform (figure 2.1). Early public buildings included Building L-sub along the southwestern end of the plaza and Building IV-sub on the northwestern end of the plaza; the walls of both buildings were constructed with huge monoliths.

Other sources of revenue could have involved the mobilization of agricultural production, tribute acquired through conquest, and the control of long-distance trade. Agricultural production available to the inhabitants of Monte Albán was probably insufficient to provision the city, necessitating the taxation of agricultural producers in communities outside the city, especially in newly settled piedmont areas (Kowalewski et al. 1989:123–126). The scale of resource mobilization to provision Monte Albán and the degree to which elites controlled and benefited from such transactions is unclear, however. Several researchers in Oaxaca have argued that the rulers of Monte Albán were also able to mobilize large armies for military conquest (Marcus and Flannery 1996; Redmond and Spencer 2006), although other archaeologists disagree and view conflict at this time as much smaller in scale (Joyce 2014; Zeitlin and Joyce 1999). Although some resources were probably acquired by Monte Albán through the establishment of tributary relationships, the evidence for tribute extraction is minimal (see Spencer 1982:246–250). Evidence also suggests that nobles had preferential access to prestige goods imported from outside the valley such as nonlocal pottery and ornaments of greenstone and shell (Winter 1984; Whalen 1988). The data do not suggest that rulers

Figure 2.1. *View of the Main Plaza of Monte Albán. (Photograph by Sarah Barber.)*

directly controlled key utilitarian resources such as land or the production of pottery and stone tools (Fargher 2007; Parry 1987; Whalen 1988). Most of the resources on which the rulers of Monte Albán were dependent therefore represented internal sources of revenue, which would have given subjects greater power in negotiating more favorable relations vis-à-vis public goods and services (Blanton and Fargher 2008).

Since sources of revenue were largely internal, following Blanton and Fargher (2008) we would expect to see evidence of public goods provided by rulers in return. Although some exotic, nonlocal goods like ornamental shell and greenstone were available, the evidence suggests that public goods controlled by rulers of Monte Albán consisted largely of religious knowledge and authority. The first several centuries following the founding of Monte Albán were characterized by major innovations in religious belief and practice suggesting a connection with the dramatic political changes of the time (Blanton et al. 1999; Joyce 2000, 2010). Associations of elite residences and burials with religious symbols, spaces, and artifacts indicate that the nobility increasingly came to control ritual knowledge and authority, although high-ranking commoners may have also achieved positions of political and religious authority (Joyce 2010:143; Urcid 2011). Religious objects associated with elaborate tombs and burials included effigy vessels and urns, sometimes depicting the Zapotec rain deity Cocijo, and a variety of zoomorphic vessels usually representing animals associated with water, including ducks, conch shells, frogs, and toads. Hieroglyphic inscriptions and iconography indicate that rulers performed acts of human sacrifice and autosacrifice (Urcid 2011). Sacrificial practices, especially human sacrifice, were particularly significant in contacting the otherworld, reenacting the cosmic creation, and renewing the world (Joyce 2000; Monaghan 2009).

An important and widely shared aspect of Mesoamerican worldview was the idea that the current world was the result of a sacred covenant between humans and the divine whereby people petitioned deities for agricultural fertility and prosperity in return for sacrificial offerings. Sacrifice was both religious and ideological. It contributed to the legitimation of political authority because nobles or priests trained in institutional contexts performed the most powerful sacrificial rituals and because noble blood was considered to be more potent in contacting divinities than that of commoners (Joyce 2000). At the time of the Spanish Conquest, common people provided resources like labor and agricultural surpluses to the nobility with the expectation that elites would reciprocate by staging powerful ceremonies through which deities and ancestors were contacted on everyone's behalf (Monaghan 2009). Thus, the economic and ritual obligations of the sacred covenant acted as a kind of social contract between elites, commoners, and the gods.

The archaeological evidence indicates that Monte Albán's Main Plaza was the political and ceremonial center for the polity (Joyce 2000, 2004; Marcus and Flannery 1996; Winter 2001). The scale, accessibility, artifacts, symbolism, and architectural arrangement of the Main Plaza indicate that it was constructed as a performance space where thousands of people participated in politico-religious ceremonies led by the elite, including human sacrifice, autosacrifice, ancestor veneration, and deity impersonation (Joyce 2010:131–155; Urcid 2011). The spatial arrangement of architecture and iconography suggest that the Main Plaza symbolized the Zapotec version of the cosmos where rituals could be performed that reenacted the cosmic creation (Joyce 2000, 2004).

Although the evidence indicates that the wealth and political power of the nobility increased considerably during the Middle to Late Formative and that elites gained greater control over religious knowledge and authority, there appear to have been limits on the power of Monte Albán's rulers that are consistent with more communal or collective forms of authority (Joyce 2004, 2010). We argue that public settings like the Main Plaza stressed the symbols of communal authority and an emerging corporate identity, while muting representations of the increasingly powerful rulers of the city. The accessibility of the Main Plaza would have allowed people to monitor the behavior of rulers during public ceremonies and the sacred covenant may have acted as a strong moral code; both are means through which Blanton and Fargher

(2008:203–206) argue that subjects are able to monitor and gain trust in the behavior of rulers and ruling institutions.

Although nobles lived near the ceremonial precinct and directed public rituals, until the Classic period (AD 200–800), the Main Plaza itself had few overt representations of local nobles and there were no high-status residences directly facing the plaza (Joyce 2004). Rulers were represented solely in the hieroglyphic inscriptions set in Building L-sub, which were probably understandable only to the literate nobility (Urcid 2011). The earliest known ruler's portrait, Monument J-41, probably dates to the end of the Late Formative. The size, accessibility, and symbolism of the Main Plaza suggests that during Monte Albán's first four centuries the plaza was a focus of public ceremonies participated in by people from multiple communities in the valley. The plaza emphasized public buildings, public spaces, and cosmic symbolism including images depicting sacrifice, warfare, ancestors, and the shared Zapotec view of the cosmos.

Collective action is evident in the nearly 400 carved orthostats originally set into the walls of Building L-sub (frequently referred to as *danzantes*). In a recent reanalysis of the monuments, Urcid (2011; Urcid and Joyce 2014) has raised the possibility that the late Middle Formative and Late Formative rulers of Monte Albán may have shared political authority with more communal institutions (see Joyce 2010:131–159). The stones that remain *in situ* consist of alternating rows of horizontal and vertical stones that differ somewhat in style. Using pan-Mesoamerican contextual comparisons he questions the long-standing view that the Building L-sub orthostats represent sacrificial victims (e.g., Coe 1962; Marcus 1992). Instead, Urcid argues that the figures carved on the vertical stones represent men in the act of bloodletting by perforating their penises, with genital scrolls interpreted as blood (figure 2.2). The only representations of human sacrifice in his view are four depictions of severed heads. He interprets the horizontal figures on Building L-sub as ancestors contacted through the act of autosacrifice. The people performing autosacrifice are interpreted as members of a warrior sodality including low-status members depicted on the lower sections of the program and high status elders and Rain Deity impersonators on higher levels. While we find Urcid's analysis to be compelling, the traditional conception of the stones as sacrificial victims would also demonstrate that early authority at Monte Albán was strongly collective. The Building L-sub imagery emphasizes the collectively beneficial outcome of political action, in this case human or autosacrifice, rather than glorifying the individuals performing those sacrifices.

Overall, the evidence from the Valley of Oaxaca indicates the emergence of more hierarchical and far-reaching forms of political authority by the

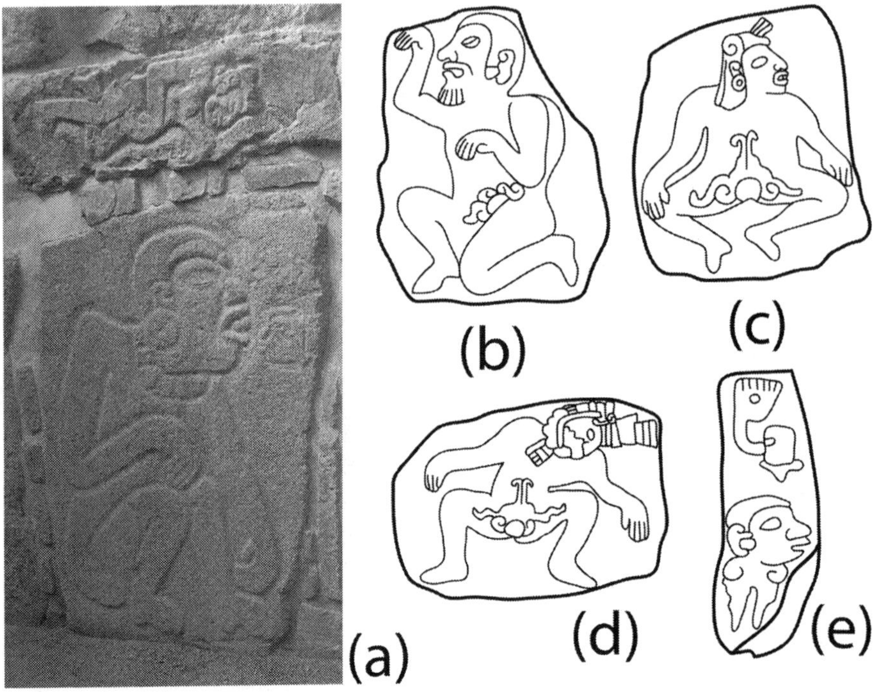

FIGURE 2.2. *The carved-stone monuments from Building L-sub. (a) Photo of in situ horizontal and vertical monuments (photograph by Arthur Joyce); (b) elder from the upper section (redrawn with permission from Javier Urcid); (c) young adult from the lower row (redrawn with permission from Javier Urcid); (d) Rain God impersonator (redrawn with permission from Javier Urcid); (e) severed head (redrawn with permission from Javier Urcid).*

Late Formative period. Yet the data also indicate that the Monte Albán polity exhibited many aspects of collective forms of political organization and economy. Commoners provided labor and perhaps agricultural surpluses and in return rulers sponsored important politico-religious ceremonies. Yet evidence suggests that the rituals carried out on Monte Albán's Main Plaza were cast as communal in emphasis and the authority of the nobility was muted in iconographic representations. If Urcid (2011) is correct in his reinterpretation of the Building L-sub orthostats, then communal forms of authority persisted alongside newer, more hierarchical ones.

Finally, although the evidence shows that the political, economic, and religious innovations of the first 400 years of Monte Albán drew thousands of people to the urban center, there are also indications that some people and communities resisted incorporation into the polity. Redmond and Spencer (2006) argue that the political seat of the Tilcajete polity located 20 km southeast of Monte Albán successfully withstood attacks from Monte Albán for several hundred years.

Río Viejo

The Late Formative period in the lower Río Verde valley, like in the Valley of Oaxaca, was a time of population growth, increasing social inequality, the development of urbanism, and an increase in the construction of monumental buildings (Barber and Joyce 2007; Joyce 2006, 2010). In the lower Verde region, the development of urbanism lagged several centuries behind the Valley of Oaxaca. During the Late Formative, the two largest sites in the valley were Charco Redondo (70 ha) and San Francisco de Arriba (95 ha). Survey and excavations at both sites provide evidence for the construction of monumental public buildings (Butler 2011; Workinger 2002). Evidence from burials, domestic architecture, and the distribution of social valuables found at sites across the valley demonstrates the existence of modest hereditary social inequality (Joyce 1991, 2010; Joyce et al. 1998). Taking these data together, we hypothesize that both Charco Redondo and San Francisco de Arriba were seats of small-scale polities in the Late Formative.

Evidence suggests that the dominant locus of authority and identity during the Late Formative was communal, rather than hierarchical and exclusionary (Barber and Joyce 2007; Joyce 2010). At both large and small sites people were creating socially meaningful places through the construction and use of shared public spaces and monumental facilities that embedded collective actions and histories in specific locations on the landscape. For example, at the 1.5-ha site of Cerro de la Cruz excavations revealed a communal cemetery beneath the floors and alongside the walls of a public building used by multiple domestic groups (Joyce 1991). Adjacent to the building was a flagstone patio that included two hearths that far exceeded the size of typical domestic ones, suggesting their use in ritual feasts. In the presence of the dead, the living defined, maintained, and recreated a social group tied to the specific histories and spaces of Cerro de la Cruz. Similarly, the monumental public facilities of Charco Redondo and San Francisco de Arriba provided a locus at which supradomestic and probably multicommunity social ties were generated through collective actions ranging from labor to ritual.

The first urban center in the region emerged at Río Viejo, which grew to 225 ha during the Terminal Formative period (100 BC–AD 250; Joyce 2010).

Like in the Valley of Oaxaca, archaeological evidence suggests that the rulers of Río Viejo depended largely on internal revenue in the form of labor and in return provided political and religious services to the populace as a form of public good (Barber and Joyce 2007; Joyce 2006, 2010; Joyce and Barber 2011; Joyce et al. 2013). Monumental public buildings at Río Viejo provide the strongest evidence for labor as a source of internal revenue. The ceremonial core of the site consisted of two monumental earthen structures. The earlier was Mound 9-Structure 4, a large rectangular platform raised over the site's Late Formative residential areas (Joyce et al. 1998). In the first century AD, the ceremonial center was shifted approximately 500 m to the west of Mound 9-Structure 4. The new ceremonial center was located on the site's massive acropolis (Mound 1; figure 2.3). The acropolis was begun prior to AD 100, but a major occupation is not evident until the following century. At this time, the acropolis consisted of a platform rising at least 6 m above the floodplain, supporting two large substructures on its northwest and eastern sides (Structures 1 and 2, respectively) both of which stood at least 16 m high (Joyce and Barber 2011; Joyce et al. 2013). To the south, a 5- to 7-m-tall set of mounds surrounded an open plaza, the use of which cannot be determined because Formative-period occupational surfaces are now below groundwater. Our conservative estimate of the volume of the Terminal Formative acropolis is 455,000 m^3.

The acropolis was an enormous building project that almost certainly would have required labor from beyond Río Viejo itself because large segments of the structure were built all at once rather than via accretion (Joyce et al. 2013). A large and well-organized labor pool is further evidenced by the labor-intensive construction techniques used to build significant segments of the acropolis. Much of the construction fill consists of a variety of earthen building techniques, including puddled adobe, adobe block, and rammed-earth (figure 2.4). The variability in earthen and masonry construction techniques indicates that the acropolis was built by multiple work groups drawn from different communities with divergent building traditions. Energetics analysis suggests that the acropolis required a minimum of 2.1 million person-days to build, a number too large to have been provided by the inhabitants of Río Viejo alone, given that the acropolis was likely built over a relatively brief period of time (Joyce et al. 2013:table 5.2).

Other forms of internal revenue, such as taxation on market transactions or agricultural goods, are less evident. For instance, the location of Río Viejo in the river's fertile floodplain makes it unlikely that the site needed to be provisioned with agricultural surpluses as was necessitated by Monte Albán's

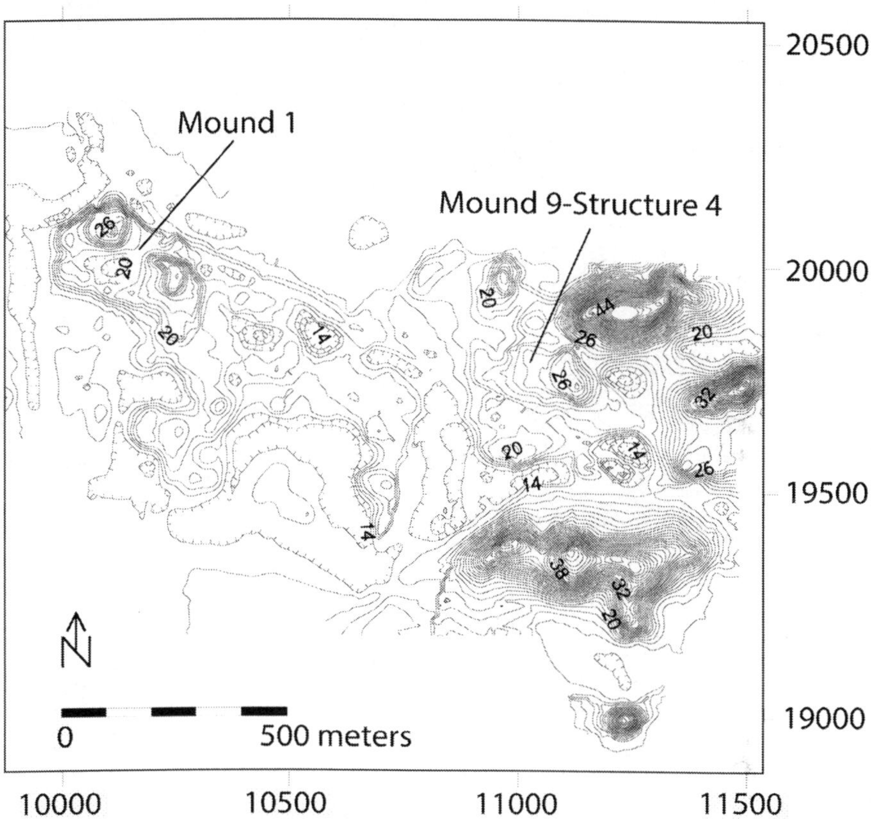

FIGURE 2.3. *Plan of the eastern half of Río Viejo showing Mound 1 and Mound 9, Structure 4.*

mountaintop location. Potentials sources of external revenue are similarly scarce. There is little evidence from the Formative period for craft specialization or the conquest of other regions that might have provided elites with external revenue. Although there is evidence that elites within the region exercised a degree of control over the importation of exotic, nonlocal goods such as greenstone, iron ore, and pottery, there are as yet no indications that these items were a major source of revenue exclusive to the rulers of Río Viejo (Barber 2013; Joyce et al. 1998; Levine 2002).

Construction and use of the Río Viejo acropolis indicate that Terminal Formative regional political authority was an outgrowth of preexisting notions of how social collectivities were defined and maintained. Both collective and

FIGURE 2.4. *Retaining wall of an adobe platform on the acropolis at Río Viejo. (Photograph by Sarah Barber.)*

exclusive activities occurred on the acropolis. Recent excavations demonstrate that large-scale food preparation and consumption took place in the Terminal Formative, most likely associated with large-scale ritual feasting. A series of large middens were deposited in deep pits dug into the fill in the southeastern and southwestern corners of the acropolis. Materials included comals and other food-preparation vessels, dense lenses of estuarine mussel and other faunal remains, as well as elaborate serving vessels, imported serving vessels, and figurines. At the base of Structure 2, we recovered the remains of a large earth oven and oven refuse, presumably resulting from food preparation associated with feasts. The oven refuse consisted of burned rock and sherds used to retain heat and measured at least 10 m in diameter. More exclusive and elaborate ritual spaces were located on top of Structure 2, a large stepped platform, which supported an adobe superstructure with remnants of the only architectural stucco ever found in the valley (Joyce 2006).

Although settlement patterns and monumental architecture strongly indicate that a ruling elite oversaw a regional-scale polity with its political seat at Río Viejo, evidence of rulers has proved remarkably difficult to find (Joyce 2010). There are no known rulers' portraits or tombs in the region that date to the Terminal Formative. Instead, we see evidence for political authority in the distribution of the population, in the coordination required to underwrite monument construction, and in the sponsorship of feasting and other rituals. It is clear that inequality was well established regionally by this time; excavations at other sites have recovered elite residences and burials with elaborate grave goods (Barber 2013). An elite residence was uncovered near the ceremonial center of the secondary site of Cerro de la Virgen. The residence covered an area of 476 m^2 and included several rooms surrounding a patio measuring 13 × 13 m, making this residence far larger and more elaborate than typical residences in Oaxaca (e.g., Whalen 1988; Winter 1986). The house overlooked and was spatially associated with a monumental public plaza that included a ballcourt. Unlike Cerro de la Virgen, the Río Viejo acropolis, however, does not seem to have an elite residence adjacent to public spaces. The most elaborate Terminal Formative interment in the region comes from a public cemetery at the site of Yugüe. This burial was a male interred wearing a plaster-backed iron-ore mirror and holding an intricately incised bone flute. The flute's incising depicts a skeletal male speaking or exhaling and likely indicates ritual responsibilities for the person with whom it was interred (Barber and Olvera Sánchez 2012).

Terminal Formative political relations in the lower Verde were dominated by collective action to an even greater extent than in the Oaxaca Valley. The evidence from Río Viejo's acropolis suggests a political strategy that was both enabled and constrained by historically embedded notions of corporate identity and practice. The tradition of geographically focused collective action in the region provided a framework through which Río Viejo's rulers were able to legitimize their authority at a regional scale. The relocation of the site center from Mound 9 to the acropolis would have made the new ceremonial center distinct from Río Viejo's earlier, more local histories and social relations. By constructing and using a regionally significant place (Mound 1) that embodied the history of the many communities in the valley that provided labor for its construction, Río Viejo's rulers facilitated a process whereby the kinds of acts that had for generations defined local places and social groups came to define a polity (Barber and Joyce 2007). People from these communities subsequently participated in ceremonial practices on the acropolis that included ritual feasting and perhaps other ceremonies (Joyce et al. 2013).

Indeed, sponsorship of these ceremonies was one of the primary public goods that rulers provided to followers. At the same time, the high visibility of such actions would have provided a means by which subalterns could monitor rulers' compliance with collective principles. The lack of rulers' portraits and of a palace on the acropolis, for instance, may represent rulers' efforts to demonstrate such compliance. Nonetheless, acts on the acropolis would have reiterated relations of domination and subordination. Its architecture was far larger than anything else in the region, providing highly visible evidence of the superordinate position of the rulers who sponsored construction and ritual action there.

Evidence for other public goods remains circumstantial. There is no evidence during the Terminal Formative for warfare such as defensive walls or a shift to more easily defended locations in the piedmont (Joyce 2010). Current data are insufficient to clarify whether this lack of conflict was a result of rulers facilitating regional safety as a public good, an outgrowth of broader social conditions at the time, or a result of archaeological sampling. Evidence for economic public goods like facilitation of market exchange or redistribution is extremely limited, although more data are needed. There is evidence for increasing standardization in fineware pottery between the Late and Terminal Formative periods (Levine 2002:167), which may indicate specialized production and perhaps regional exchange mechanisms that could either be taxed or aided by rulers.

OUTCOME OF FORMATIVE-PERIOD COLLECTIVE POLITIES IN OAXACA

Although complex, regional polities built on collective action developed at the end of the Formative period in both the Valley of Oaxaca and the lower Verde, the archaeological evidence suggests that political relations were characterized by a degree of tension and conflict (Joyce 2010). By the beginning of the Classic period (AD 250–800), both regions experienced major changes in political authority.

At Monte Albán, the Main Plaza had been a symbol of collective identity and authority during the Late Formative, but by the Terminal Formative it was increasingly controlled by and restricted to the nobility (Joyce 2004:205–207). New constructions on the plaza effectively closed off and restricted access to the ceremonial precinct. Noble residences began to be built directly on the North Platform and on the Main Plaza itself (Winter 2001). Given their proximity to the plaza, these residences may have been more "public" and their residents more closely involved in politico-religious administration. If the Main Plaza and its public buildings and spaces operated as public goods as suggested above, then by the Terminal Formative rulers were restricting access to these goods, suggesting a more exclusionary form of political authority.

There are also indications that political elites at Monte Albán may have become less dependent on followers as sources of internal revenue and instead found resources (external revenue) that they controlled directly and used to fund political administration. For example, excavations at Monte Albán have discovered 31 ovens used for pottery production, most of which were located in the elite residential area north of the Main Plaza (Markens and Martínez 2009). The ovens were used to make creamware and brownware ceramics, including creamware types C.11 and C.12. These creamwares were expensive to manufacture, often with postfire scratch incising and large hollow supports, and their distribution in the Valley of Oaxaca was markedly status linked (Elson and Sherman 2007; Kowalewski et al. 1989). Elson and Sherman (2007) argue that step-fret designs on creamware vessels symbolized Cocijo, the rain-lightening deity, and were part of a pan-Mesoamerican system of elite display (see also Kowalewski et al. 1989). These symbols may have been another indication of the increasing control of important religious symbols and ceremonies by powerful elites. Furthermore, excavations in a nonresidential architectural complex on the northwestern corner of the Main Plaza recovered evidence of the production of shell ornaments and prismatic obsidian blades (Markens and Martínez 2009). These data indicate that elites at Monte Albán were involved in the specialized production of social valuables, making nobles less dependent on revenue provided by subjects.

The evidence therefore suggests that during the Terminal Formative political authority was becoming less communal and more exclusionary. There are also indications that as the rulers of Monte Albán increasingly gained power by appropriating the Main Plaza and defeating their competitors in the Valley of Oaxaca, tensions between traditional communal leadership and the nobility intensified (Joyce 2010). Evidence from the end of the Terminal Formative suggests that these tensions may have erupted in a political upheaval at Monte Albán around AD 200. At this time the major iconographic programs of the Late Formative, including the Building L-sub monuments, were dismantled (Urcid 2011). Building L-sub was partially demolished and buried under Building L and a temple on the North Platform was burned. A defensive wall was built around parts of the site and evidence suggests that one access point onto the Main Plaza was monitored through military force (Joyce 2010:159). Regardless of how the Building L-sub orthostats are interpreted, the dismantling of these monuments may directly reflect the suppression of communal forms of authority that had existed alongside the hierarchical rulers of the polity. Evidence for the increasing formalization of status distinctions by the Early Classic period (AD 250–500) suggest that the more institutionalized and hierarchical forms of authority gained prominence (see Fargher, chapter 15, this volume).

In the lower Verde the end of the Formative period saw even greater political upheaval than in the Oaxaca Valley. Efforts to create a regional polity defined in terms of local and more egalitarian social groups were successful for a century or two, during which time Río Viejo remained the largest site in the valley and the seat of regional political authority. By AD 250, however, Río Viejo's acropolis fell into disuse and the site was severely depopulated (Joyce 2006; Joyce and Barber 2011). Several other large Terminal Formative floodplain sites with mounded architecture declined significantly in size or were abandoned. By the Early Classic, the regional settlement hierarchy decreased from five to four levels and there were as many as eight first-order centers of roughly equivalent size, indicating a period of political fragmentation. The kind of large-scale corporate social organization that had enabled the construction of Formative-period monumental buildings disappeared, and monumental construction never again matched that of the Terminal Formative.

The causes of the political collapse are unclear, although we hypothesize that one factor was tension between traditional communal forms of authority that were more local and egalitarian and the more exclusionary, hierarchical, and regional forms of rulership that were emerging at the end of the Formative. There is evidence that parts of the acropolis were heavily burned by fire prior to its abandonment. Excavations in several areas of the acropolis have revealed burned floors and burned adobe wall foundations. The burning was likely the result of termination rituals, although we cannot discount the possibility that it involved warfare or was accidental. Even though the acropolis was an important political and religious building that had required considerable communal labor to construct, it was left to slowly disintegrate over the next 250 years.

CONCLUSIONS

In this chapter we have argued that both the early Monte Albán and Río Viejo polities exemplified the kind of corporate organization and collective action relationships that has been a focus of Richard Blanton's theoretical work. Although both polities can be seen as examples of corporate political organizations and collective action, they also exhibit considerable variability in how social complexity and political authority was expressed and negotiated. In the lower Verde regional political authority collapsed after a brief and tenuous florescence, while elites in the Valley of Oaxaca were successful in institutionalizing hierarchical authority that would persist for more than a millennium.

We see a number of points of divergence that may have had significant consequences in the history of these polities, especially as they relate to the ability of rulers to extend their authority across multiple communities throughout a broader region. In the Valley of Oaxaca, evidence suggests that the rulers of Monte Albán were initially successful in negotiating shared forms of political control with more traditional communal forms of leadership. Although we see political authority in the Late Formative Valley of Oaxaca as largely communal, Monte Albán's rulers were successful in linking their authority and identity to a series of innovations in politico-religious belief and practice that served to set themselves apart from common people and provide them with a public good desired by people in the valley. An important aspect of these politico-religious innovations was the Main Plaza of Monte Albán, which was a socially significant place marked by architecture and imagery that was clearly distinct from previous ceremonial precincts. Rulers at Monte Albán were also successful in establishing a variety of sources of revenue, which increasingly through the end of the Formative included external revenue independent from the labor of subjects. Finally, polity rulers had recourse to coercive force to bring communities in the valley into compliance. For example, although we question the degree to which areas outside the valley were conquered, there is good evidence that Tilcajete was eventually defeated and incorporated into the Monte Albán polity (Redmond and Spencer 2006).

Political authority in the lower Río Verde valley appears to have been both highly communal and less successful in creating a regional political identity and extending authority across the region. Despite the scale of monumental construction at Río Viejo, the regional polity seems to have been weakly integrated and tenuous. At present, the evidence suggests that regionwide collective relationships revolved around labor as revenue and the sponsoring of politico-religious ceremony as a public good. We suspect that rulers may have been limited in their ability to extend economic and political interactions outside of Río Viejo, discouraging the development of administrative institutions that might have tied together communities. There are few indications of innovations in religious and political practice that would have distinguished rulers from followers and created pubic goods not available at the local level. Instead, what seems to have been new in terms of political relationships was limited to a scaling-up of traditional practices that had previously materialized notions of local community identity including monumental construction programs and ritual feasting. The active maintenance of strong community identities coupled with the inability of rulers to distinguish themselves from local elites, establish pubic goods distinct from those that were locally available, establish sources of external revenue, or develop a significant coercive capacity may have doomed the Río Viejo polity to collapse.

Acknowledgments. We would like to thank Lane Fargher and Verenice Heredia for inviting us to participate in this volume. We also thank the Instituto Nacional de Antropología e Historia of Mexico for granting us permission to carry out our research in Oaxaca. We thank the organizations that have funded the field research on which this chapter is based: the National Science Foundation (BCS9729763, BCS1123377, BCS1123388); the Historical Society (with support from the John Templeton Foundation), the Foundation for the Advancement of Mesoamerican Studies, Inc. (#99012); University of Colorado-Boulder CARTSS, CRCW, Norton Anthropology Fund, and Dean's Fund for Excellence at University of Colorado-Boulder; and the Office of Research and Commercialization and Department of Anthropology at the University of Central Florida. We thank the presidentes municipales of Santiago Jamiltepec and Villa de Tututepec de Melchor Ocampo who granted us permission to work in the region as well as the owner of the Río Viejo acropolis, Don Roberto Iglesia. Finally, we thank two anonymous reviewers who provided us with valuable feedback on an earlier draft of this chapter.

3

Built Space as Political Fields

Community versus Lineage Strategies in the Tequila Valleys

Christopher S. Beekman

This chapter operationalizes the corporate and exclusionary strategies proposed by Blanton and his colleagues (1996) by associating them with four contemporaneous social institutions with manifestations in built space. I use iconographic data to support and refine interpretation of the archaeological evidence in a manner similar to Small's (2009) historical-archaeological analysis in Greece. Individually and relationally, these social institutions formed a field (Bourdieu 1990) in which different forms of capital both provided admission to and constituted the objectives of competition. Once archaeological correlates for the field of power have been proposed, I consider how we might reinterpret the existing settlement pattern data for the Tequila valleys of Jalisco in the Late Formative and Early Classic periods and their relationship to the Teuchitlán polity. This allows me to bridge the gap between Blanton's work on political strategies and his regional settlement pattern research, and simultaneously contribute to the growing interest in the spatial aspects of authority (A. T. Smith 2003).

Blanton's more recent research with Fargher reorients his prior work on corporate-exclusionary strategies toward a rational choice model of human decision-making and self-organizational models of institution building. I limit my use of the model here for several reasons. Their book on collective action theory and much subsequent work (Blanton and Fargher 2008:25–32; Fargher et al. 2010) required detailed historical data

DOI: 10.5876/9781607325338.c003

to reconstruct the relationships needed to truly assess the model's utility (as used in political science; Ostrom 1990, 2003). Archaeological case studies such as Bronze Age Indus Valley society were considered and rejected as possessed of insufficient data for their purposes (Blanton and Fargher 2008:61–62). Further, while collective action stems from research in political science, the corporate-exclusionary model was designed by four archaeologists, drawing on archaeological antecedents, and defined around the interpretation of material culture. It may therefore possess particular utility in dealing with strictly archaeological data, as can be seen in Fargher's (see chapter 15, this volume) use of both collective action and corporate-exclusionary approaches as he juggles historical and archaeological comparisons. Corporate and exclusionary strategies have also proven to be applicable at levels other than the polity or the society. The study here recognizes and expects that the different strategies at play within the Teuchitlán polity of central Jalisco will not all coincide with a generalization at the level of the entire polity (see Saitta 2013 for a similar recognition of distinct social interests while still using the term "collective action," and Ferguson and Mansbach 1996 from political science). Once relationships between the different fields of power have been elucidated and recognized at the regional level, I do find certain observations from collective action theory useful for their interpretation. In this I follow Blanton and Fargher's (2008:10) advice to "study up" from local data to the broader region. In this chapter devoted to the legacy of Richard Blanton, I therefore show that his past as well as present contributions continue to inspire novel analyses.

THEORETICAL BACKGROUND: STRATEGIES AS SITUATED WITHIN FIELD THEORY

This chapter takes the position that the exclusionary/network and corporate strategies described by Blanton and colleagues (Blanton 1998a; Blanton et al. 1996) can be connected to specific institutions, often associated with formalized built space. These authors characterize exclusionary strategies as drawing upon wealth and external contacts to exclude others from power—dynastic rulership among the Classic southern lowland Maya is often cited as the canonical example, in which divine kings and queens successfully monopolized access to the symbols, tools, and positions of power. Social actors associated with corporate strategies pursue an inclusive group identity through ceremony and ritual, suppressing or redirecting competition. Teotihuacan with its apparent absence of aggrandizement for a dynastic lineage is characterized as corporate. Descent groups[1] played significant roles in both political systems, but at Teotihuacan a number of these social building blocks are argued to have shared power at some wider scale of political activity. Hence the terms "corporate" and "exclusionary" cannot be used in isolation, but only in a relational sense to describe the strategy pursued by an individual or group vis-à-vis other groups.

Social and political strategies possess a contextual and spatial aspect that defines where and when it is appropriate for social actors to pursue them. In Pierre Bourdieu's work, he pointed out that *habitus* and strategies play out differently in different *fields*, or socially defined contexts of competition for power (Wacquant 1989:38–41). His work typically interpreted field in terms of social space, so that art, business, academia, and so on each formed a field that might fit together relationally into a still larger one, such as the fields of cultural production or power or bureaucracy (see Bourdieu 1993, 1998, 2004). Fields encompass competition over one or more forms of *capital* (Bourdieu 1986), and relate to other fields in complex ways. Corporate and exclusionary strategies can be understood within Bourdieu's fields, such that social rules and cognitive codes ensure that people use and pursue different forms of capital within bounded social contexts.

Equating fields to physical spaces is not self-evident, and some interpreters of Bourdieu take pains to emphasize that fields more properly correspond to institutions and social space (Thomson 2008:74). But it is worth recalling that Bourdieu's concept of the field is not unlike Turner's (1974) contemporary idea of the *arena*, a physical space in which political conflict and competition are acted out through performance (cf. Inomata and Coben 2006; Postill 2011). Political strategies are cross-culturally bounded by proscriptions and prescriptions (e.g., Boehm 1993), and political elites are obligated to have a visible and performative component to their activities. Formally designed spaces provide an appropriate venue for performance (accession rituals, bill signing, speeches, debates, etc.) and furthermore create opportunities for political elites to reach subjects through affective means (Smith 2000). This circumscription of competition may also help to contain changes to the overarching field of power. The reproduction of one pathway to status and authority would have been partly insulated from disruption in another. Elites may seek to break down the barriers between fields and extend political activity beyond their socially accepted contexts, but these efforts to undermine cultural codes are more likely to engender resistance.

Blanton and colleagues noted the potential spatial associations of both their strategies: "Corporate and network strategies result in dissimilar and antagonistic political economies and so are likely to be temporally or spatially separated" (Blanton et al. 1996:7). Temporal distinctions have received most attention to date, but a number of researchers have pointed to evidence for the contemporaneity of the two strategies (Fleming 2004; Porter 2002; Urban and Schortman 2004). So taking the example of the Classic southern lowland Maya above, just because they are seen to have primarily followed exclusionary strategies, this would not preclude the pursuit of corporate strategies in distinct spatial settings—a council house, for example. The co-occurrence of opposing strategies forces a more direct consideration of their separate spatial needs and the fields in which they are active. Both possess significant performative aspects, with corporate strategies being more inclusive and demanding of large public spaces in which political activity can be witnessed by a greater number of people. Network approaches can derive their exclusionary aspect by limiting participation, and defining a privileged audience more narrowly through the use of smaller and access-restricted spaces (see Uriarte Torres 2011 for these and other predictive characteristics).

Complex societies rarely offer a single route to power, and there will likely be multiple examples of formalized built space that are appropriate for different forms of political activity (I suggest McGuire 1983 to be an early recognition of this point). Each of these formal architectural spaces are places in which political strategies could be acted out in the pursuit of different forms of capital. Hence to take an example from Bali that I have used previously (Beekman 2005:56), the irrigation temple networks studied by Lansing (1991, 2006) continue to be built and maintained by a collection of farming cooperatives in a self-organized corporate or group-oriented strategy. Competing political elites and their kingdoms (Geertz 1980), on the other hand, made up a second route to power more reminiscent of an exclusionary strategy, and the political elites possessed no control over the irrigation systems. Hence these two strategies with their own sets of rules and expectations are associated with and are carried out within different networks of built space. Not all forms of power necessarily have spatial loci (the work of Michel Foucault best exemplifies this problem), and some places could potentially be made acceptable for either group or individual-oriented performance. Some of these difficulties may be overcome by integrating other data sets with the archaeological record.

To summarize the theoretical argument, struggles over political power are constrained cross-culturally by defining contextual limitations on political competition. This partial encapsulation thereby limits which structures of power will be reproduced or disrupted by political activity. Participation in these arenas is dependent upon the possession of some form of capital, and capital is the medium and goal of competition. More complex societies by definition include multiple routes to power (McGuire 1983) dependent on different forms of capital, but they are kept contextually distinct by associating them with defined spaces. Once positions of power have been defined, the wider field of conflict over the relative importance of different forms of capital will become more apparent. This oscillation in the dominant form of capital is potentially the basis for the social transformations between corporate and exclusionary strategies.

FORMAL SPACES AND STRATEGIES IN LATE FORMATIVE-EARLY CLASSIC CENTRAL JALISCO

Compared to some areas of Mesoamerica, Late Formative/Early Classic central Jalisco shows evidence for a relatively limited number of recurring formal spaces (Beekman and Weigand 2008). Past archaeological research by Weigand (e.g., 1996) placed considerable emphasis upon architecture and built space, leading to his definition of a Teuchitlán tradition. It is possible to build upon this strictly material evidence to discuss the activities that took place in the architecture, and interpret them in terms of different kinds of strategies. I will discuss three forms of built space and a possible fourth, taking into account surface and excavated evidence as well as contemporaneous ceramic models that depict activities within the architecture.

Shaft Tombs

For much of the twentieth century, archaeological research in western Mexico focused on the deep shaft and chamber tombs (Fowler et al. 2006) (figure 3.1a). Various lines of evidence have been used to associate the tombs with discrete descent groups (Beekman 2008; Ramos de la Vega and López Mestas Camberos 1996). Innumerable looted tombs exist, but fortunately there are approximately 40 examples that have been excavated and published in varying degrees of detail. Shaft tombs essentially break down into those grouped into cemeteries on the rural landscape (e.g., Galván Villegas 1991), and isolated tombs occurring beneath public architecture in ceremonial centers (e.g., Ramos de la Vega and López Mestas Camberos 1996). The latter tombs are larger, deeper, and accompanied by more varied and numerous offerings. The tombs beneath the public architecture also tend to have more consistent evidence of reuse and the interment of additional individuals, while the rural tombs can often have a single occupant. The tombs in the ceremonial centers thus demonstrated greater genealogical depth for group claims to the titles or ceremonial positions associated with the public architecture, much as

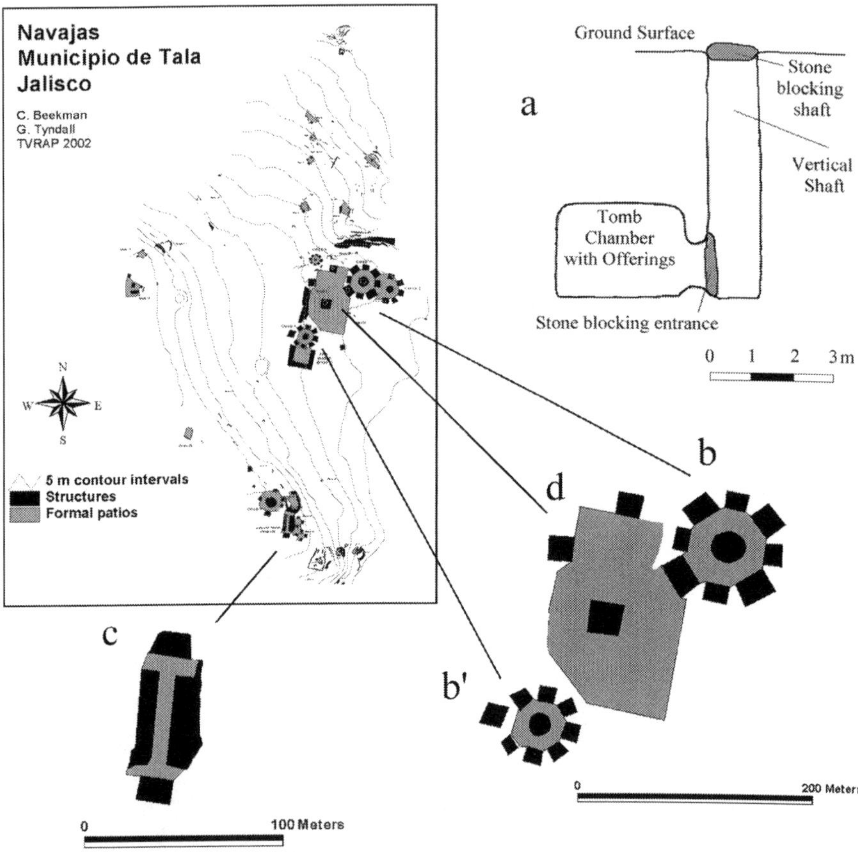

FIGURE 3.1. *Examples of each of the forms of built space proposed as associated with specific social institutions and strategies: (a) shaft tomb, (b) guachimontón (circular public architecture), (b') a guachimontón with one outsized platform, (c) ballcourt, and (d) elite residential group. (Images taken from Beekman [2005:figures 4.2 and 4.4] and courtesy of the Tequila Valley Regional Archaeological Project.*

monumental inscriptions or codices did in other times and places. The rural cemeteries emphasized the presence and cohesion of larger groups, and were probably associated with claims to economic capital in the form of land rights.

Ceramic models depict formal processions in which the dead were carried toward a structure (von Winning 1969:figures 156–157) (figure 3.2). Elite shaft tombs are found beneath buildings, and these procession scenes make it clear that burial was a public display, a performance that emphasized a family's

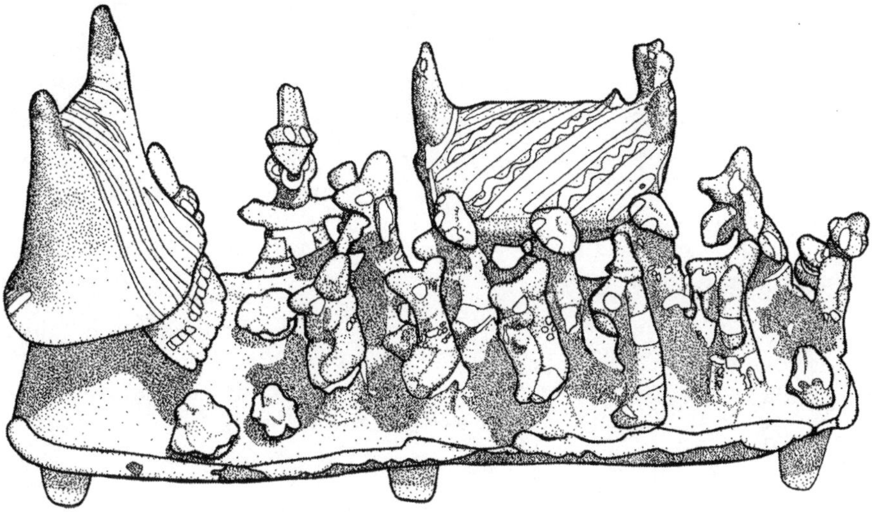

Figure 3.2. *Ceramic model depicting a burial procession with the dead being carried by pallbearers. (Drawing by Kathy Beekman, after von Winning and Hammer 1972:figure 87.)*

external connections through the exhibition of their objectified cultural capital (in the sense of Bourdieu 1986:50). Artifacts found within the tombs include ceramic vessels, shell or mineral jewelry, obsidian tools and jewelry, ceramic figurines, and hollow ceramic figures (the best published and excavated examples being in Galván Villegas 1991; Ramos de la Vega and López Mestas Camberos 1996). The raw materials used for jewelry or figurines could be exotic: jade was imported from Guatemala, and shells were brought from both the Pacific and Caribbean coasts. Fine ceramic bowls, particularly of the Oconahua Red on Cream type, usually incorporate a quadripartite division of the vessel interior, making reference to the Mesoamerican cosmological model. But there are also finely made vessels with maize symbols arranged into rows and separated by lines representing water, likely depicting farming landscapes (Beekman 2009:figuras 7–8).

The combined evidence leaves little doubt but that elite lineages practiced exclusionary strategies in association with mortuary ritual, but through the conspicuous display of wealth rather than through access-limited family ritual. Death was an opportunity to draw attention to the wealth and connections of the descent group, well demonstrated by the models of open processions carrying the dead to the tomb. This public display was used to express and reinforce the status and claims of the group (Beekman 2000). The tomb itself was a small and cramped space inappropriate for display, and the procession and mortuary rites were the true focus of the performance. In the spatial analysis that follows, I set aside rural cemeteries as potentially focused on different forms of capital and our record of their locations is in any case very incomplete.

Guachimontones

The circular public architecture (known as *guachimontones*) shows a clearly different pattern. The circular guachimontón form is composed of usually eight satellite platforms surrounding a central circular altar or pyramid (figure 3.1b). As reconstructed elsewhere, each of the surrounding platforms was built by a different descent group and occasionally had a shaft tomb beneath (Beekman 2008). Our excavations of three circles at Llano Grande and Navajas have encountered an assemblage of ceramics, lithics, groundstone, figurines and hollow figures, plant remains, and quartz crystals. Analyses are ongoing but preliminary summaries exist (Beekman 2008; Beekman et al. 2004). The ceramics include high-quality wares that nonetheless show only simple decoration (Johns 2014). Lithic remains include infrequent groundstone, manufacturing debris as well as finished products of obsidian, and limited evidence for obsidian jewelry (Hoedl 2013; Wagner 2014). No jade or shell has been identified. We are currently engaged in trying to distinguish ordinary residential tasks (particularly food preparation) from more exotic forms of the same activities, such as feasting within an otherwise sacred space. This ambiguity attests to the quotidian nature of the assemblage compared to that found in the shaft tombs, and activities within the circles placed less emphasis on communicative style. I have interpreted this elsewhere (Beekman 2000) as a decrease or suppression of open competition within the guachimontones due to cultural conventions regarding inappropriate behavior within the ceremonial circles (see Blanton 1998a:163–166).

Ceramic dioramas are a particularly rich source of information for what took place in the circles (figure 3.3). Models of the architecture often show consumption of food and drink (Butterwick 1998), musical performances, or dancing within the patio, perhaps on the occasion of a marriage (von Winning 1971:348, figures 10, 11). Other models depict a pole-climbing ceremony identified as analogous to the Postclassic Xocotl Huetzi *veintena* ceremony associated with the green-maize harvest (Beekman 2003a). The circles replicate the multilayered cosmological model so often found in Mesoamerican public architecture (Beekman 2003b; Kelley 1974). The occasional shaft tombs below represented the underworld, the patio with its very human activities was

Figure 3.3. *Ceramic model depicting a simplified form of the guachimontón public architecture. (Drawing by Chris and Kathy Beekman, after von Winning and Hammer 1972: color plate 1.)*

associated with this world, and the central altar or pyramid with a pole raised in its center made direct reference to the sacred mountain and ties to the heavens. The themes of agriculture and the cosmos were not specific to any particular descent group, and ritual performance drew together elites and subjects in an inclusive manner. Critically, none of the participating lineages was able to monopolize the link between human-built sacred space and the cosmos that it represented (Beekman 2008). The architecture replicated the Mesoamerican universe, but since the guachimontón form was divided into a series of components, each built and maintained by different descent groups, no one group was in a position to claim the role of exclusive mediator to supernatural forces. Power was thus explicitly shared among a collection of descent groups, and no single dominant family was able to emerge. At least this was usually the case. There are just a few examples of guachimontones in which one of the satellite platforms around the circle was enlarged to a degree unmatched by the others (e.g., Beekman 2005:figure 4.4; Weigand 1993:191; Weigand and García de Weigand 1997:56). These may be cases in which a single descent group was able to subvert the architecturally enshrined equality among families in some manner (figure 3.1b'), and they receive further attention below.

Overall, the guachimontón architecture was the most visible type of formal built space in central Jalisco during this period, and the activities within correspond very well with what we have come to expect from the corporate strategy described by Blanton and colleagues—ritual with cosmic or otherwise communal themes. Important ceremonial roles were shared among several descent groups that were ranked more highly than others through their possession of cultural capital in the form of sacred knowledge. Holding a privileged position within the circles and participating in these ceremonies allowed elite families to accumulate increased prestige, reproducing their position and solidifying their social distance from subjects. Opportunities existed for these families or individual members to stand out through their participation in public ritual (Beekman 2000), but the difficulty for any group to claim exclusivity as mediators with the supernatural prevented this from becoming institutionalized. Hence individual descent groups remained relatively equal when compared to one another.

Ballcourts

The third form of built space that can be associated with political strategies is the ballcourt (figure 3.1c). The Mesoamerican ballgame was a complex public spectacle in which individuals or teams competed by keeping a rubber ball in motion and moving it down an open space marked on the ground or in a formal court (Scarborough and Wilcox 1991; Whittington 2001). It has been practiced in western Mexico for some 3,500 years. Some of the earliest evidence for the ballgame occurs in the form of ballplayer figurines in the Early Formative El Opeño tombs of Michoacán (Oliveros Morales 2004:figuras 11a, 11b, imágenes 17–19), and a form of the game has been documented from the early Colonial period up to the present day (Beals 1932:113, 1933:11–13; Kelly 1943; Leyenaar 2001). The game was undoubtedly played for reasons of sport in the past, but this hardly requires a large and formally designed court, just as casual games of fútbol can be played nearly anywhere. The construction of a formal ballcourt creates a special form of built space especially conducive to special and public versions of the game.

Like the public rituals in the circles, there is an element of competition in which teams or individuals could potentially stand out through demonstrations

Figure 3.4. *Ceramic model depicting a ballcourt with a ballgame in progress. (Drawing by Chris and Kathy Beekman, after Butterwick 2005:18.)*

of their skill. Yet the ballgame once again has cosmic overtones—including sacrifice, renewal, and the dynamics of the universe, with specific variations on these abstract themes across Mesoamerica. Mediated by these higher goals, the ballgame became a controlled form of competition that reified and/or provided an outlet for entrenched social antagonisms (Blanco 2009; Gillespie 1991; Kowalewski et al. 1991; Weigand 1991). Ceramic figures and models in western Mexico portray ballplayers, the court, and their equipment (Day 1998) (figure 3.4), but the rich iconography known elsewhere remains unrecognized to date, prompting greater dependence on archaeological evidence. Ballcourt sizes in Late Formative/Early Classic central Jalisco can be arranged into a hierarchy that has been related to different scales of inter- and intrapolity conflict resolution (Blanco 2009; Weigand 1991). Blanco's summary of the excavation evidence at Los Guachimontones points particularly to the presence of human remains in pits within the court, and to the predominance of domestic wares rather than finewares (Blanco 2009:119–157). The similarity to practices within the guachimontones, and those proposed for a corporate strategy, are evident.

The intrasite placement of ballcourts in Jalisco indicates even closer links to the circles. There is no complete consideration of the ballcourts within the Tequila valleys, but of the 17 ballcourts illustrated in two major sources (Blanco 2009; duVall 2007), one court aligns perpendicularly to two circles, seven are freestanding, and nine are appended to the circles such that the end platform for the ballcourt is also a satellite platform for the circle. All are found within the ceremonial centers. The two architectural forms thus share a close relationship most of the time. More speculatively, this physical connection suggests a link between the two that was proposed for other areas of Mesoamerica by Schele and Guernsey Kappelman (2001). Those two authors have drawn attention to the Postclassic myth of Coatepec (Snake Mountain), in which the Aztecs built a temple to Huitzilopochtli atop Coatepec during one of their stops on the lengthy peregrinations toward central Mexico. They afterwards built a ballcourt at the base of the mountain. A subsequent story tells of flooding within this court, a reference to the known association of ballcourts with a watery underworld. The authors identify this symbolic relationship between mountains and water across Formative and Classic Mesoamerica, primarily in iconography but also in the layouts of the ceremonial centers of Dzibilchaltún, Uaxactún, Tikal, and others. The visually similar layout of the combined ballcourt-guachimontón complex would therefore represent the juxtaposition of the sacred mountain and the watery underworld, duplicating and reinforcing the cosmological model already present in the circle alone. The ballcourt and circle complex thus constitute a larger plan with its own cosmological meaning, cementing further their importance for group-oriented performative ritual in the Tequila valleys.

Elite Households

A final architectural form of potential importance is the elite household. It has been observed that "palaces" have not been identified in Late Formative/Early Classic Jalisco (Nelson 2004), at least not in comparison with those known elsewhere in Mesoamerica. However, Smith Marquez (2009) has defined a series of size categories in residential groups that he calls "cruciforms," since they are often composed of four symmetrically placed rectangular structures around a leveled patio. Apart from the fact that this scheme artificially separates groups of four from other residential groups ranging from 2 to 8 structures, his hierarchy duplicates an analysis that found a wide size range in residential groups as measured by structure area (Beekman 2009b). The largest groups are found in prominent areas within the ceremonial centers. For example, Group 3 at Navajas and Feature VI at Loma Alta resemble nothing so much as oversized residential groups, with three or four rectangular structures facing a leveled patio (figure 3.1d). The largest of these groups were arguably the residences for elite families, even if they are not so differentiated from commoner residences in their internal complexity to be considered as "palaces." None of these has been excavated to date, and they pose some problems for their interpretation. Did they house all the descent groups associated with a circle? Do they suggest the preeminence of a single group? Most of the ceramic architectural dioramas depict individual buildings that may be residential or places of residential-like activities. These may be ordinary or outsized residences, or one of the satellite platforms around the guachimontones, but either would suggest that the descent group is being highlighted rather than any wider social collectivity. The house models portray more private, interior views of food preparation and consumption (everyday meals, or public feasting?) and/or seated and conferring individuals (family chats, or formal reception of visitors/supplicants?) (Day et al. 1996:table 1). A lower story is often shown, perhaps depicting a shaft tomb (e.g., Furst 1975) and further emphasizing the linkage of this space to families rather than community. Only excavation across the size categories of residential groups will clarify the situation, but the large residential complexes hold out the possibility of another form of built space that may be associated with descent groups that have succeeded in lineage aggrandizement. Although this interpretation is certainly tenuous, I consider the possibility below when discussing their spatial distribution.

The field of power in Late Formative/Early Classic central Jalisco consisted of descent groups in competition with one another in multiple ways and multiple venues. Following Archer (Archer 1982:462–463, 475–477; 1995:247–293; 2000:253–305), I (Beekman 2005) have proposed that the descent groups were collective agents that served as important social actors in the Tequila valleys. Individual members aligned their interests to a significant degree with those of the group, since that affiliation is probably how they gained access to different forms of corporate property or capital, whether sacred knowledge, titles, or land. Yet the major tensions in central Jalisco were not between individuals and descent groups but between descent group and community interests. The broader institutions that sought to incorporate and dampen competition among the descent groups were the community rituals in the guachimontones and the ballcourts, and they would have been frequently challenged.

EXTENDING THE APPROACH OUTWARDS

The combination of imagery and archaeological data make it possible to imbue static architectural remains with some of the meanings associated with activities there. This presents an opportunity to unpack the social practices at different ceremonial centers and assess how they changed over time. Instead of employing a settlement hierarchy that ranks centers by their aggregated population, architectural volume, and so on (Ohnersorgen and Varien 1996), we can instead conceptualize the webs of related practices that linked some settlements and not others in heterarchical fashion (Crumley 1979). The cultural landscape takes on a different appearance, with multiple nodes where different forms of political strategies are more or less evident.

Over the course of 30 years, Weigand's (1993) survey of the Tequila valleys identified the Teuchitlán culture and mapped the central architecture of many ceremonial centers. Weigand's survey was most detailed in the southern valleys, and his work in the surrounding areas is being superseded by more systematic approaches (e.g., Anderson et al. 2013; Heredia Espinoza 2008). I focus my reconsideration primarily on the southern Tequila valleys, where Weigand recorded 38 ceremonial centers assigned to the long span from 300 BC to AD 500. Since Weigand did not collect or analyze ceramic surface collections, these sites will need to be reexamined in the near future to better situate them within the ceramic chronology (Beekman and Weigand 2008). This analysis is therefore quite rough, and my discussion of temporal change is limited to those sites that can be placed within the ceramic sequence or dated directly through radiocarbon dates from excavations.

Statics

First, the southern Tequila valleys display a cluster of 38 ceremonial centers within 26 square miles. Ohnersorgen and Varien (1996) took a standard perspective that used the total volume of public architecture to divide these centers into four hierarchical levels labeled A–D (figure 3.5). Los Guachimontones is the only A site, with 10 circles, two ballcourts, and no shaft tombs. I consider the C site of Loma Alta lying only a short distance up the hill to be part of the same site and would add seven circles, two ballcourts, a large residential group, and substantial architectural volume to Los Guachimontones as a consequence (Blanco 2009:figures 3.4, 3.12; duVall 2007:figure C.32; Smith Marquez 2009). The only B site within the southern Tequila valleys is Ahualulco, lagging far behind with a ballcourt and six circles (Weigand 1993:88). All other ceremonial centers within the southern concentration of rural settlement are much smaller C and D sites, with usually only one or two circles and perhaps a ballcourt. The map showing the settlement pattern based on site-size hierarchy emphasizes the centrality of Loma Alta/Los Guachimontones, and how sites of the next tier extend its reach into the rest of the Tequila valleys.

FIGURE 3.5. *Map of the Tequila valleys, Jalisco, with sites identified by site-size hierarchy.*

If we interpret this settlement pattern using social institutions rather than sites, we obtain a different understanding of the political landscape. The guachimontones and ballcourts are overwhelmingly concentrated in the southern valleys, and this is also where we find the greatest evidence for architectural conformity. All the ceremonial centers have guachimontones as the minimal civic architectural unit—no sites possess only the large residences and/or ballcourts. Large residential groups that were probably elite in nature and potentially associated with one or more of the very highest ranking families are only found in the primate center—Loma Alta/Los Guachimontones. The only other mapped sites with the large residential groups lie outside of the southern Tequila valleys. These are either B sites (Navajas [Beekman 2005:figure 4.4]; Santa Quiteria [Weigand 1993:87]) or a C site that is the head of its own isolated settlement cluster (Huitzilapa [Weigand 1993:191]), and it may therefore have been the head of an independent polity, something not predicted on the basis of site size alone. Furthermore, there are some places where descent groups may have been successful in consolidating their authority relative to other groups around a circle. The proposed evidence is a guachimontón with a single oversized satellite platform that dwarfs the others. These locations—Navajas circle 4, Huitzilapa Cerro de las Navajas circle A, and Santa Rosalia circle A (Beekman 2005:figure 4.4; Weigand 1993:191; Weigand and García de Weigand 1997:56)—are all outside of the southern Tequila valley settlement core and are once again B sites or the C site that was the head of its own isolated cluster. Finally, the only centers known to have the elite shaft tombs beneath public architecture are C or D centers (El Arenal, Cerro de los Monos, Huitzilapa, Resumidero, and San Andrés [Beekman 1996:159–164, figure 4.4; Long 1966:248–278, figures 8–10; Ramos de la Vega and López Mestas Camberos 1996:126–129, figures 3, 4, 12; Weigand and Beekman 1998:40, figures 8, 9]), and only one lies within the southern valleys.

The picture emerging from the distribution of distinctive spaces as opposed to sites suggests that the southern Tequila valley settlement zone constituted the most corporate area within these valleys. The south includes less indication of divergence from the corporate ideal and less evidence for descent group ceremonialism within the ceremonial centers; a large residential group is found only at the largest site in the zone (figure 3.6). Descent group aggrandizement and successful attempts to elevate one group over the others are notably limited to the more peripheral areas of the Tequila valleys. The site of Huitzilapa shows the greatest divergence from the corporate ideal of the south, but all of the significant centers in the periphery show similar evidence. This may point to the political independence of these distant centers, but it would be most appropriate to state that the corporate cognitive code (Blanton 1998a:163–166; see also Fargher, chapter 15, this volume) was being challenged on the outer edges of the Tequila valleys. Why? Was it merely the distance factor that strained the ability of the Teuchitlán polity to enforce social conformity and dampen lineage aggrandizement?

Variations in economic organization may be more likely. Blanton and Fargher's comparative analysis of the relative role of collective action across their sample of 30 premodern polities (Blanton and Fargher 2008:chapter 6, table 10.3) found that

> a measure of the degree to which the main revenue sources are produced by commoners (e.g., what we call "internal revenues" such as agrarian surpluses from a free peasantry) is highly correlated with the quantity of public goods provided by the state . . ., with degree of bureaucratization . . ., and slightly less so, but still significantly with degree of control that can be exercised over the agency of governing principals. . . . We concluded from these strong statistical results that in the more collective states, because ordinary taxpayers (including

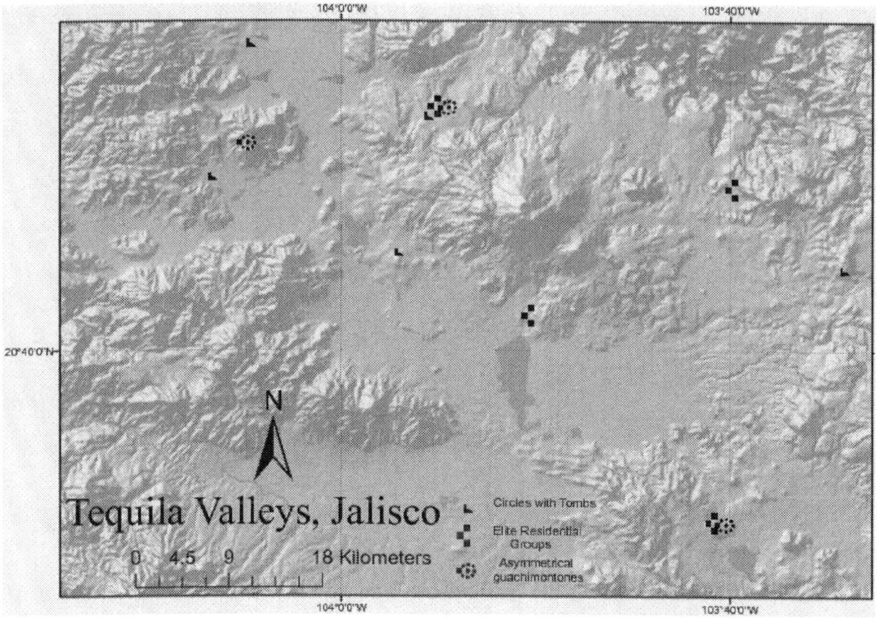

FIGURE 3.6. *Map with locations of sites with large residential groups, shaft tombs under public architecture, and guachimontones with one oversized platform.*

> peasants) are the state's principal source of revenue (what we call internal revenue . . .), they are in a stronger position to demand public goods and forms of governance that are consistent with collective political goals. (Fargher, Heredia Espinoza, and Blanton 2011:320)

Or stated in a different way, when elites are able to develop external sources of revenue, they are better able to ignore commoner demands and pursue their own agendas (see Webster 1975:465–466 for the same argument tied specifically to the spoils of warfare). This proposal may have merit for explaining variations in the political landscape of the Teuchitlán polity. The southern valleys include broad zones of quality agricultural soils today exploited primarily for sugar cane. I have argued previously that the dispersed residential pattern through the southern valleys would have been most appropriate for shifting cultivation of maize and associated crops in the precolumbian period (Beekman and Baden 2011), while the guachimontones themselves are associated through ritual and morphology with maize symbolism (Beekman 2003a, b). The northern and western valleys with most of the built space linked to lineage aggrandizement are environmentally distinct. West of the

Tequila volcano is the Magdalena Lake Basin (Anderson et al. 2013), where the presence of lacustrine resources and more limited agricultural land would surely have altered the underlying subsistence economy. The rolling and often rocky agave landscape north of the Tequila volcano is hotter, has lower annual precipitation, and cultivation is today dedicated overwhelmingly to the blue agave (Heredia Espinoza 2008; Ojeda Gastélum, Benz, and López Mestas 2008). Although this may point to different bases for the subsistence economy and are still potentially internal sources of revenue, their correspondence with the architectural evidence is suggestive. It may be that elites in the peripheral zones of the Tequila valleys successfully redefined land tenure or resource ownership in areas with different potential productivity. As survey progresses through the Tequila valleys, the distribution of residential settlement will need to receive close attention to evaluate this proposal.

Dynamics

A prior synthesis of the chronological data in the Tequila valleys found that the largest guachimontón temples are radiocarbon dated to the period prior to AD 200, after which only small additions and maintenance continued to take place (Beekman and Weigand 2008). Those circles whose construction could be dated to after AD 200 were all within Smith Marquez's (2009) smallest categories. Furthermore, the shaft tombs declined in size and in the abundance of offerings (e.g., Galván Villegas 1991), suggesting a shift in individual loyalties away from descent groups (see also Fargher, chapter 15, this volume). I interpreted this and other evidence to indicate that descent group ceremonialism and aggrandizement were failing to reproduce the conditions of its own existence, while more centers in Jalisco and further abroad were adopting the circular architecture and its corporate rituals (Beekman 2007). Smaller communities were probably swapping descent group ceremonialism for community rituals, and community or even polity membership may have successfully replaced lineage membership as a primary axis of individual identity. The existing chronological data thus show a different trend from the spatial data, and further temporal refinements will be critical for testing these proposals.

DISCUSSION AND CONCLUSIONS

Richard Blanton's research has provided archaeologists with useful tools for the study of political organization. I have chosen to highlight here distinctive contributions of the corporate-exclusionary continuum and collective action theory, both of which continue to be fruitfully tapped by archaeologists (Carballo 2013; Daneels and Gutiérrez Mendoza 2012). Each approach was developed to further the investigation of political authority and in particular the limits on that authority, though they draw inspiration from different disciplines and different theoretical bases, and may be most useful when applied to different scales of analysis and using different types of data. In this case study, the association of different political interests with archaeologically recognizable spaces provided the entry point for a more in-depth analysis of the field of power as a component of the broader landscape. Blanton's theoretical research on local strategies was combined with his interests in survey and regional settlement patterns for a more rich understanding of political action and its spatial variation. My foray in this direction was limited by the weak chronological and spatial data set for central Jalisco, but portrays the southern Tequila valleys as a strongly corporate system during the Late Formative/ Early Classic when evaluated by forms of built space that structured political interaction rather than by entire sites.

The corporate-exclusionary continuum has been applied by many (including myself [Beekman 2000]) at the scale of entire polities and societies, but the authors of the original model perceived that what was critical was the separation (be it temporal or spatial) of strategies. Tying the strategies to Bourdieu's field concept highlights the different pathways to power within polities, and not just between them. The guachimontón temples and the often appended ballcourts were the most strongly dedicated of built spaces to the enactment of inclusive community ceremonies. The ceremonies that can be associated with each of these architectural forms celebrated maize and agricultural success, reenacted cosmic myths, and built community through feasting. Descent groups or individuals would have had opportunities to enhance their status, but the overarching setting in which the activities took place would have dampened or suppressed most attempts to capitalize upon a well-performed ceremony or deftly executed ball play. The primary occasion on which a descent group could openly express their own importance and dramatize their status and accomplishments was in public mortuary ritual. Funeral processions appear to have passed through the patios of the circles on the way to their destination, so the *total* separation of these spaces is impossible. And we must note that the circles and ballcourts could encompass a degree of conflict when the actions of individuals were acknowledged within the more group-oriented rituals, just as teams and individual players each receive their accolades in modern sports. Further work with the corporate-exclusionary model will need to evaluate not only the relative presence of each strategy but also the number of pathways present that make use of one or the other, as this internal heterogeneity is a defining component of complex society (Ferguson and Mansbach 1996; McGuire 1983).

Translating these interpretations to the regional settlement pattern data identified clear variation on the political landscape that may be explained using insights from the collective action perspective. The polity based in the more corporate southern Tequila valleys was probably associated with a maize-centered subsistence economy that provided the primary source of revenue for the Teuchitlán principals. This *internal* source of funding should have made elites less independent of commoners and required principals to reciprocate with more in the way of public goods, such as feasting. The peripheral and more lightly populated corners of the Tequila valleys provided opportunities for more exclusionary strategies that aggrandized descent groups and allowed transformation in the norms of the south. Following one of the key insights of collective action theory, it is likely that there were differences in the form of resource ownership and land tenure between these areas that made them *external* sources of revenue—elites had in essence succeeded in redefining the boundaries of the political field and its basis in capital. With access to resources that provided some measure of freedom from commoner influence, elites would have had greater opportunity to aggrandize family lines, build large residences, and so on. Ongoing surveys in the northern and western valleys promise to allow testing of these interpretations in the not too distant future.

Acknowledgments. I would like to thank Lane Fargher and Verenice Heredia for their invitation to participate in this homage to the work of Richard Blanton. I would also like to thank Rich for his stimulating contributions to the discipline, whether or not he entirely agrees with the way I have chosen to use them. Lane, Vere, Rich, and two anonymous reviewers graciously provided advice that improved this chapter. This chapter draws upon research supported by the National Science Foundation, the University of Colorado Denver, and the Foundation for Ancient Mesoamerican Studies, Inc.

NOTE

1. "Corporate groups" would be a better term to encompass those groups formed through alliance as well as descent (Joyce and Gillespie 2000), but this would only cause confusion here as corporate groups do not equate with corporate strategies.

4

Complexity without Centralization

Corporate Power in Postclassic Jalisco

Verenice Y. Heredia Espinoza

Anthropological theory has turned attention from kings, palaces, and flashy remains to confederations, councils, and public architecture: in essence, more collective forms of governance to explain higher levels of integration and complexity (Blanton and Fargher 2008, 2016). These nuanced approaches have moved us beyond traditional theoretical perspectives on cultural evolution (e.g., Marcus 2008) and pointed to ways in which human agency (see also Brumfiel 1992) and the dynamic nature of social evolution combine to form a continuum of political strategies in the creation of nonhierarchical, decentralized, and more egalitarian social formations in complex societies (Blanton et al. 1996; Crumley 1995).

More than two decades ago, Blanton and colleagues (1996) brought to our attention a nontypological and nonlinear way of explaining social complexity. They presented two different complementary and non–mutually exclusive strategies used by social actors to accomplish political and economic goals, namely network and corporate strategies (Blanton et al. 1996). Since then, Blanton (1998a) continued to focus on corporate strategies such as those that suppress personal aggrandizement, including community-wide cooperation and participation, which result in more egalitarian political structures where social mobility and commoner power are possible (see also Blanton and Fargher 2008; Fargher, Heredia Espinoza, and Blanton 2011). Recently, Blanton and Fargher (2008) introduced collective action to archaeological theory. Collective action

DOI: 10.5876/9781607325338.c004

focuses on explaining variability in the organization of states as an outcome of the negotiation between rulers and ruled (Blanton and Fargher 2009:134). This new perspective shifted our attention from kings, palaces, and despotic governments to political strategies and social processes that may result in governments built around cooperation. Jalisco, an under-theorized area in west Mexico, is benefiting from this theoretical maturation.

In this chapter, I use these new developments in anthropological theory to explain the political organization of the Tequila valleys of Jalisco during the Late Postclassic. According to ethnohistory and archaeological data, multiple ethnically and linguistically diverse peoples established independent polities in the area. State formation during this period was in part a response to pressure from powerful groups, such as the Tarascans and Caxcanes. Yet, despite heightened tension and common threat, Postclassic Tequileños never unified into a single polity. Poor understanding of the nature of these societies has led scholars to reduce them into already constructed categories (e.g., bands, tribes, chiefdoms, states) or into descriptions from other regions of Mesoamerica (e.g., *altepetl*). Therefore, in this chapter, I take a first step toward filling this lacuna by adopting a more processual approach that pays special attention to political fragmentation at the regional scale and inclusive political strategies, along with the development of sequential hierarchies and corporate ideologies, at the intrasite scale. This approach lays a foundation for explaining the independent nature of these polities and at the same time their alliances in the face of common adversaries and threats.

POLITICAL ECONOMIC STRATEGIES AND POLITICAL ORGANIZATION

Constant conflict and warfare reigned at the dawn of Spanish incursion in the Tequila valleys. Polities built temporary alliances among themselves to protect their territories, but they never unified into a single political structure; instead, they retained their political autonomy. Caxcanes and Tarascans had surrounded the valleys and, as the early documents suggest, they made several attempts to take control over this highly coveted region. How did these societies manage to stand off their enemies while at the same time preserving their autonomy rather than merging into a single polity?

In response to external military and political pressure, the political architects of the Tequila valleys could have deployed network strategies to recruit the followers necessary to maintain their independence (Blanton et al. 1996). In network strategies, individuals seek to concentrate power, authority, and riches in particular families by manipulating wealth generated through long-distance exchange (commercial, political, etc.). Wealth accrued through such trade is used to recruit clients locally via elaborate consumption, gifting, and feasting, accompanied by patrimonial rhetoric. An aura of tension and competition holds sway in these cases since there are multiple competing groups fostered on personal relationships and gift-giving. In this scenario, alliances are highly volatile between people or groups; in addition, ties among the population within these polities are weak, because only a portion of the population benefits from such relationships. The overall populace has little or no social mobility and their demands can be ignored because the power and authority of leaders are grounded elsewhere or depend on external revenues (see Blanton and Fargher 2008 for a discussion of the impact of external revenues on political organization). This results in little incentive for elites to provide for voice and public goods in an effort to obtain compliance from a broad constituency (e.g., free peasants). Rituals emphasizing community integration are also absent, hence large and accessible plazas that may suggest community-wide gatherings do not figure into the overall configurations of towns and cities (see Fargher, Heredia Espinoza, and Uriarte Torres 2014). The result is a politically fragmented landscape that is in some ways analogous to the Tequila valleys during the Postclassic, as ethnohistoric sources suggest (see below). Yet, evidence of prestige goods and patrimonial rhetoric in the Tequila valleys during the Postclassic is fleeting, suggesting that network strategies were not the basis of polity-building during this period.

Conversely, people in the Tequila valleys could have responded to the external threats by using collective strategies to unify the polities in a single powerful state with the capacity to mobilize the entire population for military defense, like their contemporaries in Tlaxcallan (Fargher et al. 2010; Fargher, Heredia Espinoza, and Blanton 2011). In Tlaxcallan (the prehispanic name of the state of Tlaxcala, Mexico), an ethnically diverse group responded to external military pressure (from the Aztec Triple Alliance) by building an inclusive state grounded in collective action. In previous publications, my colleagues and I (Fargher et al. 2010; Fargher, Heredia Espinoza, and Blanton 2011) argued that Tlaxcallan was built around cooperation in which the state bureaucratized to accommodate taxpayer voice and offer public goods, especially military leadership, in exchange for revenue payments, especially military corvée, by commoners. However, the state went further in order to build the trust and confidence among Tlaxcaltecans necessary to cultivate compliance with their revenue demands. The state's political architects instituted a powerful egalitarian or corporate ideology that rewarded service to the state, especially in warfare, with promotion to "noble" status and political offices (including, possibly, a seat on the ruling council), as well as implementing a strict and austere moral code that governed the behavior of officials (cf. Fargher, chapter 15, this volume). Religion was also employed to justify and reinforce the merit-based egalitarian ethos, emphasizing Tezcatlipoca, a deity associated with merit, judicial equality, and self-sacrifice regardless of social position (Fargher 2012; Fargher et al. 2010). Building on this corporate ideology, the Tlaxcaltecas created a unified political structure based on a ruling council that consisted of somewhere between 50 and 250 officials, who made consensus-based decisions and were promoted on merit (Fargher et al. 2010:238).

Much like Tlaxcallan, wealth and prestige goods were deemphasized in the political strategies employed in the Tequila valleys during the Postclassic; yet, the Tequila valleys remained decentralized and the political landscape was fragmented, which was very much unlike Tlaxcallan. Thus, I posit that an additional pathway to complexity was followed in Tequila that differed from both network (or external revenue) strategies and collective action based on internal revenues, bureaucratization, voice, and public goods. Here, the multiple multiethnic groups maintained their political autonomy, yet they created short-term alliances to defend their territories. Specifically, I contend, based on the archaeological record and ethnohistoric information, that Late Postclassic political architects employed a form of sequential hierarchies and powerful corporate or egalitarian ideologies to build strong polities able to withstand common foes, retain political autonomy, and limit personalized power based on monopolies over long-distance exchange and ideological resources. They built a horizontal and highly collective political structure, but unlike other states high in collectivity, they did not assemble large amounts of revenue and invest in major public goods and/or a complex bureaucratic apparatus.

CORPORATE POLITICAL STRATEGIES

In the original formation, corporate strategies are those where aggrandizing behaviors are suppressed in favor of more egalitarian or group-oriented dimensions (Blanton et al. 1996). In these strategies, images of individuals are obscured and/or shown in groups or crowds. Depictions of specific personages that could be identified with a name are absent and instead groups or individuals are depicted, if at all, wearing masks. Writing, where present, tends to be dedicated to keeping records of transactions, rituals, and other important events and it is not used to document the deeds of specific individuals.

Corporate strategies used by political actors are implemented to create community-wide integration and cooperation, and can be recognized archaeologically by investment in public ceremonial and civic architecture, an emphasis on fertility rituals, community feasting, and the minimization of individualizing behaviors (Blanton et al. 1996; see also Blanton 1998a). Thus, palaces, personal monuments, elaborate tombs, and control over prestige goods are masked or absent.

As Fargher points out (chapter 15, this volume), corporate strategies are one tool that may be used in building collective states, but the institutional plan varies among polities. In the particular case of the Tequila valleys, corporate strategies involved power-sharing (probably in the form of consensus decision-making) and an egalitarian ideology but not heavy investment in the development of institutional infrastructure. Instead, I posit that consensual decision-making was achieved through the development of sequential hierarchies and an ideology that downplayed the role of wealth accumulation in gaining political prestige and power.

Sequential Hierarchies and Corporate Strategies

Johnson (1982, 1983) pointed out that the information stress created by the nucleation of large numbers of individuals within a single site or polity could be solved through heterarchical strategies, such as sequential hierarchies, instead of developing more centralized and hierarchical structures through either network strategies or bureaucratization. The development of sequential hierarchies involves grouping individuals or households into larger and inclusive decision-making units (e.g., complex households, clans, moieties, etc.) thereby diminishing their number and allowing for more horizontal decision-making (e.g., power-sharing or consensus decision-making). As opposed to dealing with a myriad of small units, the reduced number of larger aggregations decreases the number of individuals with whom consensus must be reached and, thus, decision-making stress is scaled down. At the same time, these larger units are horizontally constructed with unranked representatives who possess equivalent decision-making authority. They are also responsible for ensuring that their unit conforms and cooperates with decisions reached in council.

Sequential hierarchies have been documented in a range of societies from simple to complex (Bargatzky 1995; Johnson 1982, 1983; see also Kowalewski 2006 for similar patterns). For example, the Huron and Iroquois used sequential hierarchies to build confederacies. Huron councils originated informally from extended families. At the village scale, Huron ruling councils were built of the heads of the long houses (extended families) and larger political agglomerations (e.g., communities and peoples) were formed through supravillage councils made up of representatives from each member village. In both village councils and supravillage councils, decisions were reached through voting and consensus (Trigger 2002:89–91). The position of chief was hereditary and within a community he was "recognized as the principal chief and spokesman for the entire community" (Trigger 2002:87–88). Two types of chiefs existed among the Huron: civil or peace chiefs and war chiefs. Only the former were members of the council and hence of the national and confederacy levels. Civil chiefs were selected from their own lineages, and their positions were inheritable within a lineage. Their main duties were to act as speakers or chairmen: they assisted their group to reach agreements and represented them in foreign affairs (Trigger 2002:81). Although some chiefs were more prominent than others, there is no indication that they had more authority than the rest. In fact, chiefs could not coerce or force their will on anyone (Trigger 2002:84), their duties resided in being the spokesmen of their people. Given this nested political structure, all clan segments had representative chiefs at the Huron confederacy level.

Among the Iroquois, the structure of the confederacy of 50 sachems representing the six nations was formed through "nested hierarchies" that strongly resemble sequential hierarchies. At the very base of this political structure were households that made up tribes. A group of tribes (e.g., Onondaga, Seneca, Mohawk) constituted a nation. Tribes selected sachems, which then represented their nation and group decisions at the confederacy level. Each sachem possessed the same powers and shared equal rank and authority as all the other members of the confederacy despite the uneven number of representatives for each nation. The sachem office was hereditary within the tribe, but elective among the male members of the tribe itself (Morgan 1901).

Finally, the Cheyenne are another example of government by council where the council of 44 peace chiefs was established to prevent internal disagreement and to solve matters of war (Hoebel 1988). Those chiefs selected for the council possessed certain qualities, which included "good temper, generosity, energy, courage, altruism, wisdom" (Hoebel 1988:43). Chiefs selected for the council were experienced warriors who upon selection into the council became peace chiefs, losing their position (but not their memberships) as war chiefs (Hoebel 1988:43). Hereditary factors did not play an important part as a requisite in the selection process, but frequently, after a 10-year term in office, a peace chief could choose a son as his successor (Hoebel 1988:51). A chief was both a representative of his extended family and band; first and foremost, however, as member of the council he acted as protector of each member of the tribe (Hoebel 1988:43). Above all, the council made decisions such as "camp moving and tribal war polity" and in some cases (e.g., criminal acts) it acted as a judicial body (Hoebel 188:52). Decision-making was reached by prolonged debate until agreement was attained. All peace chiefs were equally ranked and as a council their authority was invested with supernatural skills, making it the supreme authority (Hoebel 1988:49).

The aforementioned forms of organization are quintessential examples of successful sequential hierarchies. Most important, this form of organization distributes authority and representation of the population in such a manner that people have a voice in the political decision-making process. Nonetheless, the distribution of decision-making among various segments is not sufficient for maintaining order and cohesion.

Creating a sequential hierarchy does not by itself prevent fissioning, internal conflict, or prestige-good consumption and wealth competition among groups (see for example the societies of the Northwest Coast [e.g., Ames 1995]). As Stark points out (chapter 5, this volume), there is always tension between individuals seeking personal gain (network strategies) versus those who favor group interests and welfare (corporate strategies). Accordingly, individuals seeking personal power may attempt to derail or obstruct consensus-building to thwart collective decision-making. As Blanton and Fargher (2016) point out, achieving cooperation in human societies is extremely difficult and designing functional governing structures based on collective action may be the most difficult challenged faced by humans as a species.

In order to control such behaviors, political architects must support sequential hierarchies with corporate cognitive codes. Such codes function to reinforce group identity and an egalitarian ideology that suppresses prestige goods and network strategies. However, once codes are in place, ceremonialism and ritual activity become essential for group maintenance and cohesion. In situations where leaders seek to integrate multiethnic groups into larger aggregations, such as was the case in the Tequila valleys, a common symbolic language that would cross-cut linguistic barriers and allow integration and a shared identity must be developed (see Kowalewski 2006). Ritual, in these situations, plays a significant role in creating group cohesion and allowing corporate cognitive codes to support a "corporate political structure" (Blanton et al. 1996:2).

Corporate cognitive codes also emphasize reciprocal obligations among groups of distinct origin through the enactment of rituals that "transcend local belief systems" (Blanton et al. 1996:4). Integration of different units requires the creation of new myths and spaces for the enactment of these rituals. Sizable plazas capable of bringing together large numbers of people—or conversely, multiple plazas—are indicative of such collective events (for examples, see Fargher et al. 2011; Fargher et al. 2014). The sizes of plazas or the number of them may indicate such collective codes, and public architecture, such as certain monuments, may also function as an integrative mechanism. Connectivity between different architectural groupings may also be indicative of such codes (e.g., Blanton and Fargher 2012; Fargher et al. 2014). Emphasis on shared identity through these communal events would have served as a key ingredient for the construction of strong polities.

A corporate cognitive code is also vital to restrain the emergence of exclusionary strategies such as aggrandizement, massive and luxurious tombs or burials, or monuments in honor of particular individuals or dynasties (Blanton et al. 1996:6). Such codes allow power-sharing strategies "across different groups and sectors of society" (ibid. 1996:2; Blanton 1998a:158, 159), thereby inhibiting the power over decision-making by a particular group (see also Fargher, chapter 15, this volume). Representatives of the different units, and hence the people represented, would have voice in the decision-making processes. Reduced wealth disparities are also consistent with a corporate cognitive code, where significant hoarding of material riches and their use as political currency is controlled. A corporate cognitive code and egalitarian ideology thus create unity in groups that cross-cuts ethnic and linguistic boundaries. This must have been extremely advantageous in times of warfare. Mobilization for military defense would have profited from these egalitarian strategies and ideologies as they effectively functioned to congregate an army when needed (cf. Kowalewski 2006).

Data Collection: Ethnohistory and Archaeology

A full-coverage systematic survey north of the Tequila volcano provides important data complementing the ethnohistoric information available on political organization. Using a modified version of the regional survey developed in Central Mexico and Oaxaca highlands (Kowalewski et al. 1989; Kowalewski et al. 2009; Sanders et al. 1979), the Tequila survey recorded all evidence of human occupation from small artifact scatters to sites covering hundreds of hectares. Our methodology consisted of walking every field with a distance of 20–40 m between surveyors. Site limits, identified by the fall-off frequency of artifacts and architectural limits, were drawn on the aerial photo. Our collections were of two different kinds: grab bags, or general collections, and specific collections. General collections were usually taken at small sites that we recognized as single-period occupations. Specific collections were taken at exact locations within a site such as structures, terraces, and other prominent features in order to provide accurate information on site components.

RESULTS

Ethnohistoric Data

The cultural history of the Tequila valleys of Jalisco can hardly be framed in simplistic diachronic terminology where early (or simple) societies gave way to more complex social formations. Here we see periods of stability and continuity, highlighted by rupture, coalescence, and diversity in the way that societies formed and were created at different points in time. This is the case for the Late Postclassic, when multiple independent, ethnically and linguistically distinct polities occupied the Tequila valleys (Baus de Czitrom 1982; Soto de Arechavaleta 1994; Weigand 1993:136–137; Weigand and García de Weigand 1996:31–32). Ethnohistoric documents mention Cocas, Tecuexes, and Cazcanes among those ethnic groups (Acuña 1988; Baus de Czitrom 1982). Linguistically the region was also diverse and Cora, Coano, Cazcán (closely related to Nahuatl), Coca, Vitzurita or Wixárika, Tecozquín, and Nahuatl were spoken in the region (Yáñez Rosales 1994, 2004:85–86). In addition, Caxcanes (and other Chichimec groups) and Tarascans were encroaching on the valleys and, as the early documents suggest, both groups made several attempts to take control over this highly coveted region (Coria 1937:558; Weigand 2013:59–61).

Ethnohistoric data on the polities north of the Tequila volcano are paltry and dispersed in various documents. The Spanish understated the political organization of the region and descriptions are superficial, mixing terms such as *señor, calpixque,* and *cacique*, among others. Nonetheless, it is evident that not all towns shared a single organizational structure, and the documents also mention the existence of several polities, which indicate political fragmentation, decentralization, and the presence of many rulers. In some cases, sources (Coria 1937:559) document the presence of two *calpixque. Calpixque,* however, were not heads of towns such as a *tlatoani;* yet the presence of two in one town, each from a distinct ethnic group, points to some sort of power-sharing, where each ethnic group had a representative in government affairs. This information indicates that each barrio or small settlement had a representative who ruled together and/or formed short alliances with his peers. Etzatlán, in the southwestern Tequila valleys, was a province (*provincia*) headed by a *señor* and organized "*a barrios*," each with its own representative (Weigand 2013:33–34). Weigand (2013:31) considers significant the fact that Etzatlán was organized "*a barrios*" and not "*de barrios*," the difference being that the use of *a barrios* signifies that the barrios were not continuous, but rather dispersed and *de barrios* refers to a single settlement subdivided in neighborhoods (Weigand 2013:31). A recent regional survey in Etzatlán recovered settlement-pattern data, indicating a discontinuous settlement rather than a centralized capital (Heredia Espinoza 2014).

Finally, a ceramic stamp, from Etzatlán, portrays two defleshed faces or skulls that may represent regional variants of Xipe Totec and Tezcatlipoca, respectively (Weigand 2013:28). Tezcatlipoca has been suggested as an important deity in the region (Weigand and García de Weigand 1995:23, 65–66), perhaps originating in the Classic period. He is closely identified with the desert north and nomadic or seminomadic groups, such as the Caxcanes, who were encroaching on the Tequila valleys sometime before the Conquest. Caxcanes spoke some sort of corrupted Nahuatl that was intelligible with Nahuatl from the Basin of Mexico (Yáñez Rosales 1994:60). Nahuatl has been identified as one of the many languages spoken among the resident populations in the Tequila valleys (Baus de Czitrom 1982:25). As I noted above, Tezcatlipoca is associated with merit, judicial equality, self-sacrifice regardless of social position, and egalitarianism. He is associated with Atecpanecatl, the "destroyer of royal privilege," in opposition to Quetzalcoatl, who is associated with royal lineages and privileges (Fargher et al. 2010:241; Ringle et al. 1998). As such, the ideological association with Tezcatlipoca favored social mobility over injustice, aggrandizement, and strong hereditary power. Therefore, the promotion of his cult and associated ideology would have proved the basis for the type of egalitarian corporate cognitive code that I suggest was a key aspect of the region's political organization.

ARCHAEOLOGICAL DATA

Settlement Patterns

In an area of 463 km², we recorded 242 sites that date to the Postclassic (figure 4.1), and 119 sites with standing architecture. Although population densities may have been lower here than in Central Mexico or Oaxaca, a conservative population estimate suggests that tens of thousands of people occupied the Tequila valleys at the Conquest (Weigand 1993:130) and nearly 15,000 lived in the area surveyed (including Santa María but not Etzatlán) based on a conservative figure of 17.5 persons per hectare, but debate exists on how

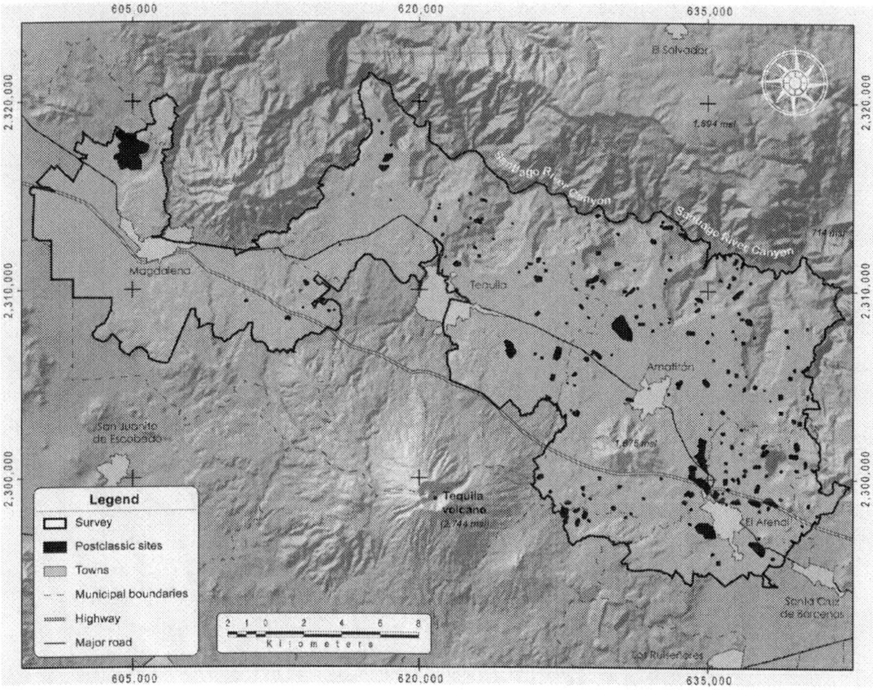

Figure 4.1. *The central valleys of Jalisco, showing the area surveyed, settlement patterns for the Postclassic, as well as various localities mentioned in the text.*

densely populated these valleys were at the time (Weigand 1993:128–130). An increase in number of settlements and in the overall size of sites from previous occupation phases along with data on residential architecture and domestic debris are in line with these observations.

Late Postclassic sites are located on hills, *lomas*, and along the Río Santiago canyon (figure 4.1), but the largest centers and political seats tend to cluster around the natural entryways into the valleys and toward the northern piedmont of the Tequila volcano. The canyon, which limits the valleys to the north, was occupied with small sites on hilltops and lomas, but no large site was recorded in this subregion. To the northwest is Santa María and to the southeast and into the Atemajac Valley is a cluster of four large sites. Etzatlán and Tala are mentioned in ethnohistoric documents as important centers in the Late Postclassic (Coria 1937; Tello 1968). The Tarascans ravaged Tala just 50 years before the Conquest (Tello 1968), and it was located on the southeastern entry. Etzatlán on the other hand, was located on the southwestern passage. Recent archaeological survey identified a large and sprawling site on a piedmont ridge above (south and west of) the modern town of Etzatlán that was probably the settlement of the same name mentioned by the Spanish. All of these are strategic locations for vigilance and to control access, themes of primary importance during this period.

Based on site size as a proxy for the political administrative hierarchy, I have identified at least four settlement tiers (figure 4.2). The first tier consists of one large site (Santa María) at the northwestern corner of the survey area (212 ha). The second tier is composed of six sites that range from 17 ha to 75 ha, located in the southern half of the survey region, including four clusters of sites in the southeastern corner. A rough population estimate of 10 to 25 persons per hectare indicates that together these centers housed approximately 7,000 individuals. The third tier consists of secondary centers ranging in size from 4 ha to 16 ha (10 people per hectare). The lowest level consists of dispersed rural settlement, artifact scatters, and isolated residences (5–10 people per hectare). Political seats are surrounded by secondary centers and rural settlement in compact clusters separated by shatter zones, or scantily occupied areas (figure 4.2).

The settlement pattern, evident in the map, suggests a highly decentralized or fragmented landscape (figure 4.2). There are two shatter zones, suggesting the presence of multiple polities (possibly three). In the southeast, four sites are clustered together, suggesting one polity, then a shatter zone of 6.5 km separates this cluster from another with two large sites at its center that may also represent another polity. The latter cluster is located near the center of the survey area. In the northwest, Santa María stands as a third polity. Unlike the two other polities, settlement at Santa María is highly nucleated, nearly everyone lived in the capital and there was little or no rural settlement. This site corresponds with the settlement of Xochitepec (home of Guaxicar), now called Magdalena, described in ethnohistoric documents (Tello 1968:128). South of Santa María at the southern end of the Magdalena Basin, the archaeological site of Etzatlán forms the seat of yet another autonomous polity. Thus, I interpret these settlement clusters, divided by shatter zones, as politically independent units, which concurs with the fragmented political landscape described in early colonial documents (summarized in Soto de Arechavaleta 1994:346).

At 212 ha, Santa María is the largest Late Postclassic site. Based on survey data, it functioned as a political seat in the northern Magdalena basin. Only two small sites (< .4 ha) northeast of Santa María were recorded in this portion of the valleys, indicating a strongly nucleated, primate settlement pattern for the polity. This cluster is separated from the nearest cluster to the east by a significant linear distance of 25 km.

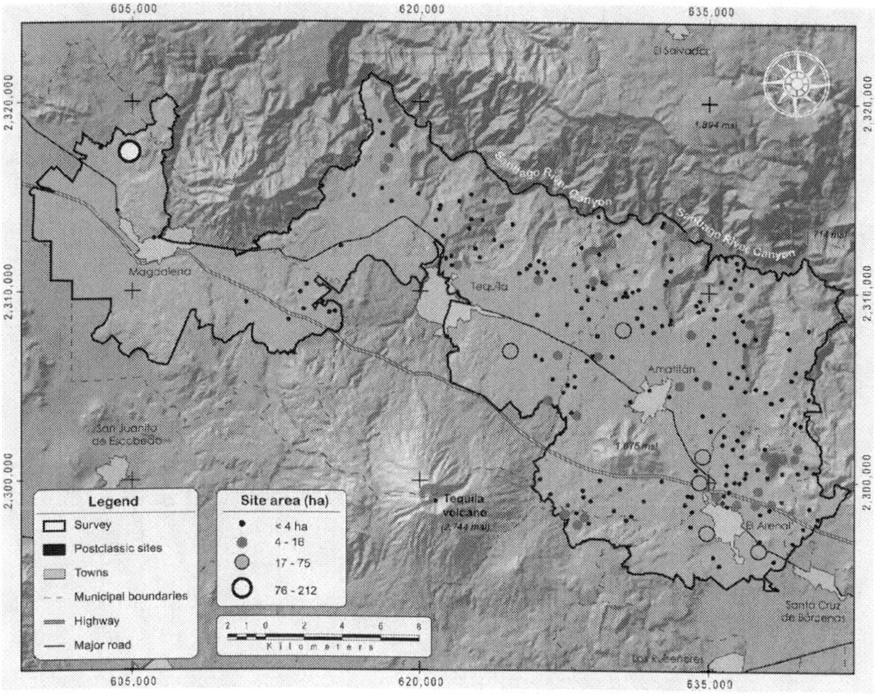

FIGURE 4.2. *Postclassic settlement tiers, based on site size.*

Of the six remaining large sites located north of the volcano, three of them do not have standing architecture due to their location in agave fields and toward the center of the valleys. It is my contention that these places were similar to Santa María, but mechanized agriculture has done away with extant architectural remains. This conclusion is based on our field observations, which recorded both construction stone and domestic debris on these sites. Besides Santa María, the only other primary site with standing architecture is located in the southeastern corner of the survey area on a smoothly sloping loma. Only two large residential platforms, superstructures, and stone alignments were visible on the surface, significantly less impressive than Santa María (see below). The loma has been affected by the contemporary urban growth of town of El Arenal (figure 4.1) and most of the site is currently planted with agave. Therefore, the possibility that many more structures existed is very likely. All of these primary sites are located within four to eight linear kilometers of one another, indicating dense populations in a politically decentralized landscape.

Intrasite Architectural Arrangements

Previous studies in various regions of Mesoamerica have tested the degree of centralization and differing political strategies through architectural spatial configurations at the site level (Blanton 1989; Blanton and Fargher 2012; Fargher et al. 2011; Fargher et al. 2014; Heredia Espinoza 2007). Generally speaking, network strategies (Blanton et al. 1996) are correlated with the presence of large palatial edifices, as well as small and enclosed (private) plazas that are, at times, directly associated with civic-ceremonial spaces. Conversely, corporate strategies are often linked with open and accessible civic-ceremonial architecture; the presence of multiple, equally ranked public spaces or civic-ceremonial architecture point to community integration at multiple levels and power-sharing strategies (see Beekman, chapter 3, this volume; Fargher et al. 2011).

For major sites in the Tequila valleys north of the Tequila volcano, it is important to identify those potential spaces where public events would have taken place as well as their location within a site, since ritual is an important aspect in the organization of sequential hierarchies. Plaza sizes are also important because they indicate the potential number of people who could have participated in various gatherings (e.g., large for public vs. small for private events). At Santa María, plaza sizes range from about 125 m^2 to 835 m^2. Some plazas are not formal, meaning they are open spaces that do not have walls nor are they delimited by buildings on all or some sides.

For the purposes of this section, I concentrate mostly on Santa María, which, I argue is exemplary of the architectural arrangement of polity capitals, and where archaeological preservation is optimal. A recurrent residential architectural form, known as *corrales,* prevails in most Postclassic settlements. As yet, there are no excavated *corrales*, but we know that they had a limited distribution following the canyons to the north (Weigand and García de Weigand 1996:51). During salvage work at Bugambilias, in the Atemajac valley to the east, Galván Villegas (1983) recorded several of these structures; and Hers (1977) reports on them in the Sierra del Nayar in Nayarit.

At first glance, corrales seem to conform to a regular construction style; two rows of stone blocks parallel to each other (between 40 and 60 cm apart), forming a rectangular C; but the survey recorded more information on their varied morphology. For example, corrales can be built directly on the surface or atop platforms and terraces. Most have a single entrance, but others have two or three entrances and staircases/steps, and therefore are of substantial size. Weigand and García de Weigand (1996:61) had already noted somewhat of a hierarchy of corrales in this area, yet they seem to be built following general guidelines. We think they were either special-function architecture or the residences of principals within a settlement (as defined by Blanton and Fargher [2008:19], principals are "the chief decision-makers and policy-makers of a polity). Sizes of corrales range from 12.5 m^2 to 772.8 m^2, following a lognormal distribution. The geometric standard deviation (1.852 m^2) indicates significant variation, and given the geometric mean (56.023 m^2), there are some very large corrales that appear to be outliers. The largest corrales are found in primary centers. Differences in size may also be indicative of stratification within the site such that larger corrales housed important, perhaps extended, families, while smaller structures served as residences for commoner households, possibly attached to the larger corrales. The importance of corrales to this discussion lies in the fact that this type of architecture is ubiquitous throughout the survey region, and it is to some extent a diagnostic feature of the Late Postclassic period.

In order to understand the architectural spatial patterning at Santa María, I drew Thiessen polygons between all standing residential architecture greater than 8m^2, which allowed the easy identification of architectural clusters (figure 4.3). The polygons indicate that architecture clustered in groups consisting of several structures, small plazas, mounds, large corrales, and other features of unknown function. These architectural groups tend to be spread evenly across the site, pointing to a decentralized patterning within the site where several formal arrangements can be detected. These architectural clusters are the material representation of larger units (e.g., such as groups of households or neighborhoods).

Next, using structure areas as a proxy for labor investment (person days), I identified those places within the site that point to high amounts of labor (figure 4.3). Cluster analysis shows several locations where labor investment is highest, indicating that labor-intensive architecture is located in multiple locations within Santa María (a heterarchical pattern in the architecture). High amounts of labor investment concentrate at the highest point of the hill, where I mapped a three-mound group surrounding a small patio atop a high platform (figure 4.4). Terraces and structures run northeast and southwest of this complex. To the west, there is an abrupt slope and from there the view of Laguna Magdalena is impressive. While we collected artifacts on the slopes around this three-mound group, we did not find a single artifact associated with it that suggested a nonresidential or civic-ceremonial function. This architectural group is located on one of the highest points of the mountain and is located in a cluster, which includes a number of large corrales, terraces, and structures. Its location at the center of a neighborhood cluster probably indicates that it functioned as the neighborhood civic-ceremonial space, as

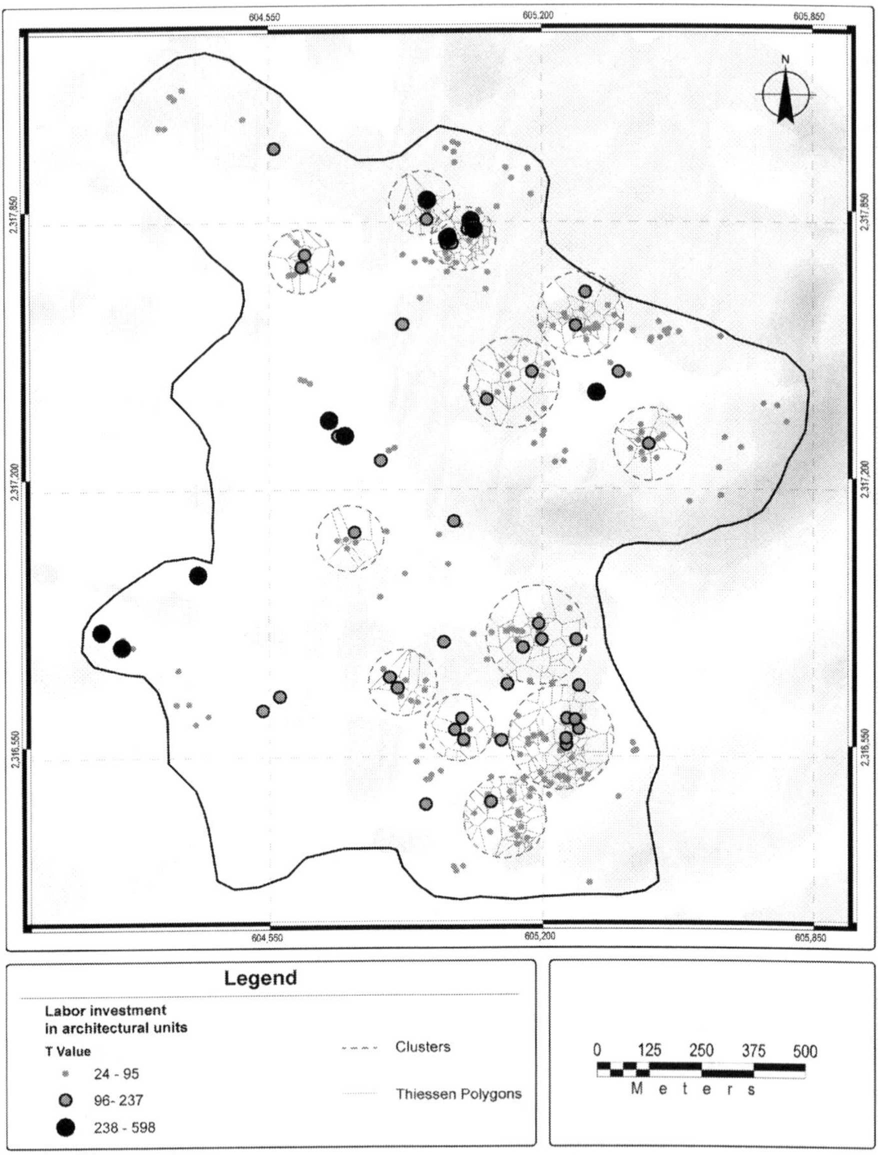

FIGURE 4.3. *Santa María architectural settlement clusters.*

well as the central architectural complex of the site. Hence, this architectural group can be interpreted as a unit formed by both residential and nonresidential architecture (e.g., civic-ceremonial).

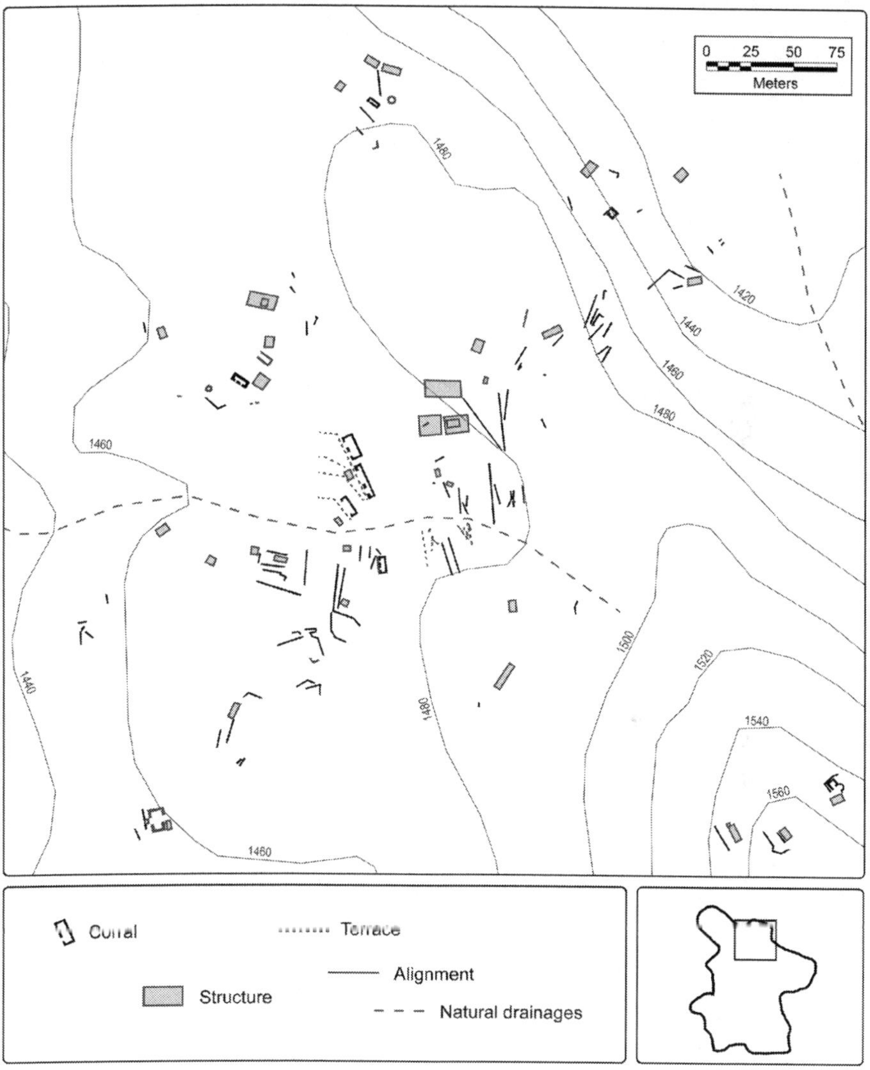

FIGURE 4.4. *Three sections of Santa María, showing various residential groups: highest point of site.*

In the middle section of the site, spatial arrangement of the largest corrales and other residential architecture follows a pattern where open spaces or plazas are surrounded by groups of three or four large corrales, smaller corrales, residential structures, and terraces (figure 4.5). These groupings are separated

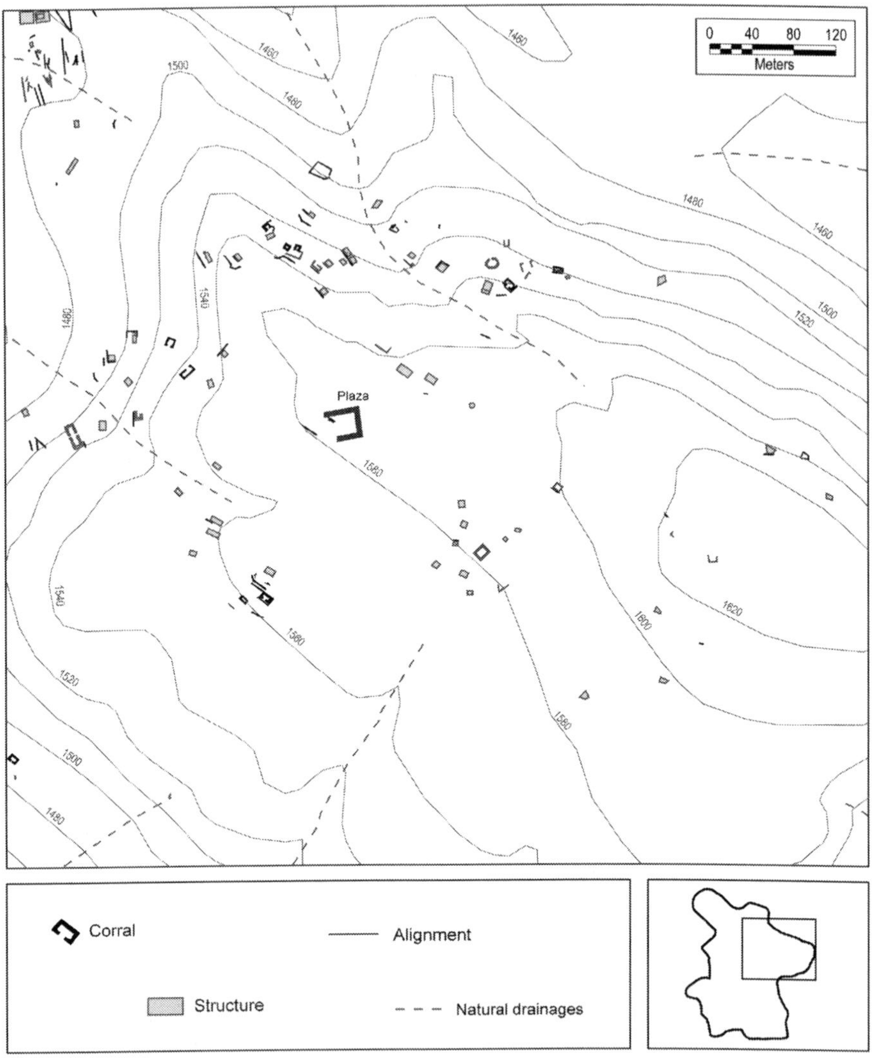

Figure 4.5. *Three sections of Santa María, showing various residential groups: hill.*

by natural topography such as drainages, constituting discrete or separate residential zones. In the northern portion of the site, several discrete groups of residences, divided by natural drainages, are arranged in circular fashion around a large plaza. The plaza is located on an open area high on the hill and at the very center of the above-mentioned residential aggregations. Given that these groups are divided by natural topography, they suggest discrete units (groups of households); thus, it appears that these units shared a large plaza as a common public space. However, no single group was significantly larger or more impressive than any other, suggesting relatively equal ranking among the groups that participated in public events at the central plaza.

On the piedmont, the arrangement varies somewhat, and groupings are not as evident, but again clusters are divided by natural drainages (figure 4.6). In this section of the site, some residences and corrales are arranged in groups of two or three. Spatial proximity to other residences is high, so the pattern is more clustered than on the highest portion of Santa María. A mound associated with a plaza and/or large corrales repeats in various places on the piedmont. Mounds have few to no artifacts on the surface, but fragments of ritual vessels (braseros, sahumadors, etc.) are found in association with these structures; thus I interpret these features as civic-ceremonial architecture. Mounds may also function to separate residential groups or create independent residential groups. In one case, we recorded a large corral and other residences associated with a mound and a small open plaza (figure 4.6). The mound and the plaza do not have any associated artifacts that indicate a nonresidential function, whereas the large corral and the surrounding residential architecture display a high density of domestic debris such as manos, metates, comals, and jars. On a different hillside, a large corral or platform is directly associated with a mound, and a "double U" shaped structure that may have functioned as a public space (figure 4.6). Finally, in another section, a large corral is associated with one of these "double U" structures.

Thus, the entire site displays multiple clusters of equally ranked civic-ceremonial and residential architecture, indicating distinct units based on spatial breaks between clusters. The ancient settlement at Etzatlán suggests variation on this pattern where there is not a single settlement focus and instead people lived in multiple separate clusters of settlement distributed along the piedmont. However, this pattern may be similar to Santa María in that these distinct groups or sites may be equivalent to the clusters. In addition, all large Postclassic sites feature debris patterns that suggest the presence of public architecture, a pattern not seen at smaller sites. Besides the three-mound group mentioned above, there are two large open plazas surrounded by residences, resulting in small groups of structures analogous to neighborhoods. Further down on the piedmont, the spatial structure of architectural arrangements follows a similar pattern; houses are situated in groups of various sizes where large corral structures were recorded, some of which are grouped around patios and yet others were constructed on terraces in the sloping terrain. Thus, it appears that the site

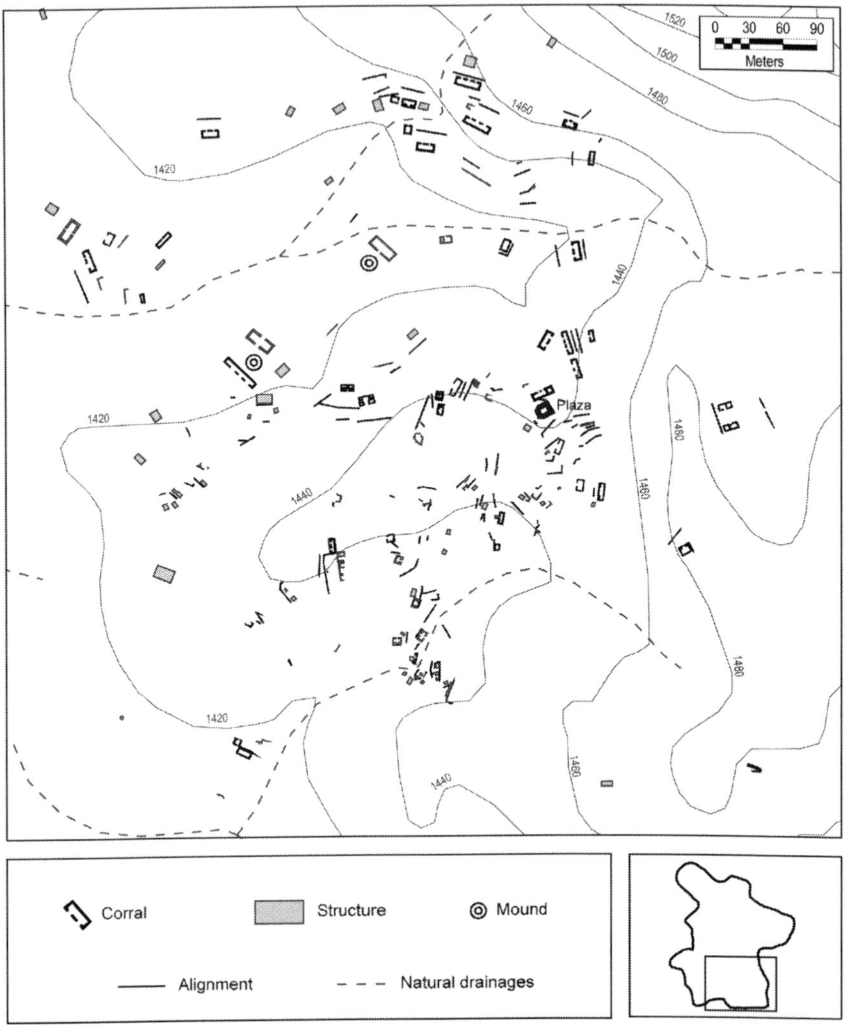

FIGURE 4.6. *Three sections of Santa María, showing various residential groups: piedmont.*

consists of unranked and relatively architecturally similar neighbors characterized by a group of one or a few large houses, small houses, and public spaces.

Spatial Distribution of Artifacts

Our field methods did not implement systematic, controlled collection points of a specific size, but our survey methodology was systematic. We made collections in specific features and in small artifact concentrations within the site in order to have better control over chronological components and over intrasite artifact distributions. Collections included only diagnostic artifacts such as rims, decorated vessel bodies, and figurine fragments. In the absence of diagnostics, and where ceramic densities were too low, we collected undecorated sherds and dated them based on paste attributes. Hence, although a statistical sampling method was not implemented, some generalizations on the distribution of artifacts for the entire survey region and within Santa María as well can be made based on these collections.

Differences in the quality of artifacts in a spatial dimension are one way to identify possible variations in wealth and status. Artifacts, especially ceramic artifacts that display high amounts of labor investment (Feinman et al. 1981; Garraty 2000; Hagstrum 1988) and specialized goods that require complex technologies or secret procedures, can be used as proxies to study differential access to wealth within and among sites. Differential distributions of these goods within settlements as well as among sites (Fargher et al. 2010; Heredia Espinoza 2007; Kowalewski et al. 2008; Uriarte Torres 2011) can effectively identify the degree of difference in access patterns to a diversity of goods and resources as well as to possible discrete functional spaces, and they shed light on the variety of strategies used by individuals or groups.

Postclassic diagnostic ceramics can be reduced, but not limited to, three main forms according to their frequency in our collections: jars, *molcajetes*, and tablets (figure 4.7a, b, c, respectively). Jars vary in size, but in terms of form they do not differ significantly. Typically, they display a distinctive, thickened rim (Beekman and Weigand 2000:68–69) and they tend to be painted in red and highly burnished, although decoration in red and white vertical bands also occurs. Molcajetes are a recurrent form. They are striated on the interior, have supports, and display diverse decorations (monochrome, bichrome, or polychrome) and vary in ceramic fabric texture. Molcajetes are very common and they do not seem to be limited to domestic contexts, as they have also been identified in interments (Glassow 1967; Smith Marquez and Herrejón Villicaña 2004). Finally, tablets are flat and rectangular with raking or striations on the upper surface; their function is unknown. They do not show wear patterns and, given their shape, they were not used for food serving or food preparation. Again, they do not seem to have restricted distribution and they are not confined to a specific context. Surprisingly, bowls and plates, among the most common forms in other parts of Late Postclassic Mesoamerica (Brumfiel 1987, 1991; Garraty 2000), occur rarely in the Tequila valleys' Late Postclassic ceramic assemblage (Beekman personal communication 2012; Heredia Espinoza 2011).

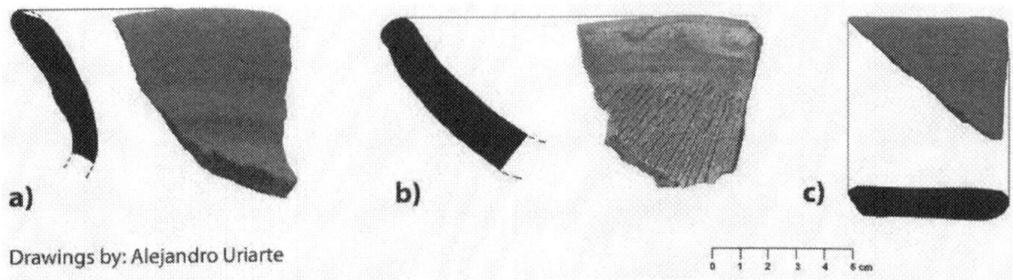

FIGURE 4.7. *Late Postclassic ceramic diagnostic forms: (a) Atemajac jars; (b) Huistla molcajetes; and (c) tablets.*

Overall, frequencies in ceramic forms and ceramic types do not show significant differences across the surveyed region or in areas within sites, suggesting a highly homogeneous distribution. Two forms dominate the entire collection: jars and molcajetes. Together, these two make up 85 percent of the entire assemblage. The proportion of jars to molcajetes is 2:1. Other forms that head the list, but that occur in significant lower proportions, are comals and, still less frequently, bowls/plates.

Competitive and aggrandizing feasting behavior is usually inferred by the presence of fancy serving ware and a wide variety of ceramic forms, which suggest consumption of diverse and elaborate foods, in a few (elite) households (Garraty 2000; Levine 2011; Smith et al. 2003:251). Specifically, the ratio of jars to bowls (11:1) indicates that serving for aggrandizement and political legitimation was not an important pursuit at this time; furthermore there is no evidence that bowls were concentrated in high-status residences (e.g., in large corrales). Cooking and storage were more important and, in the absence of highly decorated or luxury serving forms, the ceramic assemblage is more indicative of communal types of food-sharing behavior. Thus, the near absence of fancy serving vessels, the lack of diversity in vessel forms, and the low frequency of serving vessels, suggests that competitive feasting was limited. I think that such archaeological patterns indicate that cognitive codes downplayed the importance of personal wealth aggrandizement and gifting as a pathway to political power. This pattern fits with the ideological focus on Tezcatlipoca and his cult, mentioned above.

Moreover, two ceramic types dominate the assemblage: Atemajac and Huistla that together represent over 93 percent of the entire Late Postclassic collection. Both types are ubiquitous and represent the bulk of the domestic pottery in all households. Furthermore, these types do not show high amounts of labor investment in decoration, but the Atemajac type tends to be well burnished. Various pastes were used in the production of these two types, but on average pastes seem midway between a fine and a coarse clay (Beekman and Weigand 2000:68), suggesting the use of piedmont clays with little additional labor investment (e.g., they did not levigate and carefully temper clays). Accordingly, Atemajac vessels for the most part are plain, but examples of polychromes appear in small quantities. On the other hand some Huistla type vessels may have fine pastes and bichrome or polychrome decoration (Nance 2013:104–107), indicating the presence of some more finely made dishes. However, in the survey the highly decorated sherds do not constitute a large enough sample to be significant, and they were not disproportionately concentrated in and around large corrales.

Several conclusions can be drawn from the ceramic evidence. First, there are no major differences in types, suggesting a collective form of material culture and perhaps the presence of exchange mechanisms that cross-cut political boundaries, such as a regional market system (*tiangues* are mentioned for several towns in Coria 1937). The similarity in local types with limited decoration also indicates a lack of competitive behavior such as in feasting events, and it also concurs with the presence of markets. The limited ceramic forms indicate standardized food-consumption patterns and again no major differences that may indicate considerable distinctions in behavior (consumption and wealth). Finally, the data described above show that big residential units (e.g., corrales) did not monopolize access to better and fancier pottery or consume significantly larger amounts of pottery, especially serving vessels, than more modest households. The architectural clusters attest to a greater investment in architecture, but not in portable wealth.

SEQUENTIAL HIERARCHIES, COMMUNAL RITUALS, AND EGALITARIAN IDEOLOGIES: A SUMMARY OF THE EVIDENCE

At the regional scale, the spatial distribution of sites indicates a fragmented political landscape, with small-scale, relatively autonomous polities divided by shatter zones. Each polity apparently consisted of a dominant center surrounded by lower settlement tiers. Each center was internally organized around multiple semiautonomous and unranked residential groups or neighborhoods. Each neighborhood seems to have focused on a small plaza and to have shared access to a large central plaza at the center of the site. Thus, the heterarchical organization of residential groups gives the impression that they were integrated through mechanisms analogous to sequential hierarchies and not simultaneous hierarchies. Each residential group in turn had its own civic-ceremonial architecture, where important public ceremonies may have taken place, yet they shared a common plaza accessible to all residents of the site.

Consistent with a sequential strategy, evidence indicates that corporate ideologies downplayed the importance of personal wealth accumulation and networks of power. These corporate ideologies were personified in the worship of Tezcatlipoca and the adoption of his cult, which are evidenced in figures and ethnohistoric documents (in Weigand 2013:28). Furthermore, these corporate ideologies were materialized in artifact inventories, especially ceramics, which were highly homogeneous within sites and across the region. Moreover, elaborate serving vessels comparable to Tlaxcallan polychromes or Aztec III Black-on-Orange occur only rarely in the artifact assemblage and were not disproportionately concentrated in large corrales. Given the political situation in the Late Postclassic, the societies occupying the Tequila valleys adopted corporate strategies and a corporate cognitive code that provided for the effective mobilization of the population for defense against attacks from outside forces (e.g., Tarascans), but limited the degree to which warrior-leaders could monopolize material and ideological resources for their personal aggrandizement and power (see Earle, chapter 14, this volume).

CONCLUSIONS

The introduction of dual-processual theory into archaeological theory has opened a wide door for the explanation of the evolution of societies along multiple paths toward complexity. Research in Jalisco has benefited from these new insights and is already producing important evidence for the coexistence of diverse political strategies at certain points in time (e.g., Beekman 2008; Beekman, chapter 3, this volume). Whereas information was limited in the recent past, we now have assembled an important corpus of data that is throwing additional light onto the ways people organized themselves in the Tequila valleys during the late prehispanic era. Here, I have made a first attempt to characterize the Late Postclassic political organization through regional and intrasite spatial organization, architecture, and portable material culture. Rather than simply fitting these societies into known societal types or political forms elsewhere in Mesoamerica, I ventured to think about how these different ethnic groups organized effectively without centralizing power into a single polity or lineage in response to military threats from the Tarascans and the Caxcanes. The information available indicates an emphasis on differences in organizational strategies, implying an alternative pathway to complexity. In this case, sequential hierarchies proved a useful tool for disentangling these strategies, which combined corporate political strategies and collective action to achieve complexity without centralization. In addition to providing an alternative model for a more egalitarian political structure, sequential hierarchies enrich corporate theory because they help operationalize this theory for archaeological research.

Acknowledgments. This project was supported financially by El Colegio de Michoacán, A.C., the Foundation for the Advancement of Mesoamerican Studies, Inc. (FAMSI, #07012), National Geographic Society (#8711–09), National Science Foundation (BCS-1219619), Consejo Nacional de Ciencia y Tecnología (#131056), Secretaría de Cultura del Estado de Jalisco, and three municipios in Jalisco (El Arenal, Amatitán and Tequila) provided small funds for the project. I would also like to thank a very dedicated group of archaeologists and students of archaeology from different institutions in Mexico and abroad. Thanks to Lane F. Fargher and Christopher S. Beekman, who have been involved at various stages of the project throughout the years and have provided helpful advice in the development of the survey and laboratory stages. Stephen A. Kowalewski and Beekman read an earlier version of this chapter and provided useful comments. However, all omissions and errors are my responsibility.

5

Central Precinct Plaza Replication and Corporate Groups in Mesoamerica

Barbara L. Stark

In a series of pivotal contributions concerning Mesoamerican urbanism and political organization, Richard Blanton (e.g., Blanton 1978; 1998a; Blanton et al. 1993; Blanton et al. 1996; Blanton and Fargher 2008) highlighted organizational diversity in early states. He and colleagues have called attention to the roles of corporate groups in governance and the influence of subjects from the perspective of collective action. In an argument for important corporate groups signaled by central plaza layouts in south-central Veracruz, I offer a somewhat different perspective concerning the relationships between the influential concepts of exclusionary versus corporate power strategies presented by Blanton et al. (1996). My analysis suggests that cycling of dominant principles, which they proposed, while it may occur, must be understood against a backdrop in which governments commonly occupy a middle ground of competing principles. Subversion of dominant power strategies likely is part of ongoing power tensions but does not always shift the dominant emphasis or establish a dominant one.

A convergence of settlement pattern and architectural research focuses on the layouts of central precincts in ancient Mesoamerican centers, where both buildings and the spaces among them can be highly indicative of the social and cultural order. Opportunities for analysis of such layouts have increased with detailed mapping of settlements, but comparative analysis remains spotty. In this chapter, I evaluate persistent,

DOI: 10.5876/9781607325338.c005

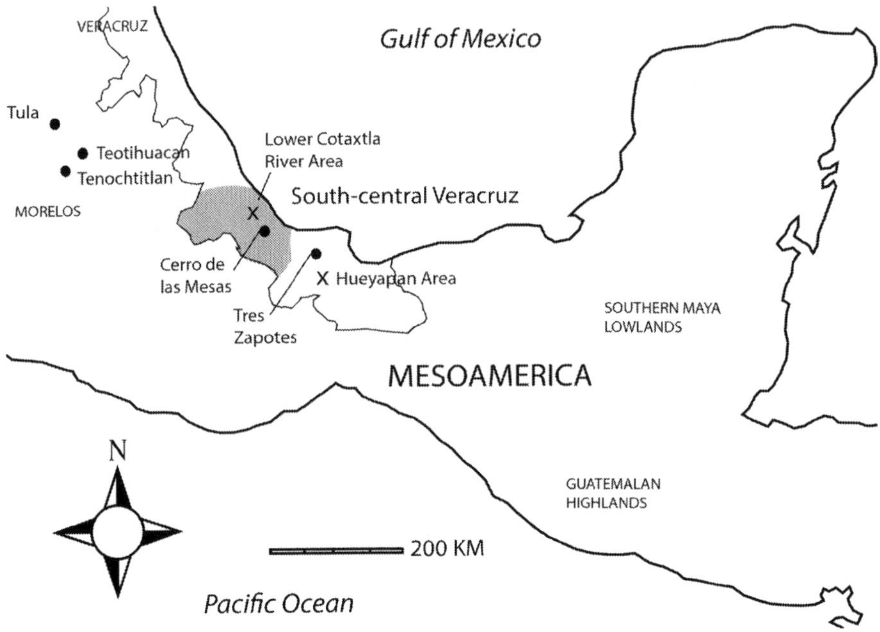

FIGURE 5.1. *South-central Veracruz within Mesoamerica, with selected sites and regions. (The Teuchitlan tradition, located west of the map area, is not shown.)*

highly repetitive arrangements, focusing on a layout in south-central Veracruz (figure 5.1), which I suggest expresses shared power involving both dynasties and corporate groups, and I compare this layout with others in Mesoamerica that have a large sample of central precincts mapped.

There are myriad reasons *not* to expect highly consistent, repeated center layouts. The built environment both shapes human actions and is created by them (Rapoport 1990), a recursive relationship similar to the general theoretical posture of structuration (Giddens 1984). The built environment is flexible, and the architecture of centers is cumulative and remodeled over time. Established forms may persist despite changing significance and activities. Even when buildings and spaces emulate prestigious places, the particular interpretations may be highly selective according to the agenda of the persons involved in planning and execution (Ashmore and Sabloff 2002). Physical constraints of locations affect execution of layouts. Prestige competition may encourage leaders to innovate within architectural and design traditions to create more magnificent or distinctive layouts. Across polities and through centuries, there is ample room for the design, construction, and remodeling processes to vary within general canons. "Agency," the capacity of diverse people to act according to variation in goals, values, means, and circumstances, allows variation within a shared cultural background.

Consequently, center layouts often express a degree of patterning, yet lack identical or extremely similar arrangements—exactly the situation commonly encountered in Mesoamerica. Likewise, variations in ancient Roman urbanism, analyzed by MacDonald (1986:5–31) through the concept of "urban armature," show a basic arrangement of streets with the central forum, yet considerable variation in the placements of other key civic buildings. Highly repetitive layouts that persist across polities and through time are unexpected. Granted, assessment of patterning is complex, requiring attention to variables, measurement, and different scales of analysis, an endeavor that will require much future work with the settlements discussed here.

The monumental construction in south-central Veracruz is almost entirely earthen, and the archaeological remains are today mounds, likely substructures of vanished buildings. Daneels (2011) has documented the use of adobe blocks in south-central Veracruz monumental constructions; perishable wattle and daub or pole and thatch could have been used in some instances, also. Stone is not locally available. The mounds have been subjected to agriculture and sometimes other damage. The central precincts discussed have been contour mapped, showing the general conformation of the structures and their positions. These conditions of preservation preclude consideration of styles of façades, which would require excavation. My focus is on the main monumental plazas at these complexes, but the larger centers may have a number of additional structures. Monumental palace platforms are acropolis-like, an important form seldom located on the main plaza, which will figure in the discussion of some plaza variants and their timing.

Two elements of planning are distinguished by Smith (2007:7): (1) coordination among central buildings and spaces (comprising traits of arrangement of buildings, formality and monumentality of layout, orthogonality or other forms of geometric order, access, and visibility) and (2) standardization among centers (comprising traits of architectural inventory, layout, orientation, and metrology). He saw these variables as ordinal, but this scaling is not yet operationalized. My comparisons utilize the arrangement of buildings and, among centers, the architectural inventory and layout. The central precincts have noteworthy formality and monumentality, along with geometric plaza groupings. I do not address access, visibility, or metrology.

The Classic period (AD 300–900) center layouts for south-central Veracruz score high in both coordination and standardization. Their pattern is so

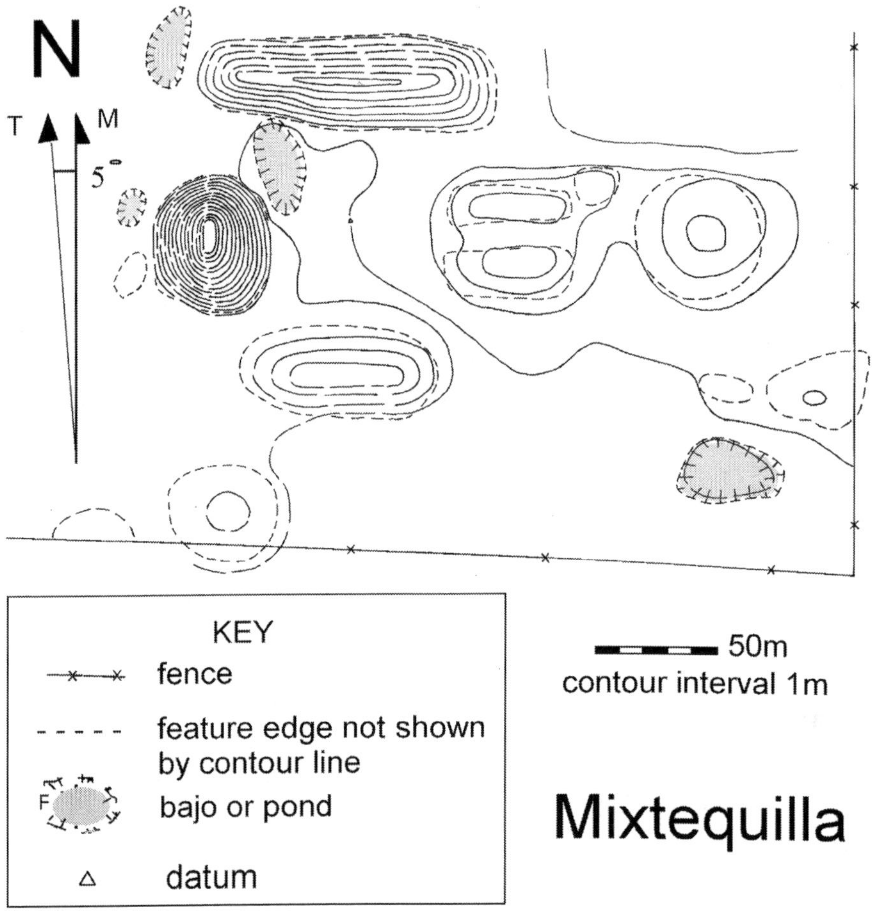

Figure 5.2. *The secondary center of La Mixtequilla (the same name is applied to the region). A monumental platform, not shown, is nearby to the northwest.*

frequently repeated that Daneels (2002:174–177) termed it the "Standard Plan." I focus on the plaza group of the Standard Plan, a roughly square plaza with a conical mound opposite a ball court, with the other sides of the plaza occupied by two (sometimes one) elongated lateral mounds (figure 5.2); the Standard Plan contains additional structures near the plaza, including monumental palatial platforms that seldom front the plaza. Standard Plan monumental complexes are the foci of settlements with dispersed residential occupation (Stark and Ossa 2007).

This chapter addresses why the Standard Plan Plaza Group (SPPG) was highly consistent up and down the settlement hierarchy and across independent polities in south-central Veracruz, as well as throughout the Classic period. Variants occur, mainly in terms of the original definition that accommodates one or two lateral mounds or with omission of one element (usually the ballcourt). This degree of consistency appears to be the exception rather than the rule in Mesoamerica, but it is part of an apparent continuum in consistency of layouts.

Despite characterizing the SPPG as highly repetitious, I analyze the occurrences of variants to better understand the social and political context. I argue that highly consistent layouts and inventories point toward multiple social interests realizing their roles through central precincts. I do not suggest that representation in the central precinct is a political microcosm (De Montmollín 1995:119–122), however, because I do not assume that all social or political segments are represented. The central precincts considered here contribute information about corporate groups and shared power (see also Beekman, chapter 3, this volume; Heredia Espinoza, chapter 4, this volume).

CENTRAL PRECINCT LAYOUTS: COMPARISONS AMONG REGIONS

To establish the SPPG at the high end of a continuum of consistency, I discuss selected Mesoamerican reports, emphasizing those with multiple centers with central precinct maps, essential for detecting consistency. In the Classic-period southern Maya lowlands, Ashmore and Sabloff (2002, 2003) proposed underlying cosmological directional principles and political emulation of major sites to account for the arrangements of buildings and plaza groups, but these patterns are enacted in a varied fashion, partly due to long histories of many sites. The cosmological implications have been disputed (M. E. Smith 2003a, 2005). Ashmore and Sabloff focused their discussion on nine centers but cite additional cases. Categories of buildings are repeated in Maya centers, along with certain subsidiary plaza groups, as well as a degree of cardinality, but the overall arrangements of buildings and groups vary considerably among centers.

In Chiapas in the Upper Grijalva Valley, De Montmollín (1995:125–135) notes 31 Classic Maya centers with the "Tenam Rosario Plaza Format," and he recognizes varying degrees of similarity. The 14 centers most closely representing the plaza format have three pyramids plus a ballcourt surrounding the plaza. Orthogonality and exact placement of structures vary. Elongated "range" structures are present at some plazas, also. The overall amount of variation is akin to the variation in a smaller sample for four Aztec-period settlements in Morelos in Central Mexico. There, Smith (2008:87–89) argues for a repeated arrangement of buildings around a central plaza following the "Tula plaza plan," an instance of emulation of a historically significant capital. In these examples, particular kinds of buildings and some placements recur, such as a major pyramid on the east side of the main plaza, but other buildings vary (e.g., two of the four examples have ballcourts, two of the four examples have T-shaped buildings on the south side of the plaza), and some sites have rows of other structures. Consequently, a plan is recognizable but not enacted in a highly standardized fashion.

The Quiché and Cakchiquel areas of the western Guatemalan highlands exhibit moderate coordination and standardization in Postclassic-period plaza groups. The central plazas are dominated by a temple mound (sometimes two or more of them) and partly framed by one or more elongated mounds (substructures) supporting long buildings (with multiple rooms and benches along the back wall in many cases); usually a ballcourt is present, and sometimes one or more small altars are preserved in the plaza (Fox 1978; Wallace 1977). Large sites may exhibit multiple plazas as well as palaces. Positioning of plaza groups varies considerably, in part reflecting topographic constraints on ridgetops. Among Fox's (1978) compendium of 43 mapped centers, 30 exhibit plazas with the temple(s) and elongated mound(s).[1] Fox (1978) discusses regional and temporal variants within this corpus, along with emulation of capitals. Despite the variations in placement of structures and in the number of temples, the plaza units at different sites are relatively consistent in their inventory and arrangement.

Quiché and Cakchiquel elongated mounds are interpreted as "council houses" or "community houses" that were foci of lineages with varying stature (Fox 1978). A considerable body of ethnohistoric data attests to the lineage structure and ruling houses of Quiché, Mam, Ixil, Cakchiquel, and other language groups in the Guatemalan highlands, several of which eventually fell under Quiché rule (Carmack 1973, 1981; Fox 1978, 1987; Nance et al. 2003; Wallace 1977). Elaborate palaces were associated with the local ruling families, while elongated structures likely provided varying functions, with some that fronted the main plazas likely supporting mainly assembly and civic activities, perhaps with some residential roles, but with other long structures likely mainly residential. The repeated long structures appear to be manifestations of the patrilineages prominent in these polities. The Quiché and Cakchiquel ethnohistoric data play an important role in linking a fairly consistent form of central precinct and a particular elongated building form with an important corporate element in political life. The Quiché and Cakchiquel information provides a key warrant for interpreting repeated elongate buildings as linked to corporate groups in Mesoamerican architectural traditions. Other examples may represent different groups than patrilineages, however. Dual organization is a possibility for south-central Veracruz.

Compared to the Guatemalan highlands, even greater coordination and standardization in layouts is evident in the Teuchitlan tradition during the Late Formative to Early Classic periods in Jalisco (Beekman, chapter 3, this volume; Weigand 1996, 2000). Sites in this tradition display a distinctive pattern of circular elements. Around a central structure, a raised circular patio is framed by a circular banquette supporting a series of platforms and buildings. Often other circular groups are contiguous or nearby but not in a set juxtaposition. The sizes of the circular units vary, as does the central element (altar or pyramidal platform), but the general layout is highly consistent. Sites differ in their number of circular units according to the settlement hierarchy. The banquettes with multiple platforms atop are argued to reflect a corporate emphasis in governance (Beekman 2008; see also Beekman, chapter 3, this volume), and palaces that might signal a ruling dynasty are absent. Maps are available for 30 of the Teuchitlan sites (Ohnersorgen and Varien 1996).

Thus, we see a range in Mesoamerica from more tenuously manifested layout principles (Classic Maya) to more consistent ones (Postclassic Quiché-Cakchiquel plaza plans) to highly consistent ones (Late Formative/Early Classic Teuchitlan). In this array, the last two instances show marked standardization in buildings and layouts in central precincts, and Guatemalan highland ethnohistory records strong patrilineage corporate group organization. The SPPG in south-central Veracruz are more coordinated and standardized than most cases discussed, matched or exceeded only by the Teuchitlan tradition.

To examine the SPPG, I stress that buildings are created and renovated for particular purposes by particular sets of people. Buildings may serve a broad range of the society but very differentially. To unlock the reasons for a repetitive pattern, we need to consider the possible roles of buildings and their sponsorship or designated users. While such information may be sketchy for south-central Veracruz and partly dependent on extrapolation from regions with ethnohistoric and epigraphic data, such information may point us toward a better understanding of south-central Veracruz centers. In particular, I argue that the repetitive plaza layout reflects a balancing of different social interests and activities expressed in a culturally recognized template. During the Classic period, south-central Veracruz was not obviously disrupted by conquest or emulation of foreign capitals, maintaining an independent trajectory of change. Thus, the SPPG was not dislodged by external factors nor based upon them. Nevertheless, the Late Classic period (AD 600–900) witnessed changes in the proportions of variants of the SPPG and an increase in the numbers of palatial platforms, both of which point to some erosion of corporate emphasis.

Background for the Gulf Lowlands

The central and southern Gulf lowlands comprise several regions geomorphologically and culturally. During the Classic period, the SPPG is one of a "family" of repeated layouts at centers. The SPPG is particularly characteristic in south-central Veracruz. In southern Veracruz, the Long Plaza Plan (Domínguez Covarrubias 2001; Killion and Urcid 2001; Urcid and Killion 2008) is alternatively labeled the Villa Alta Quadripartite Arrangement (Borstein 2001, 2005; see Lunagómez 2011 for additional plaza groups in southern Veracruz). These plazas had a dominant conical mound and elongated dual laterals—sufficiently elongated that the plaza is distinctly rectangular in contrast to the square tendency of the SPPG. Ballcourts do not occur at the end of the plaza opposite the conical mound, but, instead, parallel the plaza, using the back of a lateral as one of the flanking mounds for the court. Opposite the conical mound is another smaller conical mound, which may link to the next plaza in a linear "chain" of long plazas of diminishing proportions. Whereas the SPPG spans the Classic period, the Long Plaza Plan appears to be mainly a Late Classic phenomenon.

At the western margin of the Tuxtla Mountains, prior to the Classic period but continuing during the Early Classic (AD 300–600), Tres Zapotes exhibits the Tres Zapotes Plaza Group (Pool 2008) in which a conical mound is accompanied by an elongated or slightly elongated mound. The plaza may contain a small adoratorio, and the end of the plaza opposite the dominant conical mound usually has another conical mound. No ballcourts are associated with the Tres Zapotes Plaza Group. Tres Zapotes has four widely spaced repetitions of this plaza group, which Pool (2008) ascribes to factionalism. The laterals are presumed to be headquarters of factions, as some exhibit residential midden debris, but are thought also to play some administrative roles.

Classic-period Tres Zapotes has historical roots in the Formative Olmec era, and Olmec stone monuments at Tres Zapotes include colossal "portrait" heads, suggesting a society with both corporate and exclusionary principles in governance. Pool (2008:147) suggested the Tres Zapotes Plaza Group was also employed at other sites in the vicinity, such as El Mesón, confirmed by

Loughlin's (2012:244) research. Although the Tres Zapotes Plaza Group does not occur at other smaller monumental centers in El Mesón's vicinity, other examples in the Tres Zapotes vicinity are likely (Pool 2008:147). Because of its long Formative history, Tres Zapotes provides a possible antecedent for some aspects of the SPPG. An apogee of Tres Zapotes in size and activity occurred during the Late Formative period (Hueyapan phase, 400 BC–AD 1). La Venta, with a Middle to Late Formative span, also has been suggested as a partial antecendent for the SPPG (Stark 2007:58–59), due to the two long parallel mounds arranged north of the main conical mound (which do not form a ballcourt). The primary point is that conical and elongated lateral mound plaza arrangements have a long history in Veracruz, and more than one regional tradition exists. Longitudinal and comparative data point to a mix of corporate and exclusionary principles, the latter emphasizing personal leadership and clientage networks (Blanton et al. 1996).

I examine SPPG data primarily from a survey by Daneels (2002) along the lower Cotaxtla River and from my adjacent project in the western lower Papaloapan basin, along the lower Blanco, Guerengo, and Tlalixcoyan rivers (Stark 1999, 2003), an area sometimes referred to as the Mixtequilla (also the local name for the *municipio* of Ignacio de la Llave and of a monumental center nearby; figure 5.2). My underlying method is systematic comparison of plaza groups in multiple centers, not focusing on the layout of a single center. For example, all the capitals recorded in my survey have some unique features that set them apart. The Early Classic capital of Cerro de las Mesas is unlike the later capitals in the Mixtequilla in its agglomerative tendency, with multiple SPPGs and other buildings, such as monumental palace platforms (Stark 2003). The paired, adjacent Late Classic complexes of Ajitos and Pitos on the paleodunes likewise are unlike other capitals in south-central Veracruz. Capitals are particularly prone to idiosyncratic characteristics in part because they incorporate more structures and groups than the Standard Plan, and, as noted, they may be subject to innovation to achieve greater distinction. For many lower-order centers, a single SPPG dominates the central precinct.

The Standard Plan was defined by Daneels (2002:174–181) to include a conical mound (probably pyramidal) at one end of an approximately square plaza (figure 5.2) also framed by two elongate lateral mounds facing each other (but sometimes only one). Opposite the conical mound at the other end of the plaza, a ballcourt is formed by two parallel, closely spaced flanking mounds demarcating a game court. In Daneels's definition, the ballcourt axis matched the plaza axis bisecting the conical mound; however, rarely, ballcourts are transverse to the plaza axis in the data discussed here, which constitutes a variant. A rectangular monumental platform is nearby in Daneels's original definition, and some platforms have a conical mound on them. Monumental platforms likely supported a palatial residence (Daneels 2002:188–192; Stark 1999:209), as confirmed at La Joya along the lower Cotaxtla (Daneels 2008a, 2011). Elite or royal residences may take other forms than monumental platforms, such as a mound with an attached terrace, or, at lesser centers, a monumental platform may be less imposing to the point that its status is debatable. Variation in the Standard Plan as a whole is an important topic, but not the focus here. In this study, I concentrate on SPPGs.

In terms of arrangements, formality, monumentality, orthogonality, and access, the SPPGs are highly planned. The SPPG is a formal geometric arrangement, and, at the higher-order centers, many of the mounds are monumental; however, at lower-order centers, the structures are more modest. Access between buildings makes the plazas relatively open.

In the Cotaxtla survey, 32 SPPGs were mapped, including variants (Daneels 2002:181–183), two in a single complex. In the Mixtequilla, 41 SPPGs were mapped including variants, five of them in complexes with two or more SPPGs. Variants are of interest because they may disclose the circumstances in which a degree of deviation from a prevalent pattern occurred. Included in the counts are instances of possible ballcourts, where a mound of the appropriate size is located in the ballcourt position but disturbance (mainly plowing) has erased any surface evidence of the two flanking mounds of the court. The possible functions of the component SPPG buildings are a starting point for explaining the prevalence of the layout.

Structures and Their Functions in the Standard Plan Plaza Group

Conical Mounds

In keeping with general Mesoamerican analogies, steep conical mounds likely were pyramidal platforms for temples with ritual functions conducted by a few people but witnessed by large assemblies in the plaza.

Ballcourts

SPPG ballcourts have not been excavated, and we have no information about caches or other ritual associations; we lack associated imagery, such as accompanies some Tajín ballcourts (Ladrón de Guevara 1999), and we do not know if ballgames were occasions for public feasts, as argued for Honduras

(Fox et al. 1996). In terms of general Mesoamerican practices, the ballgame was multifaceted, with cosmic, competitive, social, and recreational functions (summarized in Stark 2012). Lines of sight and the number of people who could have directly watched the ballgame reveal that ball-game viewing involved social distinctions, as many fewer people could witness a game compared to those who could be accommodated in a plaza to witness ceremonies atop adjacent platforms and pyramids (Stark 2012).

Lateral Mounds

The functions of the lateral mounds are the most enigmatic among the central plaza structures. The elongated lateral mounds may have supported adobe or perishable multiroom structures that may have served particular corporate social groups in the society, possibly in some form of dual organization. If so, the typical inequality in height of the laterals suggests the possibility of ranked groups. Moieties are one possibility.

Some Maya elongated platforms with multiroom structures have been proposed as council houses, that is, structures used by some corporate entity(s) related to civic activities (Cheek 2003; De Montmollín 1995:66; Fash et al. 1992; Stomper 1996, 2001). This proposition is akin to the possible corporate group role of elongated Quiché or Cakchiquel platforms, interpreted as council houses for segmentary lineage affairs (Fox 1987; Wallace 1977). Corporate or council functions may involve periodic or continuous residence by some participants, and activities could include training and feasting. By way of comparison, at Tres Zapotes, Veracruz, Pool (2008) excavated domestic refuse behind three of the long structures in Tres Zapotes Plaza Groups and suggests these mounds were elite residences possibly also with administrative or ceremonial roles.

Could at least one lateral have had a royal administrative role? The likely presence of royals in government can be inferred both from monumental carvings at Cerro de las Mesas (Miller 1991; Stirling 1943) and from the occurrence of the monumental palatial platforms that represent substantial investments of labor. These palaces likely had multiple roles, including royal residence, storage, rituals, and some audience functions, perhaps also some crafts. However, the palaces do not commonly front on the Standard Plan Plaza, and use of laterals for activities related to royal governance therefore remains dubious. Royal representation is unlikely at all levels of the settlement hierarchy in which SPPGs occurred. Where monumental palatial platforms are closely associated, however, it is likely that a local noble family or a member of a royal line held sway.

Daneels (2002:179) proposed a role in astronomical rituals for laterals (or buildings on them), with solstitial sightings from one side across a line of sight transecting the edges of the opposite structure, similar to "E Groups" in the Maya lowlands. However, the use of E-Groups for solar seasonal observations is now thought unlikely in most cases, although seasonal rituals may have been important at the groups (Aimers and Rice 2006). In Daneels's interpretation, all SPPG structures likely had ritual roles. Nevertheless, I consider a corporate or governance role more likely for laterals because of the considerable variation in the azimuths of plaza "facings"—the axial direction bisecting the conical and the plaza midline and looking from the conical mound toward the other end of the plaza. Figure 5.3 shows facing percents by cardinal quadrants, using 34 measurements for SPPG facings for the Cotaxtla survey, 37 for the Mixtequilla, and 21 for Hueyapan Long Plaza Plans (Domínguez Covarrubias 2001). The Cotaxtla SPPGs have facings northward or southward much more frequently than the Mixtequilla SPPGs, which favor eastward or westward, as does the Hueyapan area. For solstitial observations an appropriate north-south alignment of the SPPG and the laterals is requisite. Thus, a solar observatory or related symbolic role for laterals as proposed by Daneels is more feasible for the Cotaxtla area, but even there it does not account for all the layouts.[2]

Summary

The SPPG clearly includes key buildings associated with ritual events at the main temple or at the ballgame, but laterals may have been used by corporate groups, possibly ones involved in civil administration, supplementing the administrative roles of major palatial platforms placed nearby. The regular presence of laterals around the main plaza assembly space attests to their importance. At larger centers additional mounds and other plaza groups may have provided administrative or ritual functions as well (see Daneels 2002:186). Single laterals could have been used on a rotational basis by corporate segments or by a single council; single laterals suggest less corporate representation than dual laterals, however.

Explaining the Standard Plan

To posit particular functions for mounds in the SPPG does not address why the SPPG is frequently repeated at different settlement levels, across multiple polities, and over many centuries, with minor variations that maintain much of the same format. Several interpretations could explain the consistency and are not mutually exclusive.

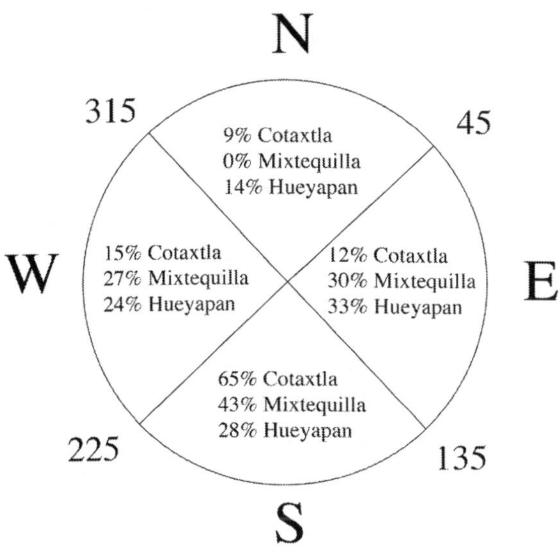

FIGURE 5.3. *Summary of facings (azimuth starting from main conical mound and bisecting the plaza) of SPPGs and Long Plaza Plans based on data from three surveys.*

Ruler or Government Control

The highest political authorities command resources and may engineer the design of buildings and their arrangement in centers. Such efforts are particularly likely if they lead to display and glorification of a ruler's power and of sacred precepts. Conceivably, then, the SPPG testified to rulers' power, with insistence on conformity in several levels in the settlement hierarchy. The AD 1573 edicts of Phillip II concerning the central plaza, streets, and associated buildings in the founding of Spanish towns in the New World are an unusual example of a top-down process that met with some success (Crouch et al. 1982). Apart from the central plaza and streets opening onto it, however, other buildings were left relatively flexible in their positioning (Crouch et al. 1982), with a church to be near the central plaza, civil buildings facing the plaza, and porticos for merchant activities. In fact, Roman Catholic churches were usually placed facing the plaza or alongside it (facing an access street), and they represent a separate corporate organization from the Spanish Crown. The Spanish edict did not result in the degree of standardization shown by the SPPG, but the position of churches suggests a role for a powerful corporate entity assuring itself a prominent plaza position.

The small states of Classic-period south-central Veracruz were not, so far as we can discern, expansionist militaristic states. Thus the resources and power to despotically mold settlements to a single plan do not seem adequate to the task. Among the many demands on state resources and efforts for maintenance of elite power, meddling in the layouts of third-order settlements or neighborhood complexes, for example, does not seem a pressing concern. We confront a problem in explaining why multiple rulers used the same plan and enforced it up and down the settlement hierarchy across centuries in the Mixtequilla and the neighboring lower Cotaxtla region (see Daneels 2011 concerning La Joya settlement hierarchy).

Emulation could create a combined top-down and bottom-up effect, with people in different community levels striving to repeat a prestigious pattern to which leaders adhered. At capitals, some types of elaboration or innovation occurred, yet we do not see replication of capital innovations in lower-order settlements, which cleaved instead to the SPPG layout. Emulation leaves considerable room for local interpretation and seems unlikely to achieve the degree of repetition across polities and over centuries that we see in the SPPG, for which most variation is minor and consists of whether one or two laterals are present. Partial SPPGs keep to the same plan but omit either the conical mound or the ballcourt (though they may be present elsewhere in the monumental core).

Cultural and Social Values and Symbols

One might argue that south-central Veracruz was subject to extensive, active communication and sharing of cultural and social values expressed in the SPPG. We see that both the highest echelons of society and commoners in small communities expressed the same design for central precincts because the SPPG appears at first-, second-, and several third-ranked centers. Extensive communication and sharing of ideas across south-central Veracruz is documented not only in settlement patterns but also in ceramics (Daneels 1997; Stark 2001). Possibly the frequency of ballcourts (Daneels 2008b) and periodic marketplaces (Stark and Ossa 2010), in addition to a ritual calendar of observances, fomented an unusual degree of visiting by people among centers. Daneels (2008b) argued that ballgames performed a key integrating function in south-central Veracruz, as the courts were widely distributed. Active communication would be abetted by a standard layout so that visitors were well acquainted with the facilities and practices at centers across south-central Veracruz.

We have little concrete information concerning symbolism of centers that might be shared. Perhaps the pyramid represented access to a celestial realm, with the ballcourt at the other end of the plaza for ritual events involving descent to the underworld (e.g., Taladoire 1981:545, 548). The space between, with one or two laterals and a plaza, could represent the domain of regular human activities in civic, political, and economic life (cf. Beekman, chapter 3, this volume).

Despite considerable promise, an explanation based on a shared template is unsatisfactory taken alone because a template may have existed without precluding considerable interpretive license, as is evident in the comparative societies discussed. Also, each building had particular functions, and buildings differentially served elements of the society, such as the ballgame in formal courts serving higher-status persons (Stark 2012), or lateral mounds possibly serving particular corporate or civic groups. South-central Veracruz society was hierarchical, with different interests and values among classes and communities, at least to some extent, which could yield considerable deviation.

Shared Power

I propose that in addition to a context of communication and emulation, the SPPG Group construction was likely related to more widespread social action than simply rulers' dictates or nearly slavish adherence to a shared concept for centers. The use of laterals by ranked, dual corporate groups would create a broader interest in construction and maintenance of the SPPG. If the laterals imply a corporate element in civic affairs, despite the presence of royal families, SPPG centers would exhibit a degree of collective action or "voice" (e.g., Blanton and Fargher 2008; Levi 1988)—that is, "bottom-up" efforts to consistently engineer architectural forms that represented diverse vested interests.

I suggest that the repetitive character of the SPPG reflects checks and balances among key social segments (e.g., royals and corporate groups), a dynamic tension that recreated the culturally accepted presentation of a center because doing so afforded an avenue for expression of shared power. The economic context of south-central Veracruz provides insight into a basis for shared power. Stark et al. (1998) noted that the region is suited to cotton production as well as sustaining reliable agriculture. Whorls and sewing gear attest to cotton production and processing. Cotton production and processing require both land and labor. Thus, important land-holding elites, royals, and commoners might have enjoyed exchange advantages trading cotton and cloth to areas that could not produce them, as well as social display advantages (fine cotton garments).

Testing the Shared-Power Interpretation Using Variants of the Standard Plan

To test the idea of shared power versus top-down control, I examine the occurrences of variations in the SPPGs to see if they are linked to innovations by central powers at the top of the settlement hierarchy or, instead, prove to be more characteristic of lower-level settlements that did not have to maintain a strict conformity (or lacked some of the resources—or permissions—necessary to create all the SPPG repertoire). In the examination of the relationship to the settlement hierarchy, I show that single lateral variants initially were associated more with lower-order settlements, a pattern not compatible with strong top-down conformity. I also show that the incidence of variants overall is higher in the Mixtequilla area than the lower Cotaxtla, so the realm that underwent unification in the Early Classic period did not demonstrate greater conformity, calling into question top-down processes as a basis to account for conformity. As an outgrowth of examining variants in relation to the settlement hierarchy, I detect that later in the Classic period, variants implying a decreased corporate role (single laterals) became increasingly associated with the upper part of the settlement hierarchy, suggesting shifts in power relationships.

Some sources of variation (table 5.1; figure 5.4) include the initial definition allowing one or two laterals. Other variation involves rare instances in which the ballcourt is transverse to the main plaza axis bisecting the conical mound. A different kind of variation is introduced by some arrangements that are "partial" because one ingredient is left out, either the conical mound or the ballcourt, usually the latter. In addition, in a single case (figure 5.4, Variant D), the two elongated laterals do not face each other but are placed instead at right angles, and the ballcourt is absent (making it debatable whether this case is usefully considered as an SPPG variant). A different variant was defined for the Cotaxtla area, discussed further below. I first examine settlement associations for one versus two laterals, which I argue does not support a top-down process of conformity. I next describe the partial SPPGs and then consider the contexts of partial SPPG and rare transverse ballcourts, which also do not support an explanation based solely on top-down conformity.

Single and Dual Laterals

The Standard Plan definition allows one or two lateral platforms. This variation is politically significant if the lateral structures represent corporate groups with civic functions.[3] Note also that with a single lateral, access to

Table 5.1. Counts of Standard Plan Plaza Groups (SPPG), partial Standard Plan Plaza Groups (Partial SPPG), and a Cotaxtla Standard Plan variant in Cotaxtla and Mixtequilla survey data. Possible ballcourts are counted as having the common orientation (follows plaza axis bisecting the conical mound). Figure 5.4 displays the Mixtequilla variants.

	SPPG 1a	*SPPG 1b*	*SPPG 2a*	*SPPG 2b*	*Partial SPPG A1.a*	*Partial SPPG A2.a*
Cotaxtla Survey	5	1	15	1	0	0
Mixtequilla Survey	6	3	14	1	1	1

	Partial SPPG B1	*Partial SPPG B2*	*Partial SPPG C*	*Partial SPPG D*	*Cotaxtla Variant*	*Total*
Cotaxtla Survey	0	5	0	0	5	32
Mixtequilla Survey	10	2	1	2	0	41

Note: Among the Cotaxtla variants, one complex, 87 (Mata Naranjo Sur), would otherwise be classified as partial SPPG-B1.

the plaza is more open, with a lower overall investment in construction. We can ask if the single laterals correlate with a particular part of the settlement hierarchy to assess top-down versus bottom-up variation or if some other factor is at work, such as change over time or geographic variation. Mixtequilla data show that the single laterals initially had mainly a lower position in the administrative hierarchy during the Early Classic period, but not in the Late Classic. Eventually during the Late Classic period, an increased frequency of single laterals in several parts of the region suggests a diminished but not extinguished role for the corporate groups or councils they possibly represent.

The Cotaxtla survey has 6 single laterals among 32 SPPGs (19%), and the Mixtequilla has 20 single laterals among 41 SPPGs (49%, partial SPPGs are included). Greater prevalence of single laterals in the Mixtequilla likely reflects more variation in the degree to which corporate groups played key roles in centers. In particular, the Early Classic unification of the Mixtequilla under Cerro de las Mesas means that the region that experienced the most developed central authority had more variation, undermining the idea that royal edicts ensured central precinct conformity. In the Late Classic, the Mixtequilla, like the lower Cotaxtla throughout the Classic period, was divided among several

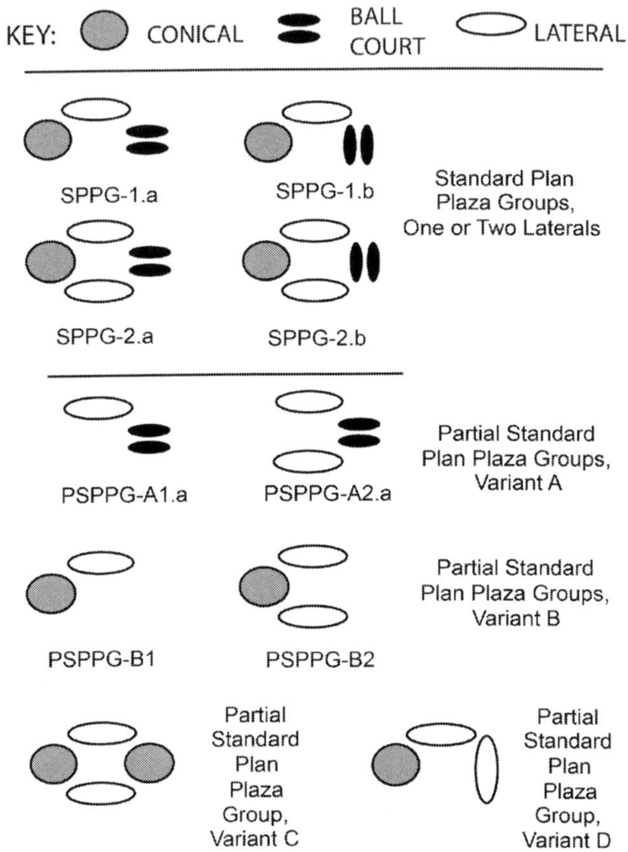

FIGURE 5.4. *Mixtequilla survey: SPPG variants.*

smaller polities. In a section that considers all types of variants, not just single versus double laterals, I return to this point.

Mixtequilla temporal analyses show that some of the higher incidence of variants is due to the single lateral becoming more common during the Late Classic period (table 5.2).

In the Mixtequilla by time periods, 5 complexes are Early Classic, while 12 are Late Classic, and 5 occur in both periods. The only top-ranked center during the Early Classic period is Cerro de las Mesas, with an extensive central precinct that includes 6 subsidiary SPPGs or partial SPPGs; among these, only one partial complex has a single lateral. During the Early Classic period, 9 lower-level centers have single laterals. In contrast, during the Late Classic

Table 5.2. Monumental complexes or subsidiary segments with single laterals, according to Mixtequilla settlement rank. For this table, possible ballcourts are assumed to follow the dominant axis.

Period	*Rank 1*	*Rank 2*	*Rank 3*	*Rank 4*
Early Classic	Cerro de las Mesas, at one of the subsidiary segments, PSPPG-A1.a	Tuzales, PSPPG-B1	Bartolo, PSPPG-B1	
			Coyote, SPPG-1.b	
			Mulas, SPPG-1.a	
Early and Late Classic		Dicha Tuerta, SPPG-1b and PSPPG-B1	Sabaneta, SPPG-1.a	
		Loma, SPPG-1.b	Nuevo Porvenir West, PSPPG-B1	
			Rincon del Tigre, SPPG-1.a	
Late Classic	Zapotal, SPPG-1.a		Lobato, subsidiary of Azuzules, SPPG-1.a	Mulas?, SPPG-1.a
	Nopiloa, PSPPG-B1		Pinchones North, PSPPG-B1	
			Tuzales? PSPPG-B1	
	Nacaste-Patarata, main complex Tio Perciliano, SPPG-1a, and 4 subsidiary segments, all PSPPG- B1	Santa Catalina, SPPG-1.a		

period, three capitals have a single lateral in their primary plaza along with 14 in lower-order settlements or in parts of settlements (e.g., Nacaste-Patarata). Single laterals are particularly common in the Guerengo area, in the interfluve between the Guerengo and Blanco rivers, and in the mangrove swamp. Thus, there are subregional and Late Classic emphases.[4]

Concomitantly, Mixtequilla data show a Late Classic proliferation of monumental palatial platforms, which points to the expansion and entrenchment of elites, probably large landholders. If single laterals represent an erosion of dual corporate group representation in civic centers, then the proliferation of high-ranking families with palatial residences may have come at the expense of wider corporate representation.

Overview of Partial Standard Plan Plaza Groups

This section describes partial SPPGs, which then figure in settlement analysis in the next section. Two circumstances apply. A few occurrences are at primary or secondary centers in which the missing component is present elsewhere. For example, two partial SPPGs at Early Classic Cerro de las Mesas lack the conical mound, but these plaza groups are in the immediate vicinity of the north part of the central precinct where four conical mounds are positioned closely around interlinked plazas (Stark 2003). The Late Classic Nacastle-Patarata settlement in a mangrove swamp provides a different example (Stark 2003). Partial SPPGs at subsidiary neighborhood groups lack a ballcourt, but the central complex, Tio Perciliano, has a ballcourt. In both of these capitals, partial SPPGs are part of subsidiary segments of construction. In contrast, partial SPPGs in many lower-order settlements do not have the "missing" structure supplied elsewhere in the same settlement (table 5.3).

Daneels defined a different Cotaxtla variant. She observed five instances in which a monumental palatial platform faced a plaza with a ballcourt; with one exception the courts are not at the opposite end of the plaza but alongside the plaza. Elongated lateral mounds are not present except in one case. In the Mixtequilla variants, laterals are always present.

Mixtequilla survey registered two complexes similar to those recognized by Daneels. Both are within the Late Classic Nopiloa settlement. Complex 6309 is near the Nopiloa central precinct and forms a subsidiary part of that center. Complex 6309 has a monumental palatial platform and a quite small ballcourt adjacent to a small plaza, with a few additional low mounds. This Late Classic phenomenon resembles the appropriation of architectural prerogatives of the center by an elite family or perhaps a collateral royal, a process of delegation of privileges noted also at the Maya site of Copán during the Late Classic period (Fash 1991:160–172; Fash et al. 1992). The Nopiloa central plaza presents a partial SPPG-B1 with a single lateral mound and a massive palace located where the ballcourt normally would appear; instead, the ballcourt is alongside the plaza. In this example of partial SPPG B1, an additional low conical mound is positioned in place of the second lateral. All of the variants encountered by Daneels are on the paleodunes, and all but one date to the Middle or Late Classic period, as does Nopiloa. Some are in close proximity to other SPPG complexes and one, complex 87, is likely part of the Mata Naranjo center. Consequently, I regard these Cotaxtla variants as distinct from Mixtequilla SPPG variants, but some may be parts of settlements, such as occurred at Nopiloa.[5]

The partial SPPG variants in the Mixtequilla that occur in parts of settlements or lower-order settlements show that laterals appear even if a ballcourt

Table 5.3. Distribution of partial SPPGs in the Mixtequilla settlement hierarchy.

Partial SPPG Variants	*Rank 1*	*Rank 2*	*Rank 3*
A1a no conical	part of Cerro de las Mesas, EC, conicals elsewhere		
A2a no conical	part of Cerro de las Mesas, EC, conicals elsewhere		
B1 no ballcourt opposite conical	Nopiloa, LC (ballcourt at side of plaza); four segments of Nacaste-Patarata (Palma Real, Nacastle, San Juan, Patarata East), LC, ballcourt at Tio Perciliano main complex	Dicha Tuerta, LC, ballcourt elsewhere; Tuzales, EC, ballcourt elsewhere	Cerro Bartolo, EC; Tuzales, LC ?; Nuevo Porvenir West, LC; Loma de los Pinchones, LC
B2 no ballcourt opposite conical	Zapotal South, part of Cerro de las Mesas, EC	Salto Norte, EC and LC	
C no ballcourt, two conicals		Palmas Cuatas, EC and LC	
D no ballcourt, laterals form "L"	Zapotal, LC, possible ballcourt elsewhere	Madereros, EC, ballcourt elsewhere	Madereros, LC?, ballcourt elsewhere

Note: EC, Early Classic; LC, Late Classic

or conical mound does not, and they constitute one of the most pervasive architectural structures. Apparently, the smaller neighborhoods or communities invested in corporate civic activity, even if their circumstances were more restrictive in resources or privileges.

Settlement Contexts of Standard Plan Variants

The Cotaxtla and Mixtequilla surveys differ in the proportions and nature of variants, which further casts doubt on a top-down imposition of the replicated format. For my purposes, it is useful to focus on the percent of variants in relation to total SPPGs, rather than to all monumental complexes in the two regions (e.g., isolated palatial platforms); on this score Cotaxtla has 12 variants among 32 SPPGs (38%), while the Mixtequilla has 21 variants among 41 SPPGs (51%.) These figures count instances of transverse ball courts, partial SPPGs, and Cotaxtla variants (table 5.1). The presence of one or two laterals is not tallied as a variant in this analysis. These statistics show less variation from SPPG principles in the Cotaxtla area than in the Mixtequilla.

These two adjacent regions have relevant organizational differences because the Mixtequilla includes Cerro de las Mesas, the largest monumental complex recorded in either region, and one with several carved-stone monuments and probably six ballcourts. During the Early Classic period it appears to have been the capital of the western lower Papaloapan basin or most of it. In contrast, the Cotaxtla area includes multiple independent polities during that time (Daneels 2002). In the Late Classic period, the Cerro de las Mesas realm dissolved into at least four separate smaller polities. Thus, in the lower Cotaxtla, with more independent polities, there is less SPPG variation than in the Mixtequilla, despite the fact that independent polities imply more separate bases of political action that could have led to variation in center layouts.

Despite the Mixtequilla's Early Classic unification followed by a breakup, partial SPPGs do not appear to be confined to the Early or Late Classic period, with five during the Early Classic and eight during the Late Classic at secondary and tertiary centers (table 5.3). Some of the partial SPPG centers appear to have been active during both the Early and Late Classic periods. (For partial SPPGs that are segments of centers, the results are similar, with three partial SPPGs within Early Classic Cerro de las Mesas, and four within Late Classic Tio Perciliano.)

To summarize, conformity to the SPPG is higher when polities are both more numerous and smaller, as we see in the Cotaxtla region, with variants becoming more noticeable during the Middle and Late Classic periods on the paleodunes. In contrast, when a dominant capital integrated a larger area—Early Classic Cerro de las Mesas—there was a higher frequency of partial SPPGs; however, that larger polity later broke apart during the Late Classic period, and the number of partial SPPGs remained similar. Because of the high level of conformity in the Cotaxtla region, with no instance of extensive integration into a single polity, it is difficult to argue that conformity was imposed by a powerful central authority. Instead, a combination of shared cultural values and the interests of multiple social segments (both rulers and corporate groups) is more promising as an explanation of the repetitive SPPGs. In general, greater deviations (partial SPPGs) are more likely at lower-order centers in the Mixtequilla or within a capital where the "missing" facilities are present elsewhere. A top-down push toward conformity, if applied, did not penetrate the settlement hierarchy very effectively compared to the lower Cotaxtla region.

SUMMARY AND DISCUSSION

The SPPG in south-central Veracruz presents a striking phenomenon with high coordination and standardization of structures. The surprisingly repetitive nature of the SPPG, with a modest incidence of relatively minor variants, suggests that recurrent considerations by multiple groups were brought to bear on the erection of platforms and on activities conducted at centers. A likely prominent role for corporate groups in these settlements points to a broader basis of authority and power than solely the royal line. If so, the extent of alluvial farmlands, reliable rainfall, and climatic conditions favorable to cotton production may have involved weaker clientage and more broad-based wealth generation than in many regions of Mesoamerica (Stark et al. 1998). The Early Classic period saw settlement growth eastward in the delta of the Río Blanco and thus at least partially represents a time of expansion of people and landholdings instead of their consolidation into fewer hands. Such patterns would be consistent with Blanton and Fargher's (2008) observation that higher collectivity or more corporate forms of governing are associated with internal revenues (that is to say revenues provided by a broad constituency of taxpayers). Conversely, the Late Classic period likely witnessed more contrast between wealthy residential groups and commoner households and more consolidation of wealth (likely land holdings) in fewer hands, with diminished corporate group representation in civic affairs. Possibly revenues shifted to more external sources or to sources directly controlled by the governing elite, which could be more readily used to exclude corporate groups from participating in political decision-making and to increase conspicuous consumption by royal lineages (see Blanton and Fargher 2008). Thus, despite marked continuity in use of the SPPGs in south-central Veracruz, there is intriguing variation over time.

Other Mesoamerican instances of repeated layouts in public centers at varying hierarchical levels also point to counterbalancing between corporate and ruling governmental power. The Postclassic Quiché and Cakchiquel highland centers, with temples and long council structures as well as palaces, are a case in point. To a lesser extent, the *popol na* or council buildings of the Classic lowland Maya point in the same direction, but lack the prominence of laterals in the SPPG and of elongated structures in the Guatemalan highlands. Instead, lowland Maya royal power is conspicuously displayed.

The Teuchitlan tradition in Jalisco displays even more consistency in layouts. In contrast to the south-central Veracruz situation, there are no indications of royal lineages and palatial residences (Beekman, chapter 3, this volume). The settlements contain banquette constructions representing multiple groups, not a dual organization such as south-central Veracruz may have had. The south-central Veracruz plans embody a tension not so much between many near equals as between one or two corporate groups and royal or other socially dominant leadership. In both Teuchitlan and south-central Veracruz the pervasive adherence to a central plaza plan, with scant variation, suggests a wider participation in the construction and use of civic and religious facilities than royal dynasties alone. A degree of "bottom up" contributions to centers was likely part of a balancing of corporate and exclusionary powers.

Increasingly Mesoamerican research is documenting variety in political systems. In Postclassic Tlaxcallan, for example, Fargher, Heredia Espinoza, and Blanton (2011) argue for a nonroyal, more egalitarian political form with councils. The site lacks any central plaza and precinct (a precinct removed from the settlement is suggested as a meeting location), and the settlement has numerous neighborhood plazas and residential terraces, only modest temple mounds, and no palaces (Fargher et al. 2011). Both Postclassic Tlaxcallan and Classic Teuchitlan are exceptional for Mesoamerica in their lack of obvious palaces (see also Heredia Espinoza, chapter 4, this volume). I suggest a common situation is a "tug-of-war" or counterbalancing among powerful social segments involving both corporate and exclusionary principles (in the terminology of Blanton et al. 1996). The long-lived SPPGs attest to endurance for a political formation in which different interests interacted and likely competed. Blanton et al. (1996:2) saw either exclusionary or corporate political strategies tending to be dominant, and they suggested cycling between dominant principles. Six hundred years of the SPPG (and its earlier precedents at Tres Zapotes and possibly La Venta) raise doubts about necessary cycling between these principles or a tendency of one or the other to dominate, but see Fargher (chapter 15, this volume) for examples of shifts in predominate governing principles.

The extremes in which we see either corporate or exclusionary principles emphasized in Mesoamerica should help us better understand a "more populated" middle ground with mixed emphases. The middle ground seems remarkably resilient in south-central Veracruz, even though the Late Classic period shows some decline in corporate emphasis (single laterals) along with a proliferation of palaces. If earlier Olmec times in the Gulf lowlands exemplified a "cult of the ruler" (Grove and Gillespie 1984), and a strong role for exclusionary patronage, the Olmec La Venta layout opens the door to additional factors in governance, provided the elongated laterals there had similar functions to those proposed for Tres Zapotes and the SPPG sites. A profusion of leader or dynastic displays and claims should not obscure the possibility that other social interests were strong in governance (even if ancient rulers would have liked for art displays to convince people otherwise). For the Classic period in south-central Veracruz, resilience of the middle ground was likely powered by the interests of landed elites, perhaps kin-linked "houses"

(Joyce and Gillespie 2000) versus "the Crown" as well as by alliances between commoners and elite factions.

Because the south-central Veracruz cultural patterns were truncated at the end of the Classic period, eventually replaced by those of intrusive highland-linked groups (Stark 2008), we cannot determine if the proposed long-lived balancing act in civic principles would have endured or undergone continued change toward more exclusionary emphases. In any case, highly repetitive planned arrangements in central precincts form an important variant in Mesoamerican urbanism. I suggest that Teuchitlan, Quiché-Cakchiquel, and the Veracruz SPPGs all point to ways in which corporate groups constrained architectural variation to bolster their civic roles.

Although the focus in this chapter has been whether corporate groups may have played important roles in governance in south-central Veracruz, this issue falls within broader concerns of how different "agents" and groups of them contribute to the diversity and history of complex societies. Blanton and Fargher (2008) show that a range of collective action emphases yields governmental variability in ancient states. Fargher (chapter 15, this volume) documents an association of more marked corporate power with stronger indications of collective action. My analysis shows considerable durability for what I argue is a representation of corporate interests combined with dynastic rulers. The longevity of both elements could indicate that wider engagement of the population in governance and provisioning of public goods yields greater durability in governmental forms (cf. Blanton 2010); alternatively, a degree of balancing of power principles may contribute to durability by checking governmental excesses. These possible temporal issues, raised by my research and other chapters in this volume (e.g., Beekman, chapter 3, this volume), remain an avenue for future research. Research provoked by Richard Blanton's attention to multiple governmental forms and processes, including a diversity of agents, will contribute to research directions far into the future.

Acknowledgments. An earlier version of this chapter was presented in absentia at the 54th International Congress of Americanists in 2012, Vienna. Fieldwork in the Mixtequilla was supported by the National Science Foundation (BNS 85-19167, BNS 87-41867, and SBR-9804738) and the National Geographic Society, with permission from the Consejo de Arqueología, Instituto Nacional de Antropología e Historia. I thank the many students and archaeologists who contributed to the field and laboratory work. I also thank the Mixtequilla communities for their hospitality. George Cowgill, Annick Daneels, Lane Fargher, Verenice Heredía, Michael Smith, and anonymous reviewers provided valuable suggestions to improve the chapter but bear no responsibility for its deficiencies and may dispute some of the arguments.

NOTES

1. I set aside 11 sites that lack this pattern or that are so destroyed that the layout is not clear, plus two sites for which the elongated mound is also rather wide.

2. In all three surveys cardinality is important for centers, but variable. To evaluate cardinality, I calculated the percent of facings within 5 degrees (plus or minus) of a cardinal direction, which allows for some measurement error. For Cotaxtla SPPGs, 50% fall within these cardinal intervals, for the Mixtequilla, 44%, and for the Hueyapan area, 24%. Cardinality has an astronomical link, but may reflect a concept of world directions rather than an interest in charting seasonal progressions.

3. The presence of a single lateral is tricky to interpret because the dual laterals are almost always unequal in height, and the "missing" lateral may have existed in the form of a multiroom structure at ground level, unmarked by a supporting platform. Although this possibility would accommodate a dual organization, for the moment I treat the instances of a single lateral mound as just that. Single laterals also could represent an interruption in construction so that a second lateral was never built, but the temporal and contextual data suggest the single laterals are patterned in their occurrences, not haphazard interruptions.

4. The minor variant of a transverse ballcourt is scarce in both subregions, but slightly more common in the Mixtequilla, mainly found in lower-order centers. Of the four Mixtequilla transverse ballcourts, three are at secondary or tertiary centers. One possible example at Cerro de las Mesas, a primary center, remains uncertain due to plowing effects.

5. Too complex to address in this chapter is the possible variation in how settlements are defined among survey projects.

6

Featherwork as a Commodity Complex in the Late Postclassic Mesoamerican World System

Frances F. Berdan

Recently Richard Blanton, Lane Fargher, and Verenice Heredia Espinosa wrote a seminal piece on Mesoamerican commodities and their economic lives (Blanton et al. 2005). Phrasing their formulation as "a goods-based approach to world-systems," they analyzed the economic dimensions, dynamics and contexts of five important commodities: obsidian, salt, cacao, cotton cloth, and pottery. For each of these commodities, they addressed penetrating questions concerning labor and time allocation bottlenecks, relationships between the good and dynamic distribution systems, the good's impact on secondary industries or markets, and broader impacts of increased production of the commodity (ibid.:262). I find their empirical treatment of these commodity issues especially attractive.

Their approach and these questions have stimulated the formulation of this chapter, which focuses on the luxury craft of featherwork. Building on the goods-based approach, I examine the complex production, distribution, and consumption lives of the several different types of featherwork in the Aztec world. Practical considerations root this discussion firmly in the Late Postclassic, although Blanton and his colleagues spend considerable time and energy on issues of change. Another time.

Much as Blanton, Fargher, and Heredia Espinosa selected a small sample of goods for their study, I limit my discussion here to just one constellation of complex manufactured products: various types of feathered

DOI: 10.5876/9781607325338.c006

adornments (luxuries, and "key commodities" in M. E. Smith 2003c). These types of objects had complicated production histories: their manufacture depended on the acquisition of a relatively wide range of raw materials from varied ecological regions, the use of a variety of tools, the application of highly specialized knowledge and techniques, and appropriate and effective organizational arrangements. The fashioning of these objects required that specific materials and human skills converge at the same time and place, some of the materials deriving from distant and specialized regions. How was all this orchestrated, with a reasonable degree of predictability?

COMMODITIES IN AZTEC-PERIOD MESOAMERICA

Aztec times in Mesoamerica were materially exuberant times. Households of all types and scales required and acquired utilitarian objects ranging from cookware to weaving implements to brooms to hoes, and there were more households, and more people, in Late Postclassic times than during any other time in Mesoamerican prehistory. There was also an upsurge in the production and use of status-linked luxuries, often referred to as "prestige goods." Shimmering tropical feathers, bright jewels, and shining gold ornaments bedecked gods and nobles alike, proclaiming their exalted status. Warriors entered battle adorned with symbolically laden feathered devices and costumes. State religious ceremonies, exhibiting a special flamboyance, claimed their share of luxurious materials and objects. In the prestige arena, a consumer-oriented and status-conscious elite enjoyed fairly restricted use of certain luxury objects made from materials such as fine jadeite and turquoise, precious metals (gold and silver), glamorous tropical feathers, jaguar skins, cotton, and cacao. That said, the designation of all luxuries as prestige goods is not entirely applicable to the highly commercialized Aztec world: some high-grade materials and the splendid objects produced from them also made their way into other levels of society, leading to an increased demand beyond royal feasting, noble fashion, battle displays, and periodic rituals.[1] As "bulk luxuries," goods such as fine salt, green obsidian, cacao, and decorated clothing (Blanton et al. 2005) traveled long distances and enjoyed considerable popularity despite their relatively high values.

There apparently was quite a bit of wiggle room in Aztec-period consumerism. A royal household needed brooms, baskets, and graters as much as a farmer—indeed, probably many more. Women in all types of households produced cotton cloth—therefore cotton fibers and the textiles woven from them were present virtually everywhere. Jade(ite), bronze, and other imported luxuries have been found in humble as well as elite contexts, "indicating that commoners and elites had ready access to valuable goods," as found in excavations in Late Postclassic sites in Morelos (M. E. Smith 2003c:250). Nonetheless, elite households enjoyed greater quantities of these types of goods. In addition, there is some suspicion that drinking cacao may not have been an exclusively noble perquisite—a sixteenth-century colonial document from Tlaxcallan refers to uppity native commoners dumping "watered-down" cacao on the ground when offered such a beverage by native nobles of the time. In this case the commoners had begun to gain some wealth through their production of cochineal and, in the eyes of the local nobles, were becoming altogether too pretentious and haughty (Lockhart et al. 1986). The disdain with which the newly prosperous commoners treated the thin cacao suggests the possibility that they may have consumed diluted cacao in the past, but that they felt that their increased wealth now placed them above this.[2]

All of this goes to say that the lines of social status, materially defined, were somewhat blurry among the Late Postclassic Aztecs. But only somewhat blurry. Social positions were unquestionably hierarchical and political power was well entrenched in the hands of a small number of individuals. Distinctions in social station were accentuated, indeed announced, by highly visible displays—especially the wearing of ornate clothing and adornments that carried specific symbolic meanings. For example, only rulers and high-ranking noblemen (perhaps only judges) wore turquoise diadems as symbols of power, and only achieved warriors were entitled to wear specifically designed martial attire in battle and rituals (Berdan 2012, 2014). The imported luxuries encountered archaeologically in commoner households (see M. E. Smith 2003b) were expensive, but not so symbolically charged.

It has been suggested that the presence of valuable goods up and down the social scale is indicative of an active and pervasive marketplace exchange system (see Hirth 1998; Smith 1998). We know that these materials and goods were available in marketplaces, and this would be a convenient and customary avenue through which both noble and commoner families, for a price, could obtain goods beyond their basic necessities. It was the most widespread means by which raw materials and finished goods, as commodities, moved from region to region, community to community, and hand to hand. These movements were effected by individual producers/retailers and regional traders, and also by long-distance professional merchants (*pochteca*), who specialized in trading relatively high-value/low-bulk commodities such as precious feathers and decorated cotton cloth. These professional merchants served their rulers directly by embarking on trading expeditions on their behalf, moving fancy goods across considerable distances and establishing or cementing diplomatic relations with foreign polities.

But long-distance trade and markets were not the only avenues and venues for the distribution of goods, whether precious or ordinary, costly or cheap. There was the well-documented tribute/taxation system imposed by conquering city-states on their subjects, and, in its largest manifestation, by the Aztec empire on its vanquished polities. Imperial tribute was demanded in food staples, utilitarian goods, and precious raw materials and manufactured objects; payments were expected on a preset schedule (usually quarterly, semiannually, and annually), or delivered on demand for the celebration of special events such as a royal coronation or funeral (Berdan 1986, 1992). In contrast to the products and goods moving through the marketplace system, tributes were delivered directly to city-state or imperial rulers for distribution according to established rules and, to a large extent, at their discretion. Rulers were expected to be generous. At specified monthly ceremonies they distributed food to their commoners from their palatial coffers, and at other ceremonies they bestowed glorious honors, such as feathered costumes and devices, on courageous warriors (see Berdan 2014:260–268). Tributes would have been available to supply these foods and regalia at least to some extent. In general, a great deal of tribute income was directed to the maintenance of the ruler's extravagant lifestyle, military expansion, trading expeditions, and the establishment of alliances (Berdan 2005).

Markets, long-distance trade, and tribute provided the most pervasive contexts for the movements of goods through the Aztec world. But there was also elite exchange, most notably through extravagant feasting whereby very fancy goods were gifted from ruler to ruler, or noble to noble, to cement friendly political relations (or, in some cases, to intimidate by flamboyance). One particularly well-documented case involved the dedication of the Mexica great temple in Tenochtitlan in 1487. The Mexica ruler Ahuitzotl invited powerful rulers (friends and foes) to this extraordinary ceremony. As host, Ahuitzotl offered his august guests extravagant gifts of elegant clothing, exquisite jewelry (golden diadems and leg ornaments, and lip plugs, nose plugs and ear plugs of gold and precious gems), finely made weapons and shields, jaguar and puma skins, and sandals (Durán 1994:340). These material luxuries unabashedly proclaimed the extent of control that Ahuitzotl exercised over his imperial domain: Ahuitzotl reminded his guests, especially his enemy-guests, that "These presents . . . are won by the strength and valor of our powerful arm," astonishing them with the magnitude of his wealth and power (Durán 1994:340). High-level events such as this moved specific gifted luxuries across regions: Ahuitzotl's awe-struck guests took their luxurious gifts home to be admired, used, and perhaps emulated in their own lands. On the flip side, there is the case of the Chichimec ruler of the city-state of Cuauhtitlan, conquered by the Mexica. When the Mexica ruler offered him gifts consisting of the costume and insignia of a Mexica warrior in reward for his loyalty, this stalwart ruler flatly refused. Instead, he preferred and accepted gifts symbolic of his own Chichimec heritage (Hodge 1984:60).

With these commercial and political exchange systems in mind, how did the materials necessary for the manufacture of feathered adornments move from their places of origin to arrive in the hands of skilled artisans? What were the manufacturing requirements? And then, how did the finished products end up in the hands of appropriate consumers?

THE CASE OF FEATHERED ADORNMENTS

We can easily speak of feathered adornments as luxuries in the Aztec world. For the most part they ended up in the hands and houses of the aristocracy, they announced the achievements of courageous warriors, and they decorated the sanctuaries and idols of the many deities. In such settings and on such persons and gods, these objects carried considerable social, political, and religious importance. But this does not mean that they were economically superfluous. Indeed, objects such as these not only reflected changing social and political dynamics, they also stimulated production and exchanges in their own and ancillary areas of the economy. Secondary industries such as woodworking, glue-making, hide-curing, twine-making, and blade-production gained from flourishing production of these fine luxury objects, which required these materials. Luxuries were prominent in many marketplaces; the great Tlatelolco market of course comes to mind. As another example, the market at Tepeacac on the eastern imperial borderlands was expressly ordered by its Aztec conquerors to welcome merchants carrying exotic materials and plying precious wares (Durán 1994:159). And long-distance professional merchants made their living (and a good living it was) from trading in the most valuable raw materials and objects throughout the land. The idea that high-end goods played dynamic and significant roles in preindustrial economies is not a novel idea (e.g., see Schneider 1977; Blanton and Feinman 1984; Kepecs and Kohl 2003). It is an idea worthy of closer examination specifically in the Aztec world, for which I have selected this category of expensive, esteemed, and complicated luxuries.

The production, distribution, and consumption of fancy feathered objects are particularly well-documented in the ethnohistorical sources. In the Florentine

Codex, Sahagún's native collaborators on the luxury crafts were particularly well informed about the featherworking enterprise and may themselves have been featherworkers (Sahagún 1950–1982, book 9). This and a wide array of other ethnohistorical sources (e.g., Alva Ixtlilxochitl 1965; Anderson et al. 1976; Berdan and Anawalt 1992; Boone 2000; Durán 1971, 1994; Quiñones Keber 1995) frequently mention the presence of feathers in markets, trading expeditions, and tribute, and their use in social, political, and ceremonial events. Unfortunately, to date no featherworking workshop has been uncovered archaeologically, and only seven preconquest Aztec/Mixtec featherwork objects sit in museums today.[3]

There were two types of featherworkers: those who worked at palaces for royal or noble patrons and those who worked independently and lived in exclusive *calpolli* or urban neighborhoods. The most notable of these latter were from the calpolli of Amantlan—such was their fame that all fine featherworkers became known as *amanteca*. I propose that the basic units of production for fine featherworking were households, especially among the independent featherworkers and probably among those attached to palaces.

Featherworkers produced ornate objects in three ways: (1) feathers were tied together into long, flowing objects such as back devices, headdresses, feathered bracelets, fans, and banners; (2) small feathers were glued to solid backings to produce intricate mosaics such as shields; and (3) feathers were spun and woven into textiles. At the present time there is nothing to indicate that individual households specialized in just one type of featherworking; conversely, there is nothing to indicate that they did not so specialize. Each of these processes required somewhat different materials, tools, and skills, although there was considerable overlap. I give some weight to that overlap, and for the present discussion I assume that any given featherworking household was capable of producing, and did produce, all three types of feathered objects.

Acquisition of Raw Materials

Raw materials required for the production of exquisite feathered objects are listed in table 6.1. They included both costly and inexpensive materials and relatively inexpensive tools.

The most expensive materials required in any featherworking enterprise were the feathers—especially the shimmering, colorful, "exotic" ones. These feathers were attached to tropical birds such as scarlet macaws, lovely cotingas, roseate spoonbills, blue honeycreepers, troupials, several types of parrots, and of course the resplendent quetzal. Hummingbirds and their iridescent feathers,

Table 6.1. Raw materials used in the manufacture of feathered adornments

Materials	*Type of Object*	*Found in Markets*	*Sent in Tribute*	*Carried by Long-Distance Merchants*
costly feathers	tied, mosaics, textiles	x	x	x
"ordinary" feathers	tied, mosaics, textiles	x		
Textiles[a]	tied	x	x	x
cloth: cotton[b]	textiles	x	x	x
cloth: maguey	textiles	x	x	
animal hides[c]	tied, mosaics	x	x	
paper[d]	tied, mosaics	x	x	
cotton thread[e]	textiles			
maguey twine	tied	x		
wood[f]	tied, mosaics	x		

continued on next page

a Textiles have been identified on "Moctezuma's" headdress, seemingly woven of different materials and in different patterns (Moreno Guzmán and Korn 2012: 73).

b Both cotton and maguey cloth were already manufactured, not raw materials. Their production required raw cotton or maguey fibers, spindles, spindle whorls, looms (wooden pieces and fibers for the backstrap), and picks. The production of cloth was the domain of women; in theory at least, all women were expected to spin and weave cloth. A featherworking household would contain one or more such women.

c These are seen on "Moctezuma's" headdress, although the animal used is not known (Moreno Guzmán and Korn 2012: 73). Deer skins were paid in tribute by Tepeacac and jaguar skins by Xoconochco (Berdan and Anawalt 1992, vol. 3: folios 42r, 47r). "Cured leather" appears in Sahagún's market list (Sahagún 1950–1982, book 8:68).

d Two types of paper have been identified on "Moctezuma's" headdress (Moreno Guzmán and Korn 2012:73). *Amatl* (*amate*) and maguey paper were both used in the manufacture of feather mosaics. Both are mentioned as present in the Tlatelolco marketplace (see Sahagún 1950–1982, book 10:78).

e Cotton thread was most likely used to attach feathers to cotton cloth or to interweave the feathers with threads. We do not know if maguey thread was used in a similar manner with maguey cloth. Maguey fiber, but not cotton thread, is mentioned as present in the Tlatelolco marketplace (Sahagún 1950–1982, book 8:68).

f Wood was found in markets, and beams, planks, and pillars were given in tribute (Berdan and Anawalt 1992, vol. 3: folio 32r). But the wood pieces required for backings (mosaics) or supports (tied objects) were none of this sort. A possible reference to the types of woods used here is "fine wood for carving" paid from one province in the present-day state of Guerrero (Berdan 1986:127). I suspect that the carpenters or woodcutters listed in Sahagún (1950–1982, book 10:81) could have provided the featherworkers with their necessary pieces. Also, Cortés (1928:87) mentions that "wood of all kinds and in all stages of preparation" were available in the Tlatelolco marketplace.

Table 6.1—*continued*

Materials	*Type of Object*	*Found in Markets*	*Sent in Tribute*	*Carried by Long-Distance Merchants*
thin reeds[g]	mosaics	x		
stout canes[h]	tied	x		
cutting boards[i]	tied, mosaics, textiles	x		
knives (obsidian?)	tied, mosaics, textiles	x		
dyes[j]	tied, mosaics, textiles	x	x	
glues	tied, mosaics	x		
bone blades	mosaics			
maguey leaves	mosaics			
spindles/whorls	textiles			
loom parts[k]	textiles			
baskets	tied, mosaics, textiles	x		
bowls/pots	mosaics, textiles	x	x	

g Reeds are mentioned as possessed by the reed mat seller in the Tlatelolco marketplace (Sahagún 1950–1982, book 10:86).

h While a market presence of stout canes is not mentioned specifically in the sources, stout-cane carrying baskets are—the seller of stout cane carrying baskets also wove them (Sahagún 1950–1982, book 10:86).

i According to Sahagún (1950–1982, book 9:90), these boards, on which feathers were cut, were made of *ahuehuetl* (bald cypress) wood.

j The Aztecs used a wide array of dyes—at this point it is not clear which dyes were used in coloring feathers, although Sahagún (1950–1982, book 9:95) mentions *zacatlaxcalli*, a yellow climbing plant. Many dyes were present in the Tlatelolco marketplace, and some dyes (especially cochineal and yellow ochre) were paid in tribute to Tenochtitlan (see Berdan 1986:127–129).

j The backstrap loom is essentially a bundle of sticks with straps attached at each end, the part encircling the weaver's back being interwoven cordage. These elements may have been found in the marketplace, but the loom as a whole may have been the weaver's own construction—it is a rather personal implement.

also popular, were found widely throughout Mesoamerica; some were migratory and would have been available seasonally. To reach the centers of Aztec power in the central Mexican highlands, most of these feathers needed to be transported long distances on foot or by canoe where possible. They arrived primarily through three of the exchange mechanisms already mentioned: markets, long-distance trade, and tribute.

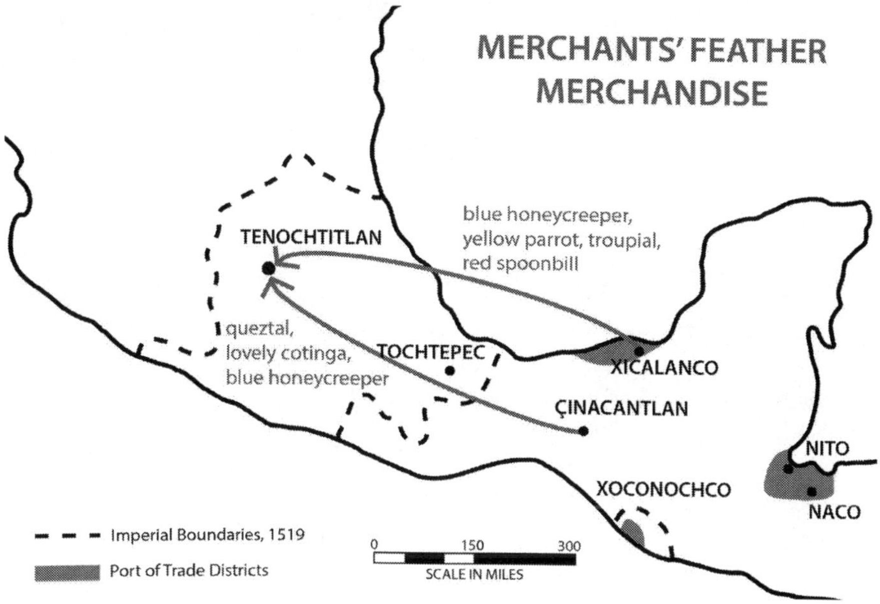

FIGURE 6.1. *Merchants' feather merchandise. (Drawing by Jennifer B. Lozano.)*

Well-heeled merchants offered fine feathers for sale in the Tlatelolco marketplace, and precious feathers were available in other regional marketplaces such as Tepeacac and Coaixtlahuacan (Berdan et al. 1996; Durán 1994:159, 182). Some of these long-distance merchants carried bundles and bunches of costly feathers from regions south of the Aztec imperial boundaries, essentially engaging in foreign trade (see figure 6.1). Some apparently also carried the birds themselves, as seen in the Codex Fejérváry-Mayer (León-Portilla 1985).

Costly feathers were delivered in tribute, on an annual basis, to the imperial capitals. These included bunches of quetzal feathers and probably handfuls[4] of roseate spoonbill, lovely cotinga, green or Pacific parakeet, and Montezuma oropendula (Berdan and Anawalt 1992, vol. 2:102–104, 110–111, 112–114, 116–118, 122–124). The feathers delivered in tribute would have been appropriate to the manufacture of tied and mosaic objects. In addition, the province of Tochpan paid an annual tribute of 20 bags of small white feathers, used to "trim" cloaks (Berdan and Anawalt 1992, vol. 4:108). Helpfully, Alva Ixtlilxochitl (1965, 2:197) tells us that the Gulf coastal province of Tzicoac (a neighbor of Tochpan) paid tribute in 20 bags of white feathers with which they made cloth. Both of these descriptions strongly suggest that these feathers were intended to be spun into textiles, and white down feathers are also mentioned by Sahagún (1950–1982, book 9:97) as elements in tied-feather adornments. Therefore, feathers for all three types of feathered productions were available through one or more exchange avenues: in markets on daily, weekly (5-day), or 20-day bases; in the caravans of long-distance merchants; and through the annual arrival of tributes from distant provinces.[5]

Other feathers, less than exotic, were locally available and widely used in the featherworking enterprise. "Ordinary" feathers such as those of ducks, crows, turkeys, and local/migratory waterbirds were used as underlayers in feather mosaics and as feathered objects in their own right. For instance, the Codex Mendoza distinguishes warrior costumes fashioned of "ordinary" feathers from those manufactured of "precious" feathers (Berdan and Anawalt 1992, vol. 4). Some of the less-expensive feathers may have included those from turkeys and crows (Sahagún 1993:270, 271, 274, 275). Rulers went about in cloaks made from duck feathers, and noblewomen similarly wore tunics (*huipilli*) of duck feathers (Sahagún 1950–1982, book 8:24, 47), the lowly duck perhaps not being so lowly after all.[6] The "feather seller" in the Tlatelolco marketplace reportedly owned the birds herself, plucking the small back and breast feathers of turkeys, geese, and ducks; she split and spun these tiny feathers into nice even threads with a spindle (Sahagún 1950–1982, book 10:92). These cozy threads could have been available in that marketplace all year long, the turkeys and ducks omnipresent in the Basin of Mexico setting.

Beyond their possible use in some warrior costumes, "ordinary" feathers do not appear in the tribute lists, nor do they seem to have interested the long-distance merchants. Feather artisans working with such feathers would have relied on the marketplaces for their supplies, with the one possible exception of the little white feathers paid by the provinces of Tochpan and Tzicoac (see above).

Other materials required in the manufacture of costly feathered adornments were maguey twine, knives, wood, baskets, animal hides, paper, glues, textiles, baskets, and possibly dyes for the tied objects; maguey leaves, knives, glues, bone blades or picks, dyes, paper, wood, ceramic vats/bowls, and baskets for the mosaics; and cloth, cotton and/or maguey thread, knives, spindles and spindle whorls, twine, baskets, a loom, and possibly dyes for the feathered textiles (see table 6.1). Any given household, whether working independently or attached to an elite palace, would have had little difficulty obtaining these materials through marketplace channels. At the very least, almost all of them are recorded for the Tlatelolco marketplace; it is not known to what extent other marketplaces might have had occasional or regular deficiencies in these materials. Very little arrived through tribute (most notably paper, bowls, cloth, and possibly the right kinds of dyes), so even the palace-attached artisans would have depended on marketplace vendors on a regular basis. The only materials not mentioned for the Tlatelolco marketplace are the maguey leaves and bone blades or picks used by the mosaic artisans; and the spindles/whorls, cotton thread, and loom parts needed to produce feathered textiles. Thin reeds and stout canes were probably available in the marketplace as part of the wares sold by the mat and basket vendors, respectively (Sahagún 1950–1982, book 10:86). Maguey, reeds, and canes were widely available in the highlands of central Mexico and could have been obtained informally. Bone blades or picks could have been acquired and regularly renewed from a recent meal: glyphs associated with featherworking images in the Florentine Codex (Sahagún 1950–1982, book 9:ill. 90) suggest that these were shin bones—although exactly whose shin is meant is not clear. Spinning was a household activity everywhere so the cotton thread would have been spun in-house. The sources of the spindles and their whorls has not been clearly established: the spindles are sticks and could have been simply fashioned by household members or been available in the marketplaces along with other wood products. Some spindles may have been made of bone (see McCafferty and McCafferty 2000). Ceramic whorls may well have graced marketplaces, but our reports of marketplace inventories largely derive from Spanish male (nonweaver) sources—these individuals may not have been able to identify such implements and therefore failed to mention them. Sahagún's informants would have certainly recognized them, but nonetheless did not mention them.

These listings of materials derive from ethnohistoric sources and inspections of existing featherwork pieces. Detailed examinations of these objects have been especially enlightening since some materials not mentioned in the documents appear on these pieces. For example, on "Motecuhzoma's" headdress, a quintessential tied object, paper and cotton backings are present, along with glues used to secure some of the feathers (Moreno Guzmán and Korn 2012). The coyote shield, a mosaic, required twine to attach the many gold pieces. And the shield in Mexico City, another mosaic, is decorated with animal skins and embellished with dangling maguey fiber tassels; its backing consists of thin reeds tied together with maguey or cotton threads (Rueda Smithers 2009:108). So, for instance, if a featherworking household focused on mosaics, it still would have required supplies of twine; if the household worked mainly on tied objects, it still would have needed to stock glues, paper, and cloth. With all of this overlap of materials, it is entirely possible, even likely that the production output of featherworking households encompassed most if not the full range of featherworking activities.

In addition to the materials listed in table 6.1, ornate feathered adornments also frequently incorporated other costly materials such as gold and precious stones. The inventories of objects sent from Mexico to Spain in the early sixteenth century attest to these additions: for example, "three shields, one the field green with some serpents of gold and blue in the center; the other, the field green with the head of an owl in the middle; the other, the field red with some fancy work of gold" and a "feather-piece, the center blue with stone mosaic-work, with other colored feathers, the border of green feathers, and lined with a tiger-skin" (Saville 1920:62, 72). Of the extant feathered objects in museums today, two especially exhibit the addition of gold ornaments: the coyote shield and "Motecuhzoma's" headdress, both in Vienna. Their distinctive and integral incorporation into the objects' designs strongly suggest that either the featherworkers themselves attached the gold pieces, or they worked in close collaboration with goldworkers (see below).

Tools and Procedures

For all the glamour and ornateness of these feathered adornments, the tool kit that produced them was singularly simple. Sharp knives, probably obsidian, trimmed feathers with an exquisite purity of line, and split tiny feathers for spinning. Maguey twines attached feathers to backings which sometimes were themselves networks of maguey twine (Moreno Guzmán and Korn 2012). Glues applied to feathers were derived from various orchid roots and pseudobulbs, and from beeswax (see Berdan 2007; Filloy Nadal, Solís Olguín, and Navarijo 2006). Flat maguey leaves, *ahuehuetl* wooden boards, and bone picks were used primarily in mosaic manufacture: the maguey leaves as a surface for preparing proper backings,[7] the wooden boards as feather-cutting surfaces, and the bone picks as an essential tool in straightening and aligning already-glued feathers. Baskets contained, constrained, and organized sometimes-unruly masses of feathers, and ceramic bowls and pots of different sizes were used in the preparation of mosaics. And spun feathers required the use of spindles/whorls and looms for the production of feathered textiles. With the possible exception of ultra-small spindles/whorls, these latter tools were women's essential equipment for the production of cloth generally and would have been found in any and all households regardless of social status, occupation, or residence.

To read Sahagún's (1950–1982, book 9:96–97) account, one might conclude that the fashioning of tied feather objects was fairly straightforward. Briefly, a frame was constructed and strengthened (with cloth and/or paper, perhaps). Quetzal feathers were laid out on the frame, their bases reinforced with pieces of cane. These feathers, neatly lined up, were bound together with maguey fiber nooses, shaken out to straighten them, and sewn to the frame. The same procedure was followed with other types of feathers (such as troupial or roseate spoonbill, and eagle down) to complete the device (Sahagún 1950–1982, book 9:96–97). However, if "Motecuhzoma's" headdress is considered, there was more to the process than that reported by Sahagún's informants. This headdress also exhibits glued feathers, a network of maguey fiber netting to which feathers were lashed, and animal hide. The object was further embellished with numerous round and half-moon gold pieces tied to the backings (Moreno Guzmán and Korn 2012).

Another set of procedures involved the protracted and painstaking process of producing feather mosaics. The creation of these elegant objects entailed several stages and operations, some of them necessarily sequential, some ongoing, and some on-call. The most detailed account of these procedures is provided by Sahagún's native associates (Sahagún 1950–1982, book 9:93–96). First, a scribe was enlisted (hired?) to draw the desired pattern. The featherworkers then carefully prepared a backing of wood or reeds, painted designs, glue-reinforced cotton, and paper supports. Thin strips of black and other colored feathers were glued to the backing to provide outlines or borders.[8] Layers of inexpensive and expensive feathers were then applied, the ordinary feathers providing a bed for the costly ones. The bottom-most layer consisted of glue-hardened feathers (feathers dipped entirely in glue) whose colors matched the costly feathers to be placed on top of them. The expensive, precious feathers were then arranged and glued on top of the glue-hardened ones to finish the object. Examination of existing feathered mosaics reveals that pattern segments were often cut out separately and added to the whole design somewhat in the manner of a jigsaw puzzle, probably by a master artisan. For example, this is the case with the coyote shield, with its numerous pieces of blue feathers bordered in gold. On this object, the gold pieces were carefully folded under, and the underside parts were then sewn to the backing. During this entire process, the carefully selected feathers were repeatedly laid out in trial designs, matched or replaced, trimmed as necessary, arranged, rearranged, and rearranged again until the master artisan was satisfied. Painstaking workmanship was a hallmark of this enterprise at every stage of manufacture. It was also complex: the fashioning of feathered mosaics involved sequential activities, each stage depending on the prior completion of other stages. But there were also ongoing activities which could be undertaken at any time, especially the trimming and dying of feathers. Still other activities, notably the making of glues, were sporadic and situational, essentially on-call.[9]

The third manner of manufacturing feathered adornments was the adding of feathers to cloth, reportedly by spinning. Sahagún (1950–1982, book 10:92) identifies the marketplace feather seller as not only an owner of birds but also a spinner of feathers. He also describes women's spinning and weaving duties, using "shallow" spindle whorls for spinning feathers (1950–1982, book 8:49); McCafferty and McCafferty (2000:47) suggest that some small spindle whorls unearthed in Cholula may have been used for spinning feathers. No feathered textiles exist from precolumbian times, but a colonial piece from Mexico is highly suggestive of these production techniques. This object, in the Cooper-Hewitt Design Museum, Smithsonian Institution, is composed of cotton warp yarns and two types of weft yarns: "The unusual white yarn consists of downy feathers and cotton spun together in a two-ply yarn" and the other variously colored yarns were "composed of a very fine animal hair." The feathers have been identified as goose, and the animal hair as rabbit (Phipps and Commoner 2006:486). Particularly interesting is the fact that downy feathers and cotton thread were spun together, which would make the spinning of feathers a manageable task. The use of white downy feathers in this type of operation is consistent with the other descriptions of white down already discussed.

The Production Process: Labor Needs and Arrangements

Luxury-feather artisans worked in three separate recorded contexts. As *tecpan amanteca*, they resided at the ruler's palace and created the feathered capes for the god Huitzilopochtli. These were intricately fashioned of quetzal, hummingbird, and blue cotinga feathers (Sahagún 1950–1982, book 9:91). Durán (1971:73 and plate 3) describes this god's green feathered headdress and feathered cape embellished with gold. His shield featured five tufts of white feathers and a border of yellow feathers (also see Sahagún 1950–1982, book 12:52–53). Yellow feathers and white down were given in tribute by conquered provinces, although hummingbird feathers were not. The production of these adornments would have been the domain of the *tecpan amanteca*, and would have required all of the featherworking skills: tying, mosaic-making, and textile production. These featherworkers also made other exquisite feathered objects that were bestowed on guests as royal gifts. Among them were probably the "three loads of cloaks of rich feather work" presented by Motecuhzoma to each of Cortés's captains (Díaz del Castillo 1963:221). Sahagún (1950–1982, book 9:91) emphasizes that at least by the time of Motecuhzoma Xoxoyotzin, the ruler settled these featherworkers at his palace, providing them with a house (*centetl calli*) of their own. There must have been several of these artisans, as the good friar goes on to say that those of "Tenochtitlan and Tlatelolco mingled with one another." The meaning of *centetl calli* is unclear. Were all the featherworking artisans housed in a single dwelling or, more likely, in rooms around a single patio workspace? Were these complete households with husbands, wives, and children (as I suspect)? Or was each artisan family housed in its own house in or near the palace? Or perhaps what was meant was the *totocalli* (bird house), an aviary with an array of captive birds including eagles, quetzals, parrots, ducks, and other waterbirds (Díaz del Castillo 1963:228–229); it contained designated workspace for artisans including featherworkers, goldsmiths, wood carvers, and stone mosaic-makers (Sahagún 1950–1982, book 8:45). The proximity of these several crafts would have provided an enhanced degree of overall productive collaboration and efficiency.

Other types of featherworkers, *calpixcan amanteca*, apparently also lived at the palace and created the finery of the ruler when he danced. This included quetzal-feather headdresses, fans, and banners (some embellished with gold); a feather arm band with gold; and a headdress of red spoonbill and quetzal feathers with gold (Sahagún 1950–1982, book 8:27–28). These featherworkers had access to Motecuhzoma's storehouse—they could use materials obtained through tribute or other royal means, and each object they made was stored and guarded in the ruler's storehouse (Sahagún 1950–1982, book 9:91). Their association with the palace and its resources suggests that their labor relations probably resembled those of the *tecpan amanteca*.

It has already been observed that women, bird-owners, spun feathers in the marketplace. These were probably commoner women. In addition, we have indications that women attached to the royal palace (who could be either commoners or nobles) engaged in weaving feathered textiles. Bernal Díaz del Castillo (1963:230), in speaking of Motecuhzoma's palace, mentions that women there produced "a huge quantity of fine robes with very elaborate feather designs." This same observer states that "chieftains' daughters" and "daughters of other dignitaries," housed near the Templo Mayor in Tenochtitlan, wore robes "entirely of featherwork." They may have woven these garments themselves, a suspicion supported by Sahagún (1950–1982, book 8:49) when he speaks of women spinning feathers while entertained by hunchbacks and dwarfs who sang and played music for them.[10] It is difficult to conceive of commoner women being so entertained.

Other luxury featherworkers, called *calla amanteca*, worked more independently in specialized *calpolli*, or neighborhoods, producing feathered objects and adornments for sale in the marketplace. This is recorded for Tenochtitlan/Tlatelolco, and it is likely that similar arrangements, both palatial and private, occurred in other city-states. It is not clear how much the *calpolli* was involved in the portioning of jobs, for example, or the regulation of standards, or the training of neophytes. The independent featherworkers would have obtained their precious feathers and other materials from the neighboring *pochteca* merchants (at least in Tlatelolco) or in the bustling marketplaces (see table 6.1).

As was usual for most Aztec specializations, featherworking parents taught their children the skills of their trade. Girls were expected to develop a keen eye for feather color variations, and boys served as apprentices, one of their duties being the making of glues. But once again, things may not be as clear-cut as they seem. Did all sons of featherworkers become featherworkers themselves? Perhaps not, as some could be dedicated to schooling in a priestly school or *calmecac*; while this education was intended to instill lofty artisanal values, it might nonetheless lead to a different life-path altogether (Sahagún 1950–1982, book 9:88; Durand-Forest 1994:173). And then, were the only apprentices necessarily sons of featherworkers? Might a featherworking father send his son to apprentice with another featherworker? We do not have definitive answers to these questions, although one might suspect that for the most part these specialized households and *calpolli* tended to be fairly insular and focus on training their own progeny to perpetuate the craft and protect their accumulated resources (including specialized knowledge and skills).

Descriptions of featherworking procedures, touched on above, indicate that the manufacture of any feathered item required a complex interplay of separate but related activities. Ongoing tasks, performed by girls and women, consisted of selecting, dyeing and trimming feathers in preparation for their use. Sequential tasks, performed by the master artisan but also surely by various associates, included commissioning the design, preparing the several types of backings, creating the many design pieces, and arranging the different layers of feathers. On-call tasks, especially the making of glues by apprentice boys, were sporadic and situational, dependent on the immediate needs of those assembling the feathered piece. All of this suggests a household division and coordination of labor drawing on men and women, boys and girls. Some of the activities, especially those performed by women, could be readily interspersed with other household duties. Most if not all of these activities undoubtedly took place out-of-doors in a patio setting, where there was sufficient space and the emerging piece could be judged in the light in which it would be admired. However, inclement conditions such as rain and wind may have periodically disrupted these daily activities.

Not all featherworking took place in-house: cooperation extended beyond the household. Much as scribes provided designs for the feathered pieces, the featherworkers made designs for the gold workers; these latter artisans "join with [and] are instructed by the feather workers," their collaboration coming as no surprise, given the nature of the multimaterial objects produced (Sahagún 1950–1982, book 9:76).

Such a household system may have worked well for the independent featherworkers, but was it the only possible form of organization to get the job done? The documentary passages describing palace artisans suggest a single "house" for several artisans but say nothing about the division of labor within that "house." It is my sense that when the settlement of artisans at a palace is mentioned, it refers to the entire household work force. It may well be that the "house" at the palace, like the *totocalli*, was large indeed, with many individual rooms, a large patio working area, and ready access to the raw materials of the trade. Also like the *totocalli*, it may be that different types of artisans were housed together in this manner, making their necessary interactions relatively easy and convenient.

It may have been that several feather artisans worked on a single piece simultaneously. Speaking of featherworking in the Colonial period, Juan de Torquemada (1969) wrote that

> if there are twenty artisans, they all make an image together, and dividing among themselves the figure of the image, into so many parts, however many there are, each [artisan] takes his piece to make it at his house, and afterwards each one returns with it [the finished piece], and they all join together, and in this way the perfect and completed image results, as if one artisan had done the work. (1969, vol. III: 210; translation and brackets, FB)

With so much continuity in this craft from precolumbian times on into the Spanish colonial world, it is interesting to entertain the idea that this might have been a strategy used in Aztec times. It assumes the action of some centralized commissioning "agent," whether a king, a noble, an overarching *calpolli* authority, or a respected master featherworker. The results of this system can be seen in a sixteenth-century colonial feathered triptych at the Metropolitan Museum of Art in New York City. The three segments of this religious piece were clearly executed by three different featherworkers:

the difficult, detailed, and meticulous work in the central section was probably completed by an accomplished master featherworker, and the artisan of the left panel was more proficient than his colleague on the right. Each of these artisans also had slightly different ways of executing similar motifs, and the finished item is not quite as ideally perfect as Torquemada describes. It is intriguing to consider that the many different pieces of the precolumbian coyote shield could have been doled out in such a fashion—in this case, Torquemada would have been proud of their perfect harmony. The consistent colors, orderly directions of the feathers, and homogeneous feather and gold sizes all point to uniform training and a high degree of quality control over the work of all of the participants.

THE BIGGER PICTURE

I have been emboldened to undertake this study of Aztec-period featherworking by the framework proposed by Richard Blanton and his colleagues in their goods-based approach (2005). That framework is fleshed out by four questions that are designed to clarify the role of commodities in their broader economic, social, political, and religious milieu. Where do those questions lead us in the matter of Aztec-Period featherworking?

1. *The presence and impact of labor and time allocation bottlenecks:* Bottlenecks in the featherworking industry could derive primarily from labor imbalances and resource availability, as well as varying consumer demand. All of these potential problems were relatively easily addressed in the Late Postclassic Aztec world. Featherworking households required trained labor of appropriate age and gender composition. One can imagine any given household, during its "usual" life cycle, lacking either male or female children, or sufficiently available or trained adults of either gender. Nonetheless, strategies were available to offset possible imbalances: households could be enlarged (as joint or compound families), and children could be apprenticed to a neighboring master featherworker. Both possibilities are suggested in the documentary record. Guild-like arrangements among the independent featherworkers, and the grouping of featherworkers and other luxury artisans in palaces, offered opportunities for shared labor among the artisan households.

   Availability of raw materials, especially costly tropical feathers, was an ever-present issue with the featherworkers. Those working in royal palaces had access to the ruler's tribute stores and, at least in the case of the Tenochtitlan ruler, a well-stocked aviary. Independent featherworkers relied on long-distant merchants, their neighbors, for supplies of exotic feathers. Both types of featherworkers could draw on markets for their supplies. Yet none of these sources was flawless. For the palace featherworkers, a rebellion in a tribute-paying province or a miscalculation (or misappropriation) on the part of a tribute collector or overseer could affect supplies of essential feathers. For the independent feather artisans, a merchant caravan might be ambushed in distant lands (a not uncommon occurrence), or the merchants may have returned from their long and perilous journeys with inadequate supplies. Ubiquitous markets with their wide range of commodities may well have offset some of these possible issues. Other materials used regularly in the manufacture of feathered objects were readily obtained in the many marketplaces throughout the imperial domain.

In addition, featherworkers relied on consumers who needed, could afford, and were allowed to acquire their fancy array. Some consumers purchased their featherwork in the marketplaces, while others, rulers and other high-ranking nobles, were directly supplied from attached artisans. Some consumers received feathered adornments as gifts or rewards (Berdan 2014:260–268). Inasmuch as a ruler supported the activities and success of his city-state, he would have provided any necessary colorful array from his storehouse to enhance his polity's image and status; this finery was stocked from tribute payments and in-house production. Consumer demand would have fluctuated with the regular or spontaneous occurrences of political, military and ceremonial activities, which benefited from the flamboyant display of feathered adornments.

2. *Relationships between the commodity and dynamic distribution systems:* Precious feathers and other materials and tools used in featherworking all moved variously through tribute, trade, and market channels (see table 6.1). A relatively small amount of the featherworkers' production needs were supplied through either tribute (for palace featherworkers) or long-distance merchants. Principal among these materials were the costly feathers native to regions distant from the imperial capitals. Nonetheless, quantities of these feathers also appeared in marketplaces throughout the realm, as did the many additional materials and tools of the featherworkers' trade. These included such mundane (but essential) objects as maguey twine, obsidian blades, glues, dyes, and baskets. Going full circle, some featherworkers took advantage of the great Tlatelolco market to sell their finished feathered objects there. The increasing

commercialization of the Late Postclassic economy paralleled the increasing demand of luxuries, including featherwork, among noble consumers. Marketplaces were especially prominent in filling the featherworkers' material needs, and also provided at least some of them with promising outlets for their labors.

3. *The good's impact on secondary industries or markets:* The complex featherwork production system drew on a large variety of producers and served a demanding cadre of consumers. Relatively low-status households manufactured many necessary components of the featherworking enterprise: obsidian blades, wooden boards and backings, glues, dyes, animal hides, paper, twine, baskets, and bowls were produced in part or in full throughout the imperial realm by households on part-time or full-time bases. Some of this production may have engaged households as intermittent crafting and/or multicrafting activities (Hirth 2009), and served the featherworkers and others as secondary industries. Such households could boost their income by selling these materials and objects in the many marketplaces throughout the realm. The output of these households would have served more than just the featherworkers, since most of these materials and objects were used in many industries, and demand in the markets may well have been quite brisk. These materials and objects hold up well. Although a few of them (such as some glues and dyestuffs) responded to seasonal rhythms, they were also eminently storable—seasonal variations in availability would not be an issue with these adjuncts to the featherworking industry. And it perhaps is a testimony to the vitality of the market system that a crucial high-end industry such as featherworking came to depend on marketplace availability of so many essential materials and tools.
4. *Broader impacts of increased production of the commodity:* With increasing numbers of nobles came increasing demands for sumptuous display objects. Among the finest, most extravagant of these, were objects made with feathers. Blanton and his colleagues speak of the "solidification of aesthetic labor" whereby decoration comes to take precedence over form to give meaning to objects, and that this process was more common in the Late Postclassic than in earlier periods (Blanton et al. 2005:280). It is worth noting that the color, texture, vibrancy, and variation of feathers provide a particularly attractive and effective medium to achieve such embellishments and convey detailed, socially charged meanings.

Throughout this discussion I have taken the position that the production of luxurious feathered objects took place in the context of individual households. This position is supported by documentary evidence for the craft's labor requirements and the style of training used in the profession. But we have seen that other slightly variable styles of production, in palace settings, may have offered enhanced conditions for the efficient and masterful production of at least some of these complex objects: the palace setting provided a stage for the collaboration of a variety of interdependent artisans, and these artisans enjoyed access to the palace's (or city-state's, or empire's) tribute stores. As the empire expanded into areas of luxury commodity availability (especially lowlands), more and more fancy feathers became available to the palace featherworkers.

This leads us to a final consideration. While much exquisite featherwork was produced in the Basin of Mexico, in city-states central to imperial expansion and commercial enterprises, much was not. Featherworking was a deep-seated craft throughout the Aztec imperial domain and beyond; feather-adorned Aztec warriors met similar warriors on distant battlefields, tired and dusty merchants encountered exquisite featherwork in "foreign" lands, and a bewildering array of feathers and feathered objects appeared in market after market throughout the realm. Many manufactured feathered items, largely in the form of headgear, back devices, warrior costumes, and shields entered Tenochtitlan through tribute. These facts indicate that such objects were being produced widely and that access to the necessary productive materials as well as the skills to manufacture them was well established in broad geographic regions. This is highlighted by the example of imperial tribute in quetzal feathers and quetzal feather devices: the feathers were derived from a few restricted areas, while the fancy devices were demanded from polities throughout the empire where they were probably manufactured (figures 6.2 and 6.3). In addition, their demand in tribute by the imperial powers suggests a need that might not have been sufficiently met by local artisans. In the end, we are left to wonder just what the lofty and demanding Mexica, Acolhua, and other Basin of Mexico nobility thought of the featherwork arriving from the provinces. After all, as Richard Blanton, Lane Fargher, and Verenice Heredia Espinosa so cogently say, the Aztec elite displayed "little evidence of consumer reticence" (2005:280), and undoubtedly set high standards for their personal, stately, and godly adornments.

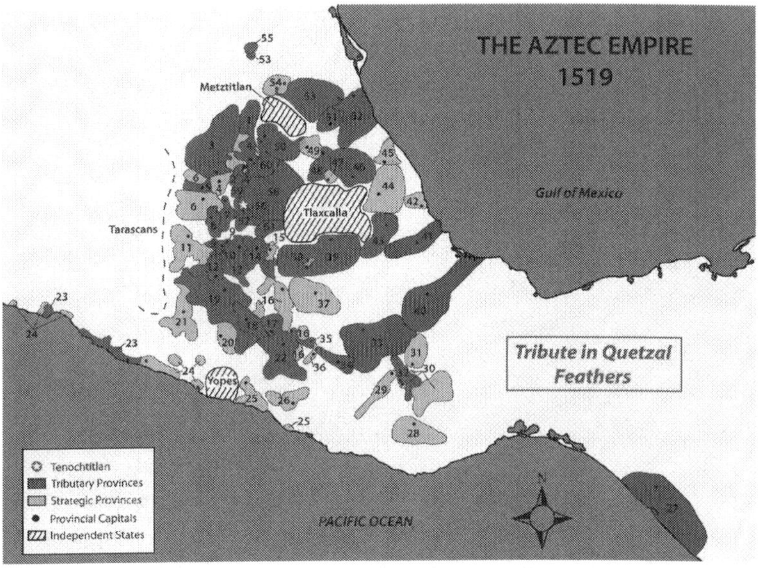

FIGURE 6.2. *Aztec tribute demands in quetzal feathers. (Drawing by Jennifer B. Lozano.)*

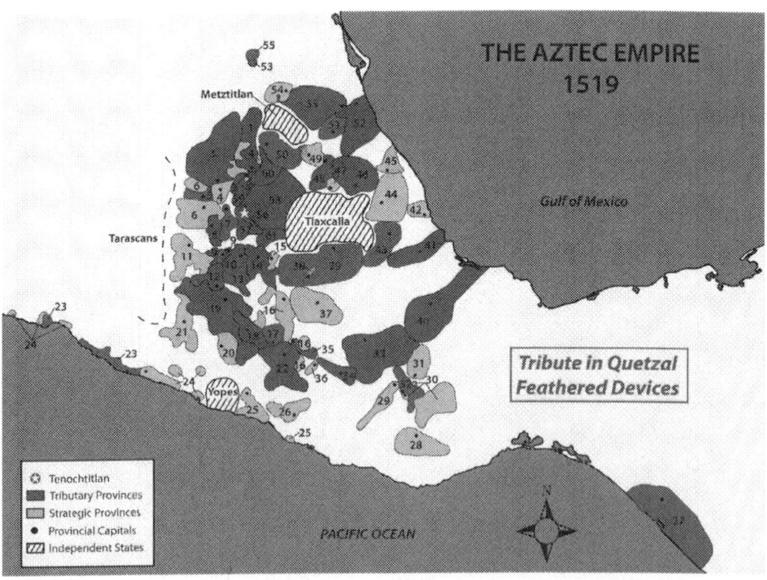

FIGURE 6.3. *Aztec tribute demands in quetzal-feathered devices. (Drawing by Jennifer B. Lozano.)*

NOTES

1. Michael Smith (2003c:123) distinguishes the different roles of luxuries in commercialized economies and prestige-goods economies. In the former, acquisition of fine goods was open to all consumers (who could afford them) with few status restrictions; in the latter, the production, distribution, and consumption of exotic goods were controlled by and restricted to elites.

2. If cacao were indeed available to commoners in preconquest times, the watering down may have been an economic as much as a social response: with less wealth, commoners could make their supplies of cacao go further, sacrificing richness.

3. There are two in Mexico City (a mosaic shield and a mosaic disk), two in Stuttgart (two mosaic shields), and three in Vienna (a mosaic shield, a headdress, and a fan). Some questions have been raised about precolumbian origins for the disk in Mexico City and the fan in Vienna. However, I believe that both of them derived from pre-Spanish times.

4. These are glossed as handfuls in the Codex Mendoza but just as feathers in the Matrícula de Tributos. I used to prefer the interpretation that these came in units of 8,000 feathers (from Tochtepec and Xoconochco provinces), but now am more inclined to think of these as the very small, soft feathers from the necks, backs, and breasts of the birds, in which case deliveries by handfuls would make more sense. These are the types of feathers used in fashioning feather mosaics.

5. Tributes reportedly arrived on a quarterly, semiannual, or annual basis. With the possible exception of distant Xoconochco, feathers were always paid annually.

6. Sahagún (1950–1982, book 10:92) mentions domestic and wild ducks (probably Muscovy ducks, nonmigratory avians), and Peru ducks.

7. Carded cotton (thin as a "cobweb") was stiffened with glue, the procedure taking place on the maguey leaf. Experiments have revealed that the glue does not stick to the maguey leaf, and the stiffened cotton piece peels off easily (Laboratory for Ancient Materials Analysis, California State University San Bernardino).

8. Sahagún (1950–1982, book 9:95) identifies the black outline feathers as those of the grackle. From even a short distance, these borders are so finely executed that they appear to be painted.

9. Sahagún (1950–1982, book 9:97) additionally describes the manufacture of small animals from wood, dried maize stalks or paper, glue, cotton, and feathers. Rivero Weber and Feest (2012:48) offer the intriguing suggestion that these figures may have been toys made for sale in colonial Mexican markets.

10. These statements are made in the context of discussions of palace life.

7

Classic Maya Marketplaces and Exchanges

Examining Market Competition as a Factor for Understanding Commodity Distributions

Lisa J. LeCount

Mounting evidence indicates that large, Classic Maya Peten capitals such as Tikal and Calakmul supported permanent marketplaces (Carrasco Vargas, Vázquez López, and Martin 2009; Jones 1996; Masson and Freidel 2012) and smaller provincial centers in the upper Belize River valley (Cap 2011; Keller 2006) may have sustained periodic markets as well. Nonetheless, archaeological signatures for marketplaces and commodity exchanges continue to be difficult to find across the lowlands (Shaw 2012). To understand why this might be the case, I used Richard Blanton's ideas that link the relationships between interest groups to organizational aspects of states. Like Blanton, I suggest that rulers promoted the development of marketplaces as a means of consolidating authority, generating taxes, and stimulating craft production. In the Classic Maya case, embedded sociopolitical interactions between market participants led to underdeveloped commercial market systems, especially in provincial centers where polities were small in scale and politically bounded. The small size of market zones and volatile relations between Maya kingdoms resulted in fluctuating supply and demand of commodities, dampening investments in permanent market infrastructure and allowing interest groups to manipulate exchange values.

A key assumption here is that the degree to which ancient market economies conform to archaeological expectations for marketplace development and market exchanges is conditioned not only by the primary

DOI: 10.5876/9781607325338.c007

economic forces of supply and demand, but also by the nature of sociopolitical interactions among interest groups who have a stake in markets. Combined, economic and sociopolitical forces affect the scale of competition within market systems and govern the outcome of market exchanges and commodity distributions.

This perspective is informed by Blanton and Fargher's recent research into the role of collective action in state formation and market development (Blanton and Fargher 2008, 2010). They suggest that organizational aspects of states can be understood based on "outcomes of bargains struck between those in power and non-ruling groups" (Blanton and Fargher 2010:211). In situations where interest groups maintain considerable resources and authority, rulers or other key decision-makers negotiate equitably with them on matters of public goods and services; in contrast, when interest groups are endowed with few resources, leaders hold the upper hand in political decision-making. These negotiations directly affect rulers' decisions to invest in infrastructure and administrative institutions to guard them, as well as to conform to social contracts, moral codes, and norms concerning personal accessibility and reciprocal obligations, including redistribution of goods provided to the state by taxpayers.

In their comparative study of 30 states, Blanton and Fargher (2008, 2010) found support for the theoretical expectations of collective action theory, but more to the point of this chapter, they present positive statistical correlations between market development and three societal factors: degree of collective action, scale of agricultural intensification, and size of population and urbanization. They postulate that markets provide alternative sources of incomes that allow commoners to specialize in the production or sale of goods and break away from patrimonial control over land and resources, while rulers benefit by leveling taxes on marketplace exchanges. Households located in prime agricultural areas tend to specialize in the production of staples, a pattern that may explain the correlation between agricultural intensification and markets. Markets also are positively correlated with the polity size and urbanization. Population growth and aggregation into centers may increase the potential consumer base, especially among non-food-producing urbanites, making specialist production in craft goods and food more feasible.

Based on these studies, it can be predicted that many large ancient polities with dense urban centers were more likely than smaller polities to have supported market economies, unless small-polity leaders encouraged market exchange and the consumer base was relatively large. Part of the problem for the development of markets in smaller polities is generating sufficient supply, demand, and price information for transactions to become predictable (Garraty 2010:7). These requirements can be met in open geopolitical landscapes that accommodate the movement of market participants including merchants and local sellers and buyers. In politically fragmented landscapes, markets may develop but transactions are limited to people embedded in existing social relationships. Blanton (2013:25) calls these types of markets "restricted" because they replicate existing social structures and exchange relations within the polity. As such, market transactions resemble reciprocal gift exchanges more so than commodity transactions, and elites are prone to meddle in market development. In more open geopolitical environments, buyers and sellers are free to engage in atomized economic transactions, especially between merchants strangers (Granovetter 1985). Geopolitical and societal factors therefore shape market forces and must be taken into account when modeling ancient market systems and marketplace exchanges.

Four approaches—configurational, contextual, spatial, and distributional—provide independent and complementary lines of evidence for identifying marketplaces and market exchange in ancient societies (table 7.1).

A configurational approach focuses on identifying marketplaces directly from infrastructural features, while contextual, spatial, and distributional approaches focus on the indirect effects of marketplace exchanges. Kenneth Hirth's (1998) distributional approach is particularly useful because it provides archaeological correlates for identifying exchange modes from the perspective of household and settlement provisioning (table 7.2).

The major tenet of Hirth's model is that, in markets, individuals may buy and sell basic commodities regardless of their social rank because provisioning networks operate independently of sociopolitical relationships. One line of evidence that supports unfettered exchange in Mesoamerican markets is ethnohistoric accounts that document the fact that the majority of individual sellers were producer-sellers unregulated by authorities (Hirth 1998:455). These kinds of highly competitive marketplace exchanges can be identified archaeologically as homogeneous household artifact assemblages in terms of sources and quantities of commodities. Once commodities enter competitive marketplaces, their exchange values are determined by the forces of supply and demand, which conventionally refers to the behavior of buyers and sellers who engage in strictly commercial transactions. Where there are many buyers and sellers, none of them, on their own, can affect market prices. This self-regulating behavior, commonly referred to by economists as the "invisible hand of the marketplace," keeps the value of commodities competitively priced once a product or service has been on the market for a long time and there are many substitutes or sources for the product. Although purchasing power should affect, at least to some extent, the distribution of high-value

TABLE 7.1. Archaeological correlates for approaches to market systems (after Hirth 1998)

CONFIGURATIONAL	Indices of spatial and architectural features of market behavior include presence of centrally located plazas, market infrastructure, transportation arteries, administrative precincts, and large walled compounds.
CONTEXTUAL	Indices of cultural features required for large-scale provisioning of commodities including urbanization and full-time specialists.
SPATIAL	Indices of regional distribution of commodities or marketplaces including fall-off curves of commodities reflecting distance from production source to market centers and the arrangement of sites predicted by central-place theory.
DISTRIBUTIONAL	Indices of commodity provisioning based on the differential distribution of commodities across households, especially long-distance commodities of obsidian goods and imported ceramics.

TABLE 7.2. Outcomes for Hirth's distributional model of exchange modes

RECIPROCITY	Dyadic exchanges result in low volume, small spheres of exchange, and heterogeneity in household consumption that reflect different social networks and procurement patterns. Household production results in less-standardized commodities.
REDISTRIBUTION	Centralized circulation of commodities results in significant differences in commodities across statuses and multiple, parallel circuits of exchange.
MARKET	Nonhierarchical provisioning results in homogeneous assemblages across households and statuses in a community. Specializations increase scale, segmentation, and efficiencies of production, including standardization.

goods in low-status households, the disparity between elite and common household assemblages should be negligible in comparison to the outcome of other exchanges modes (Hirth 1998:456).

Late Postclassic Mesoamerican markets were highly competitive, but I contend that most ancient marketplaces were not, due to the limitations of transportation, communication, and influence of sociopolitical forces on participants (also see Garraty 2010:7). Interest groups can easily manipulate the consumption of goods, especially if goods communicate information concerning social affiliations or political position (Appadurai 1986:31). Sumptuary laws limit use of symbols of royal office, but more common symbols of ethnic, clan, or other sociopolitical affiliations are also subject to regulation through more informal social mechanisms. Further, market participants can engage in formal price setting to encourage consumption by artificially holding down exchange values or control consumption by limiting supply and inflating exchange values (Block and Evans 2005:509). Trade goods are especially prone to price fixing, fluctuations in supply, and monopolization because middlemen are few in number and travel across polities is perilous without political connections.

Locally made items are not exempt from manipulation of exchange values because their supply and demand are affected by embedded social relations between producers and sellers. When producers and consumers negotiate directly for commodities, pricing includes consideration of personal social relationships and notions of fairness, especially in economies typified by numerous small-scale producers, sellers, and buyers operating with limited means (Mintz 1961). In small-scale markets, people who buy and sell are often on friendly terms and attempt to sustain regular business dealings by establishing long-term reciprocal relationships (Mintz 1961:55). In offering and accepting concessions on commodity prices, sellers acquire a group of steady customers and good customers get lower prices. In marketplaces where middlemen vend wares, rather than the artisans who make them, sellers have more autonomy in setting the exchange value of the commodity (Dilley 2005). Nonetheless, price concessions develop when sellers have relatively large stocks of goods (Mintz 1961:58).

Market competition therefore should be considered as a scale that ranges from strong to weak based on the number and social embeddedness of buyers and sellers. The scale of competition directly affects exchange values, which ultimately conditions the distribution of commodities in households. Blanton's comments on the varied nature of ancient market economies are salient here.

> Hirth's finding that imported pottery and obsidian were available to households of varying socioeconomic status, evidently through market purchase, is an important one, but I am worried that other researchers not finding the same patterns in their distributional data will infer from it an absence of markets. The particular circumstances found at Epiclassic Xochilcalco may make it unusually well suited to the kinds of analysis Hirth emphasizes; other market situations maybe more complex and less easy to decipher. (Blanton 1998b:464, comment on Hirth 1998)

In order to understand why Classic Maya market economies are not easy to decipher, I investigate marketplaces and commodities to explore the forces that shaped them. I start with an application of Blanton and Fargher's collective action theory to the Classic Maya lowlands to examine the predicted degree of state involvement in marketplaces and then compare this model to the actual evidence for marketplaces, including that from the upper Belize River valley at the sites of Xunantunich, Actuncan, and Buenavista del Cayo (referenced simply as Buenavista below). For data concerning commodities, I present the distribution of local pottery types across Late Classic households at Xunantunich and diachronic shifts in access to obsidian through time at Actuncan. The uneven distribution of obsidian and some, but not all, types of local pottery documents manipulation of exchange values within a moderately competitive but restricted market system.

CLASSIC MAYA POLITIES, MARKETS, AND COMMODITIES

Classic Maya Polities and Collective Actions

Xunantunich, Actuncan, and Buenavista sit on the Mopan River only a few kilometers from each other near the western border of Belize (figure 7.1). At times, these sites were volatile members of a larger multipolity network centered at Naranjo, an expansionistic Peten capital located less than 20 km to the west. The degree to which Mopan River centers differed economically and politically from larger lowland Maya capitals, therefore, had more to do with scale and setting than cultural practices.

One of the most pervasive characteristics of Maya politics is the cyclical nature of political power (Marcus 1993). These cycles occurred at regional scales when strong capitals incorporated provinces into a multipolity state and at local scales when provincial capitals subsumed hinterlands into their political sphere of influence. As the power of capitals waned, previously subordinate groups regained their independence forming more decentralized polities or autonomous centers. While marriage alliances and war were the two most common strategies for incorporating centers into political networks, wars between Tikal and Calakmul and their allies (including Naranjo) escalated in the Late Classic period (Martin and Grube 2008), as did raiding. In this political milieu, polity boundaries and loyalties were never settled and almost always contested.

These dynamic cycles can be seen within Mopan River centers at the regional and local scale. Actuncan was the political center of the upper Belize River valley in the Terminal Formative (or Preclassic) period (100 BC–AD 250) as evidenced by many hallmarks of early civic construction, including a triadic pyramid group, an E-Group, and a ballcourt (Mixter et al. 2013). Buenavista and Xunantunich overshadowed Actuncan during the Late Classic period (AD 600–780), when leaders at both sites initiated new civic construction projects and competed for influence over minor centers located between them in the countryside (LeCount and Yaeger 2010). By the early eighth century, Xunantunich, with the aid of its Naranjo overlords, became the provincial capital of the valley, surpassing Buenavista in the scale and elaboration of

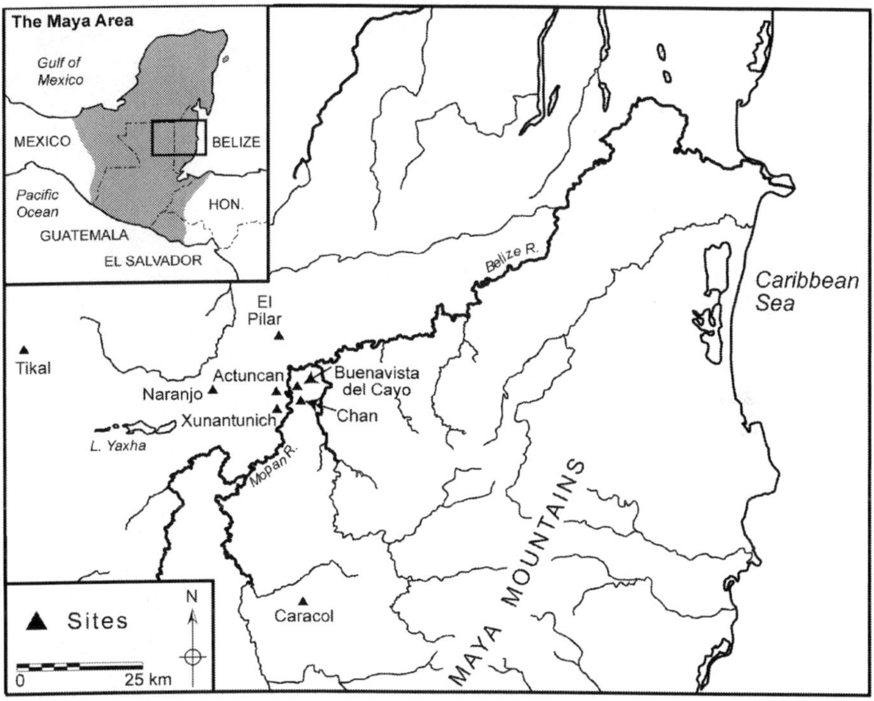

Figure 7.1. *Location of select archaeological sites in the eastern periphery of the Maya lowlands.*

its monumental architecture (LeCount and Yaeger 2010). During this time, Actuncan, and possibly Buenavista, were subsumed within the Xunantunich polity. At Actuncan, an ancient palace was remodeled befitting a vassal noble (Mixter et al. 2013), while at Buenavista settlement clusters appear to have become increasingly complex with the addition of administrative buildings that may represent the imposition of a new level of bureaucracy (Yaeger et al. 2010:165–167). However, Xunantunich's control over sites in the upper Belize River valley lasted only a short time. In the eighth century, textual references on Xunantunich Panel 2 refer to "flint and shield" events undertaken by the site's ruler and allies at an undisclosed location within the Mopan region (Helmke et al. 2010: 103). These skirmishes appear to be harbingers of events that led to the desecratory termination of Xunantunich's palace in the Late Classic period (Yaeger 2010). By the Terminal Classic period (AD 780–1000), claims of authority were increasingly made by neighboring centers. None of these sites, however, regained lasting authority.

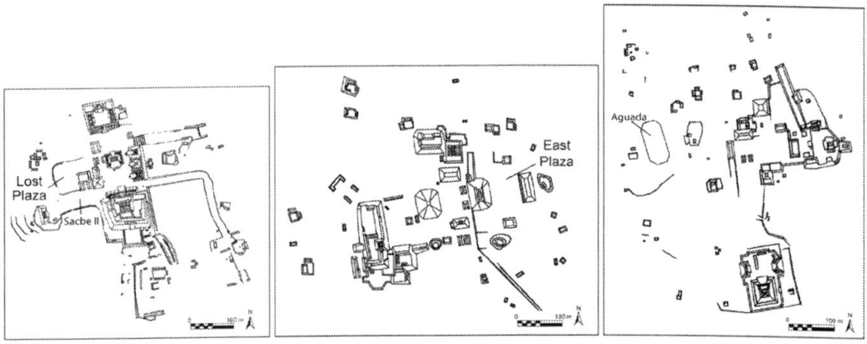

FIGURE 7.2. *The civic centers of (left) Xunantunich, (middle) Actuncan, and (right) Buenavista del Cayo, Belize.*

At the apogee of their political power, Xunantunich, Actuncan, and Buenavista were relatively mid-sized centers in comparison to Tikal and Calakmul, the two largest superpowers in the Maya lowlands (figure 7.2). Arlen and Diane Chase suggest large Classic Maya polities were approximately 8,000 km^2 in size with hierarchically ordered centers, the largest of which contained upwards of 150,000 people (Chase and Chase 1996:805). In contrast, Jason Yaeger (2003:131) estimates that fewer than 1,600 people lived within 1 km of Xunantunich in the Late Classic period, a number that is slightly smaller than that estimated for Actuncan and Buenavista populations. During their reigns as paramount capitals, each may have controlled a fairly small hinterland, no larger than 20 km in diameter. Site hierarchies in the region contain three kinds of sites—paramount center, major center, and minor centers—indicative of their classification as city-states (Webster 1997) or secondary states (Marcus 2003).

It is not surprising to find that Classic Maya states are rather low on the collective action scale. Based on Blanton and Fargher's (2008:tables 7-1, 8-1, 9-1) measurable variables for states without extensive ethnohistoric texts, small states consistently rank lower than large empires in the degree to which they engage in (1) building public goods such as transportation infrastructure and water control, (2) bureaucratization, or more specifically the nature of officeholder recruitment, and (3) moral responsibility, or more specifically the nature of ideological resources and standards of living. Classic Maya states built a moderate amount of state-supported infrastructure within centers, but their investment in public works outside them is more limited. For example, elaborate road systems (*sacbeob*) at Caracol, Calakmul, and Coba link large civic plazas with residential groups at the peripheries of centers (Shaw 2001); however, in a survey of Maya roads, Justine Shaw (2001) found that only 9 out of 190 road segments connect separate sites together. A similar pattern can be seen in water-control features. Almost all Classic cities maintained centralized reservoir systems near the largest civic architecture, and in some cities, such as Tikal, centralized water systems recharged residential tanks (Scarborough 2003). But it is difficult to assess the degree to which the state sponsored the construction of waterholes, raised fields, and other water systems found throughout the hinterland. These features could have been built by commoners alone or during the Formative period when large-scale state project directed toward the public good appear to have been more common. The nature of officeholder recruitment also indicates that Classic Maya "bureaucratic" institutions were mechanisms that protected the interests of royal dynasties more than public interests. Epigraphic studies document that Maya polities were governed by a line of rulers who bore the title of "divine lord," or *k'uhul ajaw* in Classic Mayan (Martin and Grube 2008:17). The rule of kings in Maya society was absolute, combining control over symbolic systems, such as writing and religious rites, with their regulation of economic resources (McAnany 2004). Their position at the top of the sociopolitical pyramid is evidenced by their splendid, elevated lifestyles in palace acropoli. As the Late Classic period progressed, kings increasingly privatized and segregated their lives from the rest of Maya society (LeCount 2001; Robin 2003b).

Provincial polities may not have engaged in exclusionary and hierarchical relations to the same degree as large Classic Peten polities, but it is difficult to assign them higher ranks on the collective action scale. At Xunantunich, Panel 2 contains a full emblem glyph including phonetic complements and a main sign toponym translated as "divine mountainous place lord" (Helmke et al. 2010:106). This title, as well as the construction of a Peten-style royal compound, indicates that Xunantunich's Late Classic ruler participated in the same sociopolitical kingship system as that found in larger lowland sites (Yaeger 2010). Evidence of state-funded public works designed to provide access to critical resources outside political centers is present but limited. Although a short segment of road runs between Xunantunich and Actuncan (Keller 2006), many of the agricultural terraces and *aguadas* were likely built by local communities without state intervention (Wyatt 2008).

Given the nature of Maya collective action, market participation is predicted to be limited in scale and scope. Volatile relations between provincial centers in the upper Belize River valley and large Peten states may have prevented safe passage of market participants across political boundaries. If so, the size of the consumer base was relatively small, especially in provincial polities, dampening supply and demand for local goods and restricting the influx of regional and long-distance trade goods, especially obsidian, jade, and marine shell. During times of greater regional consolidation such constraints would have been lifted, and markets would have flourished. During the mid-eighth century, when Xunantunich was briefly incorporated into the Naranjo state, the larger site's influence over the region would have opened up travel across previously contested ground. Nonetheless, given shifting politics, long-distance traders would have found traveling across political boundaries within the lowlands unpredictable, making the flow of imported goods from highland Guatemala and Mexico reliant on negotiations with a series of Maya kings or other principals. Maya kings, therefore, had greater access to imported items than intermediary buyers or lithic specialists; they could have made those items available in administered solar markets or through redistribution. Under these circumstances, exchange values of imports were susceptible to price fixing to expand commercial activities or to restrict the imports for political purposes.

Classic Maya Markets

Physical evidence of Maya marketplaces has been found at 10 ancient Maya capitals, including the Classic sites of Tikal, Yaxha, Calakmul, Coba, and Chunchucmil (Masson and Freidel 2012; Shaw 2012). In these large sites, marketplaces were located in nodal plazas that contained stall-like alignments or arcades. The configurational indices for these Classic Maya marketplaces compare favorably with other Mesoamerican markets.

At Teotihuacan, the Great Compound marketplace, located at the junction of axial causeways near the Ciudadela, was a large rectangular compound surrounded by raised platforms that could have supported workshops (Millon 1973). Likewise, the centrally located Tlatelolco marketplace in Aztec Tenochtitlan featured a square plaza surrounded by long arcade-like structures with portals and limited entryways (Feldman 1978). Inside, sectors of the plaza were devoted to the selling of particular commodities and administrators oversaw market activities from small ancillary platforms. At the Maya capital of Tikal, Grant Jones (1996:86–87) suggests that the East Plaza displays similar marketplace characteristics, including the plaza's location in the principal ceremonial area near the junction of *sacbeob* and the presence of a large double-gallery compound that contains arcade-like structures, stalls, and ancillary platforms. Calakmul (Carrasco Vargas, Vázquez López, and Martin

2009) and sites in southeastern Peten (Houston and Inomata 2009:252) have similar architectural arrangements near the junctions of *sacbeob*. Calamkul's Chiik Nahb complex is particularly noteworthy, not only because of its linear arrangements of low, long structures within a large rectangular compound, but for the murals painted on the exterior walls of Structure 1 near the center of the complex. The murals depict groups of ordinary people, some of whom are engaged in preparing and dispensing foodstuffs and other commodities. Hieroglyphs that accompany the scenes describe individuals using an agentive term, *aj*, followed by the name of a particular commodity, for instance maize-gruel person, salt person, or clay-vessel person (Carrasco Vargas, Vázquez López, and Martin 2009:19248). While these commodities may have been exchanged at festivals or public feasts, Kerry Hull (2010:251) notes that vendors at modern highland marketplaces are similarly referred to in Ch'orti' Mayan by the type of item they sold.

Ancient Mesoamerican markets also were located in multiuse plazas and vendors sold from temporary stalls, similar to modern Mesoamerican periodic markets today. To test for the presence of ancient open-air market activities at Chunchumil in Yucatan, Bruce Dahlin and colleagues (2007) systematically sampled soils from the modern Antigua, Guatemala, marketplace to establish chemical signatures for ancient marketplace activities. At Antigua, they found that the spatial distribution of extractable phosphorus (P) and zinc (Zn) mapped onto areas of vegetable and fruit vendors, food service, and food-preparation areas. Highly elevated levels of extractable P and Zn also were found in the central portion of Chunchucmil's plaza parallel to a *sacbe* and rows of small rock alignments suspected to be market stalls, supporting their interpretation of an ancient open air marketplace in a multiuse plaza. Chase and Chase (1987:52) also suggest that unrestricted, centrally located plazas at Caracol were the locations of marketplaces based on low rubble features suspected to have been vendor stalls.

In the upper Belize River valley, Xunantunich and Buenavista provide architectural, artifactual, and chemical data consistent with open marketplaces located in multiuse plazas. At Xunantunich, Angela Keller (2006:388) suggests that the Lost Plaza, located north of Sacbe II and east of Ballcourt 1, may have been a Late Classic marketplace bounded on the west side by low linear features. Here, excavations revealed higher than expected frequencies of chert debris and tools, obsidian debitage and blades, and spindle whorls that Keller (2006:389, 615) infers are the remains of point-of-sale finishing of craft goods by artisan-vendors. Similar patterns have been found at Buenavista's East Plaza, a large plaza flanked by pyramids at the endpoint of two *sacbeob*

(Cap 2011). There, Bernadette Cap found evidence of wattle-and-daub structures and two low platforms built atop the Late Classic plaza. Like Keller's discoveries at Xunantunich, Cap's (2011:248) plaza excavations revealed *in situ* lithic production associated with the final stages of chert biface shaping and resharpening and obsidian blade production. Strong spatial correlations in ceramic sherd and soil phosphorous concentrations indicated that food, likely held in pots, was sold in specific marketplace locales.

Given the stark contrasts in investments in permanent marketplace infrastructure between large capitals and smaller centers, market systems and the kinds of exchanges that took place within them varied greatly across the Maya lowlands. As Leah Minc (2006:83) points out, "different market systems create very different contexts for production, exchange, and consumption according to the structure and scale of their regional organization." The scale of exchange interactions, the amount of commodity flows between market centers of the same size and those at different levels of the settlement hierarchy, and the political geography of the landscape all factor into the organization of market systems (Minc 2006; Smith 1976a:314–5). Minc (2006) characterizes Carol Smith's (1976b) four different market system models—administered solar, noncentralized network, dendritic, and complex interlocking—using the dimensions of scale, networking, hierarchy, and political congruence. For the ancient Maya, solar markets are considered a good fit for Classic Maya market systems in provincial polities, although more complex hierarchical systems may have existed in large Maya capitals. Local markets where commoners exchanged basic goods and pilgrimage market fairs tied to calendrical events also may have operated periodically within the larger system (Masson and Freidel 2012:461; Scarborough and Valdez 2009).

Solar market systems consist of a market center serviced by small subsidiary markets located within a single political entity. Each solar market system is self-sufficient and independent of the other because they "appear to be the result of the dominance of political forces over economic forces" (Smith 1974:177). The administrative capitals of political kingdoms are the economic hub of the greater community and exert strong controls over the movement of producers and consumers (Smith 1974:176–177). Local goods move from production sources outside the center to the central marketplace where they are exchanged, with little or no goods flow across political boundaries. Therefore, households sharing a market zone have similar distributions of basic craft goods, whereas households belonging to a different polity have functionally similar yet stylistically distinct items (Minc 2006:84).

These markets were more likely administered by elites because they were closed and bounded systems (Smith 1976a, 1976b), in which market exchange is restricted, in Blanton's terms, to buyers and sellers who are socially embedded in hierarchical relationships. Long-distance imports were prone to control by elites because leaders provided safe entry into politically contested territory to traders. If interpolity commerce rests in the hands of elites, then elites are free to use long-distance commodities as "network capital" and commoners are constrained in their access to them (Blanton 2013:25–26). Although commoners may negotiate locally made items for fair market value because buyers and sellers are not strangers, they are at risk to challenge the price of long-distance distance goods when market adjudication is in the hands of powerful interest groups. These sociopolitical factors condition the exchange value of goods and result in uneven distributions of high-value commodities that diverge from expectations for highly competitive market economies. Data to support this idea are presented below.

Classic Maya Marketplace Exchanges and Commodities

Marketplace exchange may best be understood by examining basic commodities likely exchanged in large quantities in marketplaces. The interchangeability of some high-value commodities, such as jade, cloth, and marine shells, to satisfy reciprocal exchanges and tribute obligations, as well as function as market currencies, presents problems for models that presuppose dominant exchange modes (Masson and Freidel 2012:460). Given the scope of this essay, I concentrate on two high-demand commodities—local ceramics and imported obsidian—to explore their exchange values and, ultimately, their household distributions in upper Belize River valley sites.

At large Maya lowland capitals, there is good evidence to suggest that ceramics were made available through marketplace exchanges in the Late Classic period. At Tikal, access to simply decorated polychrome serving vessels and plainly finished utilitarian vessels appear to have similar distributions across households (Fry 1980), a pattern also found at Copan (Beaudry 1984), Palenque (Rands and Bishop 1980), and centers in the Petexbatun region (Foias and Bishop 1997). Compositional studies of locally made ceramics indicate that, although Maya capitals may not have been the loci of production of most paste groups, capitals were the hubs of regional exchange systems that regulated periodic markets in rural communities. In the Palenque region, distance-decay curves of four paste groups indicate that the center had the most diverse paste groups, while sites further afield had fewer groups (Rands and Bishop 1980). Nonetheless, hinterland sites have similar proportions of those paste groups found at the center indicating that hinterland sites participated in intraregional exchange of ceramics. Further, the distribution of the local paste groups coincides with that of the site's emblem glyph suggesting a politically bounded exchange system. In a more recent compositional study, Antonia Foias and Ronald Bishop found similar patterns at sites in the Petexbatun region (Foias and Bishop 2007). Hinterland sites exhibited relatively homogeneous clusters of compositional groups, indicative of local production of monochrome and polychrome pots, but they also contained high frequencies of compositional groups from other sites in the regions, indicative of interregional exchange. Like the exchange system at Tikal, the Petexbatun capitals of Aguateca and Dos Pilas had the most diverse assemblages, indicating that exchange was centered on larger centers rather than smaller ones.

Of the three study sites in the upper Belize River valley, the most systematic ceramic data come from the site of Xunantunich. There, plain and monochrome pottery make up slightly more than 75 percent of the total assemblage and are found in relatively equivalent frequencies within elite and commoner household assemblages as expected if they had been sold in a marketplace (LeCount 1999; LeCount et al. 2002). Krista Garcia's (2008) petrographic study of the two most common types and forms (Cayo Unslipped jars and Mount Maloney Black bowls) from Xunantunich, Actuncan, and the nearby community of San Lorenzo found that these types were made in four paste recipes during all phases of the Late and Terminal Classic period, lending evidence to suggest that they were fabricated in the same, highly stable household- or community-scale workshops. Her analysis of stylistic and formal attributes also found a high degree of standardization in formal attributes, including rim diameter and vessel wall thickness, another pattern expected of pottery made by specialists in quantities for sale in a market. Monochrome slipped types, such as Mount Maloney Black, Garbutt Creek Red, and Rubber Camp Brown, are also found concentrated in regionally discreet areas that correlate with provincial capitals (LeCount 2010). These data indicate that utilitarian pottery was widely exchanged but politically bounded, as predicted by a solar marketplace model.

These patterns, however, are not borne out in the distribution of high-value polychrome volcanic ash ware types. There are two groups of ash ware: the Chunhuitz Orange Group, which contains polychrome types, and Belize Red Group, which contains monochrome slipped types. Both were made predominately as serving vessels, and both make up about 11 percent of the total Late Classic Xunantunich ceramic assemblage (LeCount et al. 2002). Xunantunich elite contexts have significantly higher statistical frequencies of Chunhuitz Orange types than commoner contexts at nearby communities such as San Lorenzo (LeCount 1999) and Chan (Kosakowsky 2012), where these types make up between four and two percent of the total assemblage. Interestingly, Belize Red Group frequencies are roughly similar across elite and commoner household assemblages. Previously, I have suggested that cost was a factor in the distribution of ash ware polychrome ceramics (LeCount 1999). In general, pottery made from volcanic ash would have been more costly for families in the upper Belize River valley than similar calcite tempered wares, given the scarcity of volcanic ash temper versus more widely available calcite tempers. Detailed stylistic analyses also indicate costly differences in production techniques of the two groups. Chunhuitz Orange types are far less standardized, more stylistically diverse, and required more production steps than the standardized forms and styles seen in the Belize Red Group (Chase and Chase 2012; LeCount 1996). Higher production costs may have translated into higher exchange values for Chunhuitz Orange types compared to Belize Red types, thereby limiting access to certain ash ware types for those at the lower rungs of the socioeconomic ladder. Although decorated types may have been redistributed or exchanged reciprocally through kin relations, their presence in even the most humble homes indicates that they were widely available in markets. In marketplaces, sellers could have negotiated a range of exchange values on pottery vessels based on production costs, resulting in the differential distribution of types by socioeconomic status.

Obsidian, imported from highland Guatemala and Mexico, was in high demand as fine cutting tools, projectile points, and esoteric cache objects by lowland Maya populations. The sources of obsidian available to Maya centers depended on trade relations, politics, and exchange modes that shifted over time. According to Geoffrey Braswell (2010:135; Braswell and Glascock 2011), Tikal controlled interregional trade in the lowlands, preventing rivals such as Calakmul from receiving large quantities of obsidian during the Classic period. Within the Tikal polity, the abundant nature and widespread distribution of obsidian supports either a network or administered solar market system (Braswell 2010; Moholy-Nagy 2003, 2008). At Copan, Braswell (2010:136) suggests obsidian was both redistributed by elites and exchanged in solar markets. There, urban elites obtained higher frequencies of black Guatemalan and Honduran obsidian than rural households, and royalty received rare and highly prized green Mexican obsidian, which they redistributed to lesser members of the nobility (Aoyama 1999). Evidence of production, in the form of cores, is only found in elite contexts, indicating that they controlled the production and distribution of blades. However, Ixtepeque obsidian, which comes from a source 80 km away from the center, is ubiquitous in households, lending evidence to suggest that this kind of obsidian was more widely available within the Copan polity, possibly through a market system (Braswell 2010:136). Yet rulers residing in the Copan Principal Group had far greater amounts of Ixtepeque obsidian than elite and common households (Aoyama 2011:42). Given the differential access to different sources of obsidian, Braswell (2010:136) suggests that "elites rather than the forces of supply and demand monitored the value of and access to obsidian." The value of each kind of obsidian was based on procurement costs, use, and esthetic qualities, much like ash ware pottery types within the Xunantunich polity.

In the upper Belize River valley, access to obsidian was orders of magnitude smaller than that found in Copan or Tikal. At Actuncan, Sara Shults (2012) reports on the distribution of 594 obsidian pieces, an amount similar to that found at Xunantunich (Keller 2006:530), but far below the tens of thousands of pieces found at Copan (Aoyama 1999, 2011) or millions found at Tikal (Moholy-Nagy 2003, 2008). Shults utilizes a diachronic approach to understanding consumption of obsidian because, based on Hirth's model, the historical development of markets should homogenize the distribution of obsidian across households through time. By charting the obsidian-to-sherd weight ratios within elite and nonelite households over Formative (or Preclassic) and Classic periods (figure 7.3), she found that before the Late Classic period, elites and commoners appear to have had relatively equal access to obsidian, but by the Terminal Classic period elite households had as much as three times more obsidian by weight than did common households. But similar to patterns at Copan, obsidian is found in all Actuncan contexts, albeit in varying quantities, indicating that households had relatively open access to the amounts of obsidian they required for daily activities and, possibly, craft specialties (Masson and Freidel 2012:468–471).

At Actuncan, obsidian is overwhelming recovered as small processed bladelets. The mean cutting edge to mass ratio for blades is 7.84, with a mean width of 10.58 mm and mean thickness of 2.74 mm, indicating that blades were being consumed highly efficiently (Shults 2012:72). Prismatic blades from which these bladelets derive do not appear to have been produced in workshops at Actuncan. Besides the impromptu production loci in the plazas of Xunantunich and Buenavista, only one obsidian workshop has been found in the region (Hintzman 2000). Therefore, both the consumption and production data indicate that obsidian arrived in the upper Belize River valley in relatively limited quantities through interregional exchanges.

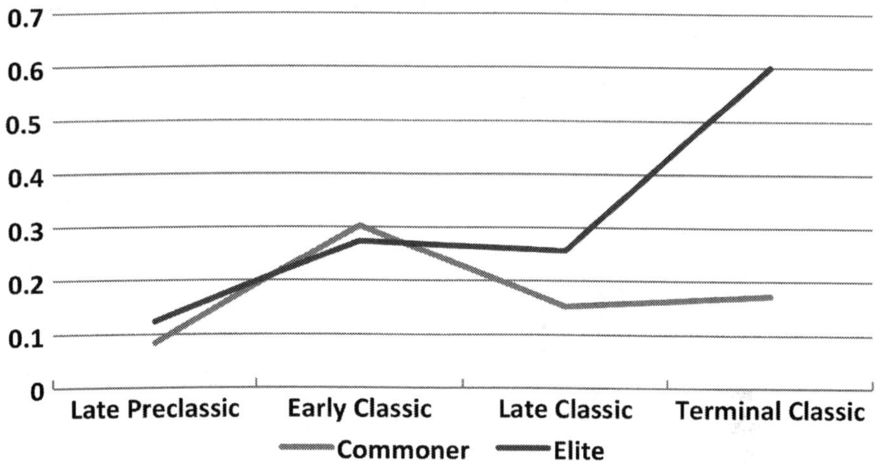

FIGURE 7.3. *Obsidian-to-sherd ratio over time in elite and commoner contexts. (Graph from Sara C. Shults 2012:table 5.4.)*

Obsidian from the highlands reached the upper Belize River valley either from Caribbean or overland trade routes. El Chayal obsidian arriving from overland trade would have flowed through Tikal, whose kings negotiated access to obsidian with subordinates at low-level centers or permitted merchants to travel and exchange obsidian across its borders. Ixtepeque and possibly some El Chayal obsidian would have made its way up the Belize River from the Caribbean coast through exchange relations between trade partners or long-distance traders. Mexican green obsidian is very rare at Actuncan, and although it too is found in both elite and commoner households, elites had more than commoners. The percentages of these sources vary over time, with Ixtepeque becoming more common in the Terminal Classic period, but access to each source by elite and common households remained similar over time (Shults 2012:91).

These data indicate that long-distance exchange of obsidian was likely overseen by elites throughout the Formative (or Preclassic) and Classic sequence in the upper Belize River valley. At Actuncan, blades were likely available in local markets at least by the Late Classic period, if not much earlier, but rulers likely reserved some obsidian for redistribution to loyal supporters. The great disparity in the distribution of obsidian by household status in the Terminal Classic period may have as much to do with changing consumption patterns as exchange relationships. Struggles between Xunantunich's paramount rulers and Actuncan's nonroyal elites for power may have concentrated obsidian in nonroyal elite households where it was used for termination rituals, veneration practices, or production of items requiring fine cutting, such as featherwork (Berdan, chapter 6, this volume).

CONCLUSIONS

Multiple lines of evidence indicate the existence of Classic Maya market systems. Based on Blanton and Fargher's collective action scales, the cultural conditions required for large-scale markets were rather weakly developed during the Classic period. Although a survey of the configurational evidence for marketplaces supports their existence in some large Maya capitals including Tikal, Calakmul, and Chunchucmil, as well as provincial centers such as Xunantunich and Buenavista in the upper Belize River valley, the distributional data for high-value goods indicates that marketplace exchanges were not highly competitive except at Tikal. Based on these data, it is unlikely that the forces of supply and demand were the sole mechanisms operating in Classic Maya markets. Like Blanton and Fargher's (2010) conclusions that the development of markets was the result of multiple factors, the evidence presented here illustrates how indices for marketplace exchange also must account for the scale of competition and the embedded sociopolitical relationships between interest groups. This is especially critical in politically fragmented regions, where market transactions are "plagued with uncertainty and opportunism" (Blanton 2013:25–26). For example, in restricted markets, commoditization of goods is restricted to locally made items, while long-distance items are used as network capital by elites.

In the upper Belize River valley, polities were relatively small, regionally bounded, and fiercely independent except during a short interval in the Late Classic period when they were incorporated into the Naranjo state. At that time, markets flourished at Xunantunich and Buenavista, where vendors sold goods out of temporary stalls situated in multiuse plazas. However, the distributional evidence for fully commercialized goods is mixed. Locally made plain and monochrome pottery types are found evenly distributed across households indicating that they were sold in marketplaces, but ash ware types, although present in nearly all households, are more variable in quantity. This pattern indicates that ash ware vessels were also sold in markets, but their exchange value was fixed using a sliding scale based on production and, possibly, transportation costs. On the other hand, the distribution of obsidian, while ubiquitous, is more than three times more common in elite households than commoner households. This pattern indicates that obsidian blades were made available through both elite-administered solar marketplaces, where their exchange value was fixed at a high rate, and centralized redistribution, in which elites received more blades than commoners.

The obsidian data from the upper Belize River valley do not support the dual-economy model in which luxury items circulated in a separate exchange sphere from ordinary goods; rather, it lends evidence to suggest that long-distance trade goods were available in markets, at a price, and through redistribution. Classic Maya commodities, such as obsidian, cloth, and decorated pottery, circulated within articulated, not separate exchange modes—reciprocity, tribute, and markets—where they were assessed along a continuum of value (Masson and Freidel 2012:458). Future research on investigating how exchange values were established for commodities will greatly enhance our understanding of market economies in Classic Maya society.

Acknowledgments. I would like to thank Lane Fargher and Verenice Heredia Espinoza for the opportunity to participate in this volume. The ideas presented here grew out of many conversations with colleagues—Angela Keller, Jason Yaeger, Bernadette Cap, John Blitz, and Steve Kosiba—and my graduate students—Krista Garcia and Sara Shults. I also appreciate the comments of two anonymous reviewers who pointed me in the direction of Richard Blanton's work on restricted marketplaces. Borislava Simova, Bernadette Cap, and David Mixter were instrumental in producing the illustrations. Data from the Actuncan Archaeological Project were gathered with support from a National Science Foundation grant (BSC-0923747) and University of Alabama College of Arts and Sciences grants.

SECTION 2

Old World Cases

8

Enduring Nations and Emergent States

Rulership, Subjecthood, and Power in Early Scandinavia

T. L. Thurston

NATION, STATE, AND POWER

Until recently, the archaeological study of past sociopolitical organization was characterized by a search for the origins and rise of elites, whose quest for power could be characterized as seeking ever-increasing capacity for coercive force. This scholarly preoccupation with power as autocracy or oligarchy created a failure to recognize other forms of authority—more symmetrical distributions of power or nonelite self-governance—even when they were relatively apparent. Over the course of the mid- to late twentieth century, this led to the overall mischaracterization of all power as asymmetrical power, from its advent within small-scale communities to its expression in politically complex societies, ancient and modern (Thurston 2010). While the power of coercive force is amply evident in the recent and distant past, it is only one type of instrument within a spectrum of organizational potentialities recognized by contemporary social theorists. At worst, most archaeological theorists long denied the existence of politically complex societies with egalitarian ideologies; at best they have displayed indifference or doubt that alternative power structures can be detected and studied. As with all orphan domains there have been notable exceptions, among the most important the work of Richard Blanton and his distinguished coauthors, through the initial innovation of dual-processual theory (Blanton et al. 1996) and the subsequent development and application of collective action theory

DOI: 10.5876/9781607325338.c008

to historical and archaeological data sets (Blanton and Fargher 2008, 2009, 2011, 2012; Fargher and Blanton 2007; Fargher et al. 2010; Fargher, Heredia Espinoza, and Blanton 2011).

It is not easy to find archaeological cases highlighting the power of a non-elite majority, but a useful instantiation of subjecthood and rulership within such a system is the subject of this case study, an example found among the array of ancient North Atlantic nations. By "nation," I mean a group with historically evolved or socially constructed shared ethnic identity; in northern Europe these sometimes can be documented as coherent entities for as much as a millennium. Over time, these enduring nations were absorbed by emergent states, political constructs whose authors sought overarching legal jurisdiction over defined geographic territories and control of large-scale political economies, based on enforceable obligations for taxation and labor. Such states usually display their own socially constructed identities, often pluralistic, frequently manifest through a politicized conquest mythos of state origins.

The case study below focuses on a sequence in which the early Svear state, expanding outward from central Sweden between the tenth and sixteenth centuries AD, incorporated a region to the south, the Småland Plateau, an environmentally marginal upland: high, cold, rocky, and heavily wooded (figure 8.1). Plateau dwellers lived in a group of small neighboring polities (kingdoms or chiefdoms), had a strongly pastoral economy, and formed tightly knit kin and community groups for cooperative labor, using slash and burn to create artificial clearings for habitation, pasture, and "garden" cultivation in densely forested uplands. The case study reveals that the negotiation of power between nation and state was often violent, yet ultimately led to the formation of a corporate, or collective, state.

Traditional Archaeological Narratives of the State

Older yet still pervasive archaeological models of state formation (e.g., Flannery 1972; Trigger 1974; Webster 1975) tacitly imply or actively proclaim the notion that increasing political complexity is always associated with an elaboration of hierarchy, and always associated with increasing centralization and displays of status and legitimation. This view of state origins (i.e., Knapp 1993; Roscoe 1993) has more recently been challenged (Blanton et al. 1996; Blanton and Fargher 2008, 2009, 2011; Crumley 1995, 2003; Fargher et al. 2010; Fargher, Heredia Espinoza, and Blanton 2011; Feinman 2001; Feinman et al. 2000; Pauketat 2007; Thurston 2001, 2010; Yoffee 2005) by the proposal of alternative organizing principles.

FIGURE 8.1. *The Småland Plateau.*

The case study presented here focuses on late prehistoric through early historic Sweden. As in many world regions, beginning in the 1980s, attempts were made to force Scandinavia's late prehistoric societies into a the traditional "elite hierarchy" model, in which virtually every complex society worldwide was interpreted as harboring centralizing elites on a steady trajectory of attaining more and more power over subjects, who were never explicitly described but implicitly presumed to have given up traditional practices and privileges when faced with variously hypothesized forms of elite "legitimation." For Scandinavia, while a culture historic framework persisted, materialist-influenced, processualist interpretations were embraced by some (Hedeager 1992; Parker Pearson 1984; Randsborg 1980) and changes in burial treatment, property, and implied social organization were all seen as representing movement toward "progressively unequal social relations" (Parker Pearson 1984:69).

While there is little doubt that the region's extant nations, their peoples and resources, were compromised by expanding and broadening states, historical and archaeological approaches to this process have long been dominated by narratives presenting an exclusively top-down perspective, interpreting the record as a trail of "events" initiated by "great men" or families. Acknowledging the role of such elites, yet adding to it a bottom-up viewpoint, balances our understanding of such sequences. The inevitability of ever-increasing inequality through time is also questionable for Scandinavia—understandable yet fraught with oversimplification. The region presents a productive case for the application of collective action theory, as Scandinavia's protohistoric and early historic sociopolitical systems are generally well understood, facilitating the development of more appropriate perspectives.

Historical Narratives of Modernity: How to Build a State

Rojas (2005) has discussed the changing historical outlook on the political and economic development of the modern nation-state of Sweden, where recent debate is intellectually situated in studies of the creation and breakdown of the "Swedish Model" industrialized welfare state, with a high standard of living and a caretaker mentalité. Rojas, along with a small number of revisionist historians, notes that there is a traditional historic narrative that stresses individual historic "personalities" and "discontinuities" within this history. These discontinuities include especially the purportedly abrupt nineteenth-century industrialization of a formerly impoverished agrarian state, and the "sudden" invention of a unique type of balanced engagement between government and the governed. The notion of *Folkhemmet*, "the people's home," as a metaphor for the Swedish state, is frequently tied to the pre- and post-WWII attempt to create a bridge between communism and capitalism through a modern welfare society (Rojas 2005:7). From this perspective, it is often asserted that this structure was the invention of sociologists during the mid- to late twentieth century, as a social experiment that lacked any basis in reality and was thus malaise-ridden and overdue for collapse by the end of the twentieth century (e.g., Enzensberger 1982; Wolfe 1989). Over the last few decades, this narrative has been challenged by the perspective that the "new model" state was far from the invention of mid-century sociopolitical engineers, or nineteenth-century industrialists, but "the result of a centuries-long historical evolution with no European counterpart" (Enzensberger 1982). New analyses that attempt to follow the state's trajectory much further back in time examine social traditions, practice, and structuration, and emphasize continuities:

> *Folkhemmet* certainly drew ideological strength from modern industrial utopias. However it was based as deeply on the egalitarian and solidaristic ideals of the rural estates, the paternalistic order of the old mining and manufacturing community known as the *bruk*, and on the idiosyncratic relationship between rulers and the ruled which lay deep in Sweden's history. (Rojas 2005:7–8)

In this newer model, long-term ethnic homogeneity, the strength of the state itself, industrial growth, and technology, all stretching back well into the medieval era, were critical to the twentieth-century *Folkhemmet*, and it was only the ensuing forces of "immigration, globalisation, prolonged economic stagnation and the information revolution [that] have undermined them" (Rojas 2005:8). With the exception of economic troubles, these forces were largely embraced and tackled head-on by the Swedes, who are attempting, with varied success, to adapt their sociopolitical structure to these new conditions.

Rojas begins his exploration in the fourteenth century, with the so-called Swedish Magna Carta, or Charter of Liberties, and the creation of the Riksdag, a parliament where municipalities and districts had a representative voice in the governing of the state. In the Riksdag, not only nobles but also peasants were present to witness the king's pledge to obey the law and refrain from excessive taxation. It continued through that century when Sweden, Denmark, and Norway were unified through a negotiated century-long settlement called the Kalmar Union, created by the Danish-born Margareta Valdemarsdotter, and governed by her from 1387 to 1412. This is admirable for a historic view, as such views are so often plagued by shallow time-depths and poor understanding of the nature of social structure, human agency, continuity, and change, and is even more remarkable for Rojas, who is not a formally trained medievalist, but a Chilean immigrant to Sweden and political economist serving as a member of that self-same Riksdag at the time of this publication.

Rethinking Traditional Narratives

The period in Sweden's history identified by Rojas, and the historians he relies upon (e.g., Österberg 1993), as critical in the creation of the Swedish model begins in 1319 with the Charter of Liberties. A socially situated archaeology tells us that the majority of ancient and modern secondary states are created from a unification of autonomous entities. Some consolidate by passing from alliances, through "consent" or negotiated hegemony (Gramsci 1971) and into a centralized polity with relative ease.

Superficially, then, 1319 seems a good place to seek Sweden's political origins: an era of negotiated hegemony. But other states begin as unrelated polities or loose confederacies that undergo integration not through consent but via "power" or "force" (Chomsky 2003; Scott 1989, 1990). In fact, Sweden emerged from an expansionist conquest sequence and postconflict reorganization more than 300 years earlier than the fourteenth-century period identified as critical, and it was a process that was torturous for all parties.

The historian Österberg (1993) states that a *lack* of regional elites and the presence of "ethnic homogeneity" enabled the Swedish state to develop rapidly;

here I believe she errs significantly. The strength and perceived *autonomy* of regional elites and the marked ethnic *differences* between the Svear and the local authors of eventual rebellion against increasing absolutism are precisely what led to the *restoration* of the corporate ideology that appears to have originated in the Iron Age and stretches forward to the *Folkhemmet*.

Rojas notes that throughout Sweden's sociopolitical development, a "common link is the tension between freedom and submission that runs clearly across 500 years of Swedish history" (2005:10). This is insightful, yet the time frame must be extended to at least approximately 1,000 years. If we push the origin of this "freedom/submission" tension back to its real beginnings, we find it in the later prehistoric era, before Sweden's unification, and its earliest portion studied through archaeologically-framed research.

THEORIZING POWER IN A HIGH-CONFLICT SOCIETY

Courses of political unification often produce sets of unequal power relations within emergent states, when inside their borders they subordinate groups with irreconcilable cultural institutions, religious differences, or ethnic identities, as well as incompatible economies. Such amalgamations often lead to the formation of "high-conflict" societies (Ross 2007) that can experience instability and internal unrest from the legacies left by expedient colonial borders and opportunistic but short-sighted political decisions. Prior to forced unification, such disparate, self-identified groups frequently have coexisted in relative peace, yet as the state begins to enact itself, implementation of ill-considered or purposefully divisive economic or political policies pits them against each other, and hostilities with both material and ideological origins can last for centuries. Frontiers that previously divided once-autonomous regions often fade slowly, or not at all—in some cases becoming more cemented, as borders can simply divide people and territories, but can also "initiate or accelerate the identification or construction of collective identities both in the past and in contemporary societies" (Klusáková and Ellis 2006:xiii). During Sweden's initial unification, a number of dissimilar groups were brought under a single government, drawing preunification societies into conflict with a centralizing polity that expected them to fall into line as others already had (Thurston 2015). The polities at issue here were not "tribal" entities, as are frequently the object of study in conquest sequences, they were kingdoms, organized in more egalitarian yet politically complex modes. This created tensions equal to but different than those that developed between so-called tribes and states (Ferguson and Whitehead 1992). The clash in early

Sweden, between two kinds of complex political systems, both in terms of their structure and their ideology, is especially interesting.

Most archaeologists working on the topic of states and their development over the last few decades remember the puzzling nature of the "differently organized" society or state (Thurston 2010) and the endless debates surrounding them. Were they states? Were they actually differently organized? Could they be divergent and still be states? Were we just missing something that would reveal them to fall within the normative expectations of the times? When the dual-processual theory of Blanton et al. (1996) was first published, many researchers working with the remnants of differently organized states welcomed it, as it explicitly dealt with the possibility that state structures and modes could vary significantly, and that ideologically egalitarian complex political systems were present in significant numbers in the past. Praised for its ability to model atypical systems, the theory proposed an infinite variety of political modes stretching between the corporate and the network ends of a continuum. A purely network strategy was hypothesized to be dominated by elites with monopolies on power, supported by creating a network of relationships with other elites, and advertised their status with all the trappings of wealth and/or prestige. At the other end lie societies where power is devolved away from rulers toward a more equitable distribution among groups or institutions within society. Such societies are characterized by less elite power, or at least less emphasis on its appearance, display, and manifestation, often with institutions that encourage some level of popular power from below.

Blanton et al. (1996) further posited that most societies are in a constant state of tension between these poles as various groups or interests within society constantly pull its characteristics and ever-evolving traditions toward one side or the other in a tug of war. In this way, envisioning as it did a reshaping and reforming of the "normal" state in a continual negotiation, it resonated with the reexamination of state formation by scholars of the 1990s who were trying to rewrite parts of the archaeological theoretical corpus through the structuration theory of Giddens (1984), the heterarchy concept introduced into archaeology by Crumley (1994), and the resurrection of older ideas in new forms: the house society of Levi-Strauss as reinterpreted by Joyce and Gillespie (2000), and the hegemony concept, both the original "consent" model of Gramsci (1971) and the resistance model of Scott (1990). In this way, at a critical time in the development of political theories on state formation in archaeology, dual-processual theory permitted some convergence between more traditional materialist and processualist views and the increasingly theoretical interpretations influenced by the postmodern turn. It did so by allowing that ordinary people, using their agency from below, can wield considerable power through their evolving practice; that societies with absolutist and shared power structures can both exist; and that both types of principles can coexist simultaneously within the same society, and that they can shift over time.

In order to understand the mechanisms that might trigger such shifts, using dual-processual theory as a starting point, Blanton and Fargher (2008, 2009) developed collective action theory to more deeply investigate these issues. The mechanisms within ideologically egalitarian states, as well as their counterparts with more absolutist ideologies, are also key to understanding the high-conflict state, and I specifically call these "ideologies" because they are not always realities. As noted in both dual-processual and collective action theories, rulers and subjects often are out of step with their understanding of each other. Heads of state may imagine that they have much more power than they actually do; conversely, so may an autonomous and self-governing subculture overestimate its own ability to hold off a state army. On the other hand, farmers may imagine they have little power when they have enormous collective capacity for fomenting change. These mismatched perceptions may result in long periods of stasis where such assumptions are untested, or in episodes or extended periods of intense and violent conflict, as ordinary people may go underground or defy authority, may conform to or confront authority. The state may despise its own populace, or fear them, or both. The fact that the lower classes are excoriated, ridiculed, mocked, or diminished in ruling transcripts is well understood by historians and historical ethnographers as an indication not only of conflict but of dread (Scott 1990).

In developing collective action theory, the authors move from theorizing a continuum between differently characterized states (autocratic/absolutist networks vs. corporate/collective forms) to investigating the infrastructural qualities that render them into the entities that display such characteristics. Using premodern yet historical states as a proxy for prehistoric or nonliterate cultures, historical clues to a more grounded view of the past are found in the records of law courts, property dispositions, and economic transactions that give us a sense of how such forms might "look" in terms of patterning, material culture, and archaeologically discernible (or theorizable) social processes.

Central to many disputes within high-conflict states is the public perception of elite demands: too many obligations for labor, service, and taxation, which diminish the citizens' abilities to work on their own behalf, allocate time to their own convenience, and make the most of their skills to stave off poverty or hardship. In addition, a state's mandate that its people conform to a preferred set of legal constructs and ceremonial obligations that favor state actors and conventions can clash with traditional local values.

The Bureaucratized State

Blanton and Fargher (2008, 2009) propose, supported by extensive quantified analysis of premodern states, that the most effective way a government can assert its authority is to build organizational structures that accommodate *voice*, distribute *public goods*, and uphold *just and fair taxation*. This is accomplished through what Blanton and Fargher call *bureaucratization*, a process that creates effective infrastructural power for the state, rather than giving its rulers heavy-handed personal or network-style power. Infrastructural power has the incidental effect of limiting elite ability to coerce or command with impunity (see also Fargher, chapter 15, this volume; Mann 1986; Weber 1947).

Taxation

Without deference to formal Marxist theories of political economy, it is the case that all states must acquire the means of funding their own activities, operations, and the maintenance of their leadership. This political economy can take many different shapes and forms. Some income may stem from raiding or appropriation of war plunder. In others, corvée labor may be used to see that the citizenry produces a yield of consumables or trade items for the sole benefit of the state, or directly labors with bodies and minds in civic activities in service to the state—labor that may be materially uncompensated. Many states, however, either supplement or replace such asset- and income-generating strategies with the collection of specific tribute or taxes from free taxpayers. There are different methods for collecting such revenue, some more successful than others.

To some, the notion that fairness is more efficient than autocracy may seem counterintuitive. After all, tyrants, despots, and absolutists are the most powerful and successful at appropriating the wealth of the taxpayer, are they not? Through their extensive cross-cultural analysis, Blanton and Fargher show that in reality, the answer is actually no. Rather, such regimes create motives for taxpayers to resist them, by cheating the collector, hiding assets, hiding themselves to avoid detection, or engaging in outright rebellion.

Thus, while it might be assumed that success in revenue collection is measured by the amount of tax money or tribute a state manages to collect, this is only a part of the picture. The collection should also be easy—in other words, rulers should avoid extraordinary efforts to support tax collecting personnel, and the method of collection should not lead to dissent, which must be dealt with, or disturbances like protests and uprisings, which are expensive to control, and which ultimately may destabilize and overthrow the state—all counterproductive strategies. The costs of collecting revenue already reduced by tax evasion, and of putting down symbolic and/or violent protests by unwilling taxpayers, result in less efficient use and lower total income for the state.

Public Goods

Ostensibly, assets collected for funding a political economy also support the provision of public goods. These services to the public such as law and order, defense, maintenance of physical infrastructure, and in some states, efforts toward social welfare, all have their costs. Conversely, if taxpayers see the offer of public goods, they much more willingly pay the tax, tithe, or tribute. Public goods ease or enhance their way of life and livelihood, and also provide some degree of confidence that they will have recourse against the abuses of petty bureaucrats or systemic unfairness through a legal system or a grievance process. Providing these assurances eases revenue collection and requires far less expenditure of labor and funds on the part of the state.

Voice

Voice entails a forum in which grievances can be aired, local/state obligations can be negotiated, and policies can be challenged. It can be, for example, in the form of large public meetings or smaller audiences with authorities, formal legal challenges to perceived violation of precedent or tradition, or the ability to cast a vote. The mere presence of such institutions goes a long way toward reducing conflict and violence between the state and its constituents.

We may see various scenarios play out either through direct detection of the process, such as an assembly place or structure serving as a legal forum, or law-related activities (specific to the region discussed here: Iversen 2013, Oosthuizen 2013, Riisøy 2013, Smith 2013), or perhaps only by the resulting form of society—revealed through the materiality of large-scale spatiotemporal patterns, and the organization and composition of landscapes, sites, and households that archaeologists are skilled at revealing.

Rationales of Resistance and Rebellion

Where states do not bureaucratize in an effective manner—by offering voice, public goods, and fair tax collection—collective action theory predicts that free taxpayers will evade tax collection, migrate or relocate to avoid surveillance and physical appropriation of their goods or money, organize protests against perceived unfair practices, or violently rebel. The quantitative analysis of premodern states indicates that the theory has great predictive power. Does this also mean that detection of such practices and actions can be indicators of poor bureaucratization and its inherent conditions?

Blanton and Fargher (2009:134) note that "the form taken by a state depends in large part on the outcome of bargains struck between those in positions of state authority . . . and non-ruling groups, especially taxpayers." We easily recognize this negotiation and the implicit agency of nonelites when we think of famous examples of "power from below"—those such as Wat Tyler's fourteenth-century English Peasants' Revolt, or the French Revolution, or any number of disruptive, violent, and sometimes failed, sometimes successful uprisings in historic and current times. Yet when we contemplate prehistoric or largely unrecorded protohistoric periods, our ability to predict—or even imagine—the unseen, undocumented common people of our study areas, we seem far less likely to concede that they had the will, agency, and power to organize and take action. For those of us explicitly studying the relationship between rulers and subjects through time, and the development of each state's unique set of compromises around "power," collective action theory gives us theoretical access to the groups, classes, or factions below the ruling class.

Egalitarian (and Other) Ideologies in Late Prehistoric Europe

Because the emergence of European Iron Age political systems is a repeated cycle of balance between power from above and power from below, our understanding of this era has benefited inordinately from all theories, past and current, that deal with varied and alternative sociopolitical organization. The Iron Age is an era variously described as illustrating the "Germanic mode of production" within a Marxian framework (Gilman 1995; Hunt and Gilman 1998), as being heterarchic (Crumley 1995; Kristiansen and Larsson 2005; Wailes 1995), and as producing "corporate" polities within the "corporate/network" continuum (Bentley and Shennan 2003; Thurston 2001, 2009, 2010) or without reference to a specific theory, a kind of ideologically egalitarian, checked and balanced system (James 1999, 2000; Collis 1997; Hill 1989, 1993). A tradition of deemphasized status and class differentiation was intermittently impacted by the aspirations of self-promoting elites who used "personal prestige, wealth, power accumulation, aggrandizement, highly individualized leadership . . . long-distance exchange, exotic wealth, princely burials, and . . . status craft goods" (Feinman et al. 2000) to advance their fortunes. Collective action theory is the most recent and sophisticated effort to characterize and situate the social, political, and economic conditions inherent in such frameworks where institutional hierarchies are not necessarily centralized, increasing political powers are not always marked by consolidated control, and ordinary people can figure prominently in political development and sociopolitical ideation.

This case study begins in prehistory and progresses to a protohistoric time when a partial elite transcript is available—a historic record not very detailed, yet indicative of the concerns of rulers. The record of subjecthood, of what it was like to be an ordinary person under the rule of the state, is found only in what James Scott calls the hidden transcript—an oblique indication of the "peasant problem" as inscribed in elite complaints, the creation and enactment of laws around "peasant trouble," and actions taken by the state to exploit, suppress, or control them (Scott 1990, 1998). Few of these, if any, are written by or described through the perspective of the lower or middle levels of society, yet the indicated state of affairs, the balance of power, is usually not as elites describe for themselves and each other. Elite-inscribed records must always be read with the understanding that they present an instrumentalized version of events and processes (Ferguson and Whitehead 1992, Scott 1989, 1990, 1998; Sewell 2005). Only at the end of the study era do we finally have a clearer direct transcript of nonelite issues and concerns.

The Svear State: A High-Conflict Society in the Throes of Organizational Transition

Before the development of the Svear state over the course of the Iron Age, Bronze Age society in northern Europe (1700–500 BC) exhibited clear distinctions between elite and commoner in both material wealth and symbolic status: large houses, cattle sheds, monumental grave mounds, distinctive long-distance and local wealth items, and restricted access to metals and weapons characterized high-status individuals. Across northern Europe, including Sweden and the rest of Scandinavia, the period closed at around 500 BC with a general collapse of visible elite culture, often interpreted as not simply failure but outright rejection of sharply differentiating organizing principles, followed by an era of social and political flattening, the Early (pre-Roman) Iron Age, marked by the disappearance of obvious status markers, large impressive dwellings, and monumental tombs with "wealthy" goods (Giles 2007; Hodos 2006; Kristiansen 1994; 1999; Morris 1999; Oubiña 2003; Thurston 2009, 2010).

If material culture alone is considered, one would think that leadership itself had vanished. When dual-processual theory was first introduced in the 1990s, one of the first applications was to such contexts, using their newfound ability to postulate forms of organization different than those rooted in neo-evolutionary ideas. It was suddenly possible to argue that leaders did in fact still exist, but that they had shifted from network-style status aggrandizers to corporate-style status deemphasizers as society took a more egalitarian-minded turn. As noted, other theories important in this reinterpretation were house society theory (Beck 2007; Chesson 2003) and the heterarchy concept (Crumley 1995, 2003, 2005).

The conclusion that a major sociopolitical shift occurred during the Bronze Age/Iron Age transition, or perhaps that this underlying shift was a large part of what *caused* the visible material transition (Thurston 2010), is supported by evidence from the historic record. As the Iron Age progressed and contact was established between the Mediterranean and western/northern European peoples, literate Romans and Greeks such as Julius Caesar, Cassius Dio, Diodorus Siculus, Pliny, Strabo, Tacitus, and many others described "primitive democracies" in the Celtic- and Germanic-speaking spheres of the late centuries BC and early centuries AD. Further textual data from the post-Roman era is found in Continental authors such as Jordanes and Procopius.

The persistence of such sociopolitical traditions is discernible in the archaeological record throughout the course of the Iron Age, most strongly in the early Iron Age but continuing into the later Iron Age and the so-called Viking Age despite the redevelopment of more visible and more powerful rulers. Archaeological evidence for this more collective and heterarchical system is seen in an egalitarian or transegalitarian material culture, reduced stratification in sociopolitical and settlement indicators, more egalitarian burial rites (Axboe 1999; Barrett et al. 2000; Earle and Kristiansen 2010; Kristiansen 2005; Smith 2004), and the appearance of assembly-places seen through textual records, place names, and archaeological evidence (Smith 2004; Sanmark and Semple 2008; Semple and Sanmark 2013). These reveal an Early Iron Age society with invisible leadership, developing slowly into those dominated by a political-military elite, yet with strong and sophisticated leveling mechanisms: "checks and balances" from below.

Some social codes and behaviors observed by the Romans continued, albeit in shifting form, in the intervening centuries when there are no textual records, only to emerge again in the documentary sources when literate traditions were adopted at the introduction of Christianity around AD 1000. These include the heterarchic Iron Age sociopolitical organization frequently discussed by archaeologists (Bondarenki and Nemirovskiy 2007; Crumley 1995, 2003, 2005; González-Ruibal 2006; James 1999; Moore and Armada 2011; Thurston 2009, 2010) with its "stand-alone" power structures: legislative, warrior, and religious. Military leaders required support from an assembly, and leaders and followers had reciprocal obligations: a warlord sustained his fighters (Christophersen 1982; Lindow 1976; Vestergaard 1979), but their support evaporated if they perceived arrogance or avarice (Thurston 2010). Allies elected a paramount to lead collectively against outside threats (Wells 1999:57), but refusal to relinquish power led to sanctioned overthrow or assassination. As clarified by collective action theory (Blanton and Fargher 2008, 2009; Fargher and Blanton 2007), the attempt to layer a "strong" form of rulership atop a long-time corporate society can inadvertently lead to the rise of a state with many internal conflicts.

After a more corporate, socially egalitarian, and status-flattened beginning, over time a slow return of more network-style traditions can be traced in the heartland of the later Svear rulers during the so-called Vendel period (ca. AD 550/570 to 790/800), directly preceding the Viking Age. For example, slight status distinctions in central Swedish inhumation and cremation traditions, with pits and small earth heaps or within modest boat-shaped stone settings, developed after AD 550 into more differentiated forms (Ljungkvist 2008). New burial rites include monumental burial mounds of imposing height and over 20 m in diameter, incorporating a central cairn with whole-animal sacrifices and rich, rare material culture. In the Svear heartland, at and around Gamla Uppsala, the protohistoric and historic seat of the Svear dynasty, are three high (ca. 10 m) burial mounds constructed in the mid-sixth through early seventh centuries, dwarfing earlier structures; several others lie in nearby regions.

Yet it is notable that while some elites constructed such monuments, others adhered to the more traditional burial context, a continuation of the smaller mounds and ship settings of the past, some cremated and some inhumed. Ljungkvist (2008) notes that osteological and artifactual analysis of both types of burials—new and traditional forms—show that the number and types of animals sacrificed and the grave offerings were identical, indicating that families of the same sociopolitical and/or economic class were selecting different rituals. This may signal that a newly emerging group, or a splinter group of elites, adopted a more symbolically uneven presentation, emphasizing individual status with unprecedented monumental construction and eventually an increasingly wealthy material culture, reflecting a more divided, uneven, and asymmetrical reality.

By the end of the seventh and eighth centuries AD, not one but two polities existed: the Svear and Götar kingdoms, documented archaeologically and through mythopoetic texts. These saga entities emerge as real political entities in the early historic era. Significant archaeological work has targeted the emergence of the Svear "core" near Stockholm in the Mälar valley, but scholars are only now focusing on the origins of Götaland, which consisted of Västergötland and Östergötland (figure 8.2). At the Skänninge site, militaristically symbolized elites of the early first millennium AD kept compounds near cult-places whose names link them to a *götar* people. Nearby gravefields at Högby show social stratification developing between the first and fifth centuries AD, culminating in several "levels" of elites with warrior trappings (Helander and Zetterlund 1998; Kaliff 2003). Late Iron Age rulers, sharing power with religious specialists and a public assembly, may have emerged through interactions between important families in Väster- and Östergötland.

Just south of Östergötland, lies what is today called Småland. This area was not initially part of Götaland, rather its borders contained several small-scale political units. Småland literally means the "small lands": 12 independent sociopolitical aggregates (figure 8.2), some mentioned as "peoples" in Jordanes's ethnohistoric *Getica* of around AD 550 (Mierow 1908). This is reflected in different runestone and mortuary styles following still-extant boundaries between Småland's *härads*, administrative units whose borders lie along yet earlier ones. Although there is no textual documentation, historians often assume that Småland's polities were consistently allied with the Götar, and in their fight to resist Svear domination, the outcome of continuing autonomy suggests that this is plausible.

Ethnohistoric traditions describe hostility and warfare between Svear and Götar, with eventual Svear domination, claimed to have begun with "conquest" around AD 1000. Yet Svear claims of "unification" at AD 1000 are improbable. Swedish historians (Sigurðsson 2006) admit ignorance of how or when unification began or proceeded. In light of the work of Blanton and his coauthors, unification, rather than comprising an "event," likely was attempted earlier and proceeded longer.

From the tenth century on, Scandinavian legal codes were transcribed to written texts, from oral traditions in which a lawspeaker recited in public to ensure fairness and adherence to the code, leading to historically recorded protests against radical interpolations that increased the ruler's power and decreased the rights of sub-elites and common people (Brink 2002, 2003, 2004). By the eleventh and twelfth centuries, the *ting/thing*, or assembly, where leaders and later kings were voted in by those eligible to cast a lot, became less electoral

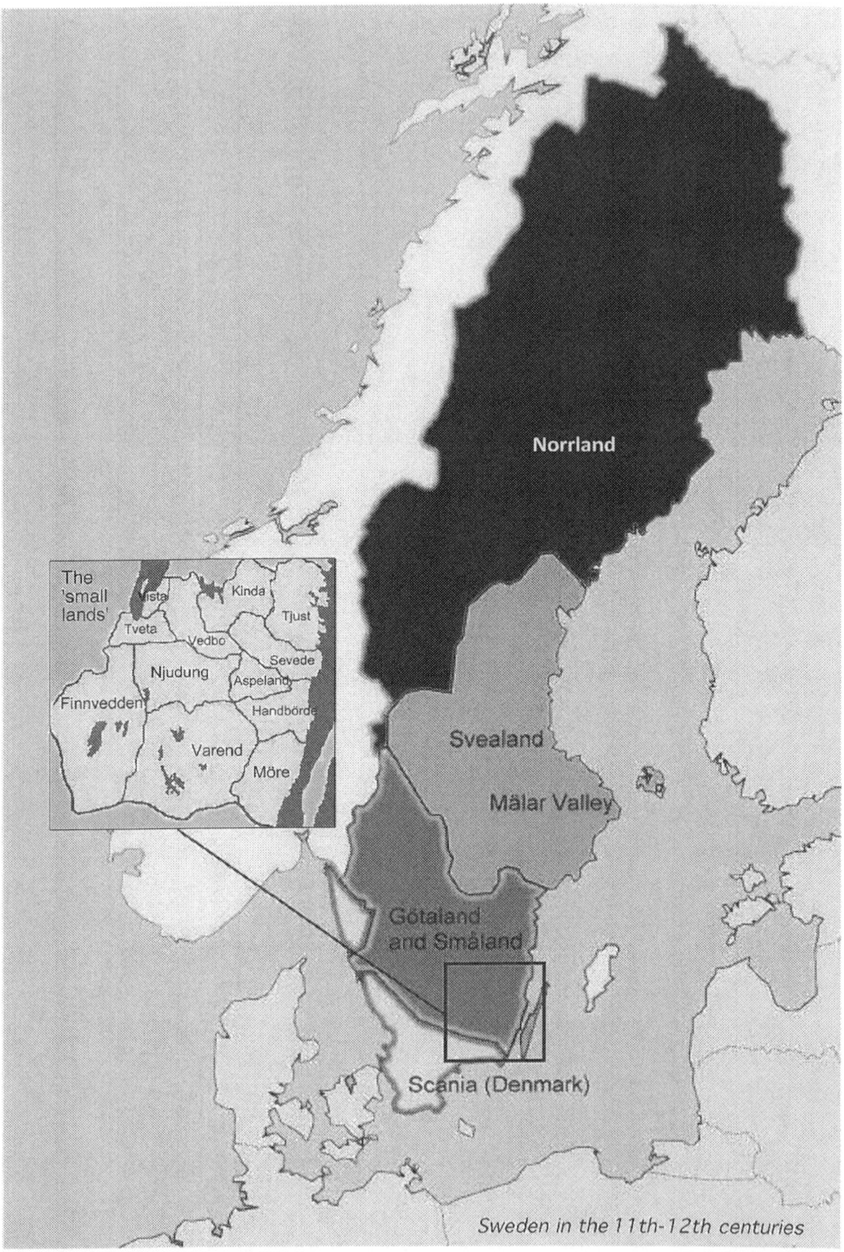

FIGURE 8.2. *Sweden in the eleventh–twelfth centuries, showing the major regions (Norrland, Svealand, and Götaland-Småland) and the "small lands" of Småland.*

and more oriented toward "approval" of heirs, and the selection of eligible men from a large group of hopefuls was replaced by royal patrimony via brothers, then sons. Urban and rural elite-run courts eventually superseded local assembly places (Myrberg 2008; Sanmark 2009; Sanmark and Semple 2008).

The drama that played out between the Smålanders and their ostensible kings has usually been cast in traditional historic terms, identifying individual people or small groups of "aristocrats" as the primary agents of conflict and change. It can alternately be seen as a struggle between two forms of political organization, older more egalitarian forms, and newer attempts at hierarchic, centralized organization, over an entire millennium, as the Iron Age system suppressed quick changes, leading to long periods of tacit or active resistance against *attempted* unification through "creeping" centralizing change, followed by short, violent struggles when rulers challenged tradition, often unsuccessfully. To understand Småland's incorporation into Sweden and its role in state formation, we must primarily use archaeology and its allied disciplines to reveal prestate conditions and later courses of change.

Change in Economic and Political Landscapes

The Svear and Götar kingdoms developed out of heterarchic Iron Age societies, where we understand that the infrastructural elements called bureaucratization were well established. As they grew less collective and more network-oriented during the later Iron Age, other regional polities did not necessarily follow suit; many retained a strongly collective nature. During the long stretch of time when the increasingly network-style Svear polity began to conquer and incorporate its neighbors, there were many continuities but also numerous shifts in the organization of the cultural landscape. Some landscape changes are local byproducts of other decisions, while "institutional landscape changes" are planned or encouraged components of political and economic strategies, or responses to broadly experienced demands or opportunities.

Småland is often characterized as marginal and impoverished, but this is largely based on the perception that wealth lies in cereal agricultural production. The Swedish state saw many profitable and productive possibilities on the Plateau. A border region, Småland was the launching pad for Swedish military expeditions against Denmark. Military musters, while drawing from across the state, unequally availed themselves of local manpower. Husbandry provided meat and dairy, horses, wool and leather with regular and military uses, and was also taxable. Large iron deposits destined for use in weaponry lay in ore-rich hills and in *malm* (bog-iron) found in ubiquitous lakes and swamps.

Forest industries included iron-smelting with blast furnaces fueled by charcoal, manufactured in large charcoaling pits. Tar manufacture was vital to rapidly expanding royal navies.

While the Svear rulers desired the Plateau dwellers' products and tax revenue, obtaining them was much more difficult. It first required some level of territorial control, which during the state's coalescence was fictional—the state had no formal presence there before the twelfth century. To collect revenue in an area where little direct control is exercised and the population is spread across a landscape that is well-suited to concealment, collective action theory would suggest that the best way to facilitate state economic policies would be to construct a vigorously bureaucratized infrastructure.

Was bureaucratization the Svear strategy? The official transcript of the medieval era indicates that tax evasion was rampant and state mandates were more often protested or ignored than followed. Farmers frequently murdered tax collectors and other officials, seen in the legal records of "blood money" paid to acquit them of these crimes, as was possible under the laws of the era. There were tax rebellions and uprisings. Complaints were made about the hiding of assets and the difficulties of prosecuting lawbreakers scattered through the deep and dark forests—tactics that might be rational and expected behaviors in the face of network-style lack of bureaucratization. This trend continued, and escalated over the next few centuries.

It might be inferred from both the textual and archaeological records that rather than building bureaucratic infrastructure, the Svear state strategy appears to have taken the path of building a physical infrastructure for the installation of top-down, authoritarian power. In a region where no towns or cities were previously known, centers were established from which authority was exercised. Around 1140, a royal castle was constructed on Visingsö island in Lake Vättern in the heart of the Plateau region, marking the Swedish king's authority. Visingsö—not the Svear homeland—became the primary royal residence for the next two centuries; it was not a temporary local residence.

A local Iron Age marketplace had long operated at the confluence of the June river and southern Lake Vättern, but in 1282, the state founded a market town, Jönköping nearby, superseding the traditional hub, and soon installed a large garrison. Access south along river roads enabled trade and moved armies downstream to attack the Danes, but also brought them north, and the city was torched and plundered several times, deeply embroiling Småland in warfare in the fourteenth through sixteenth centuries.

The Swedish Crown forbade trade between the two states, and commercial activity across the border, a centuries or even millennia-old tradition, was rendered illegal, leading to the Smålander's refusal to recognize the Swedish-Danish border at Småland's southern edge (Andrén 2000). The niche of "forest-dweller" extended from Småland far into the Danish territory of Scania, and allegiance to either state and respect for a largely arbitrary border was far weaker than bonds between forester-pastoralists on either side. During the many Danish-Swedish wars, "farmer's peace treaties" were enacted. These *bondefred* were formal written agreements to continue trade, and provide mutual warning against incursions of troops from their own homelands. Actual collusions are documented for both sides (Andrén 2000:317–318; Cederholm 2007). Eventually during the sixteenth century, the "magnate" Brahe lineage controlled the region, building castles like royalty on the mainland and on Visingsö. In the sixteenth century, Catholicism, still cherished by the uplanders, was forcibly replaced with the Lutheran Protestantism adopted by the Svear.

DISCUSSION

Traditionally, in Sweden, the *bonde* or free farmer had voice in government, but not because power was devolved by enlightened or merely clever rulers, and not because of any state-initiated institutions designed for smoothing the collection of taxes and the orderly control of society. Rather, it was because at the time the state emerged, the free farmer *was* the state. As noted above, the heterarchical nature of both prestate and early-state Scandinavia, left two-thirds of the governance of society in the hands of nonrulers, or conversely, two-thirds of society that were nonmilitary elites, ruled together, through representation, with their war leaders. While small, private-retinue armies of 200–300 full-time soldiers were kept by rulers, the actual armies of the day were simply levies of farmers and other free men. The kingdoms were divided into administrative units called *härads/herreds* (Swedish/Danish) or "hundreds"—referring to the number of men, with a proper complement of ships, that was owed to the king in times of war. The ruler would call a levy and the hundred-man would arrive at a levy-place led by local elites. Similarly, the assemblies where local and national leaders were elected or acclaimed, saw the participation of all free men, also rallied and led by local leaders. Until AD 1000–1100, the "yea or nay" was real, in the sense that rival candidates for king were put in or voted out of the job by the votes of ordinary citizens, organized of course by those with influence in the community.

Changes in this system occurred in the manner of a punctuated equilibrium over the course of the Iron Age, including the early medieval Viking Age or long Iron Age, with long periods of relatively imperceptible change interspersed with violent eruptions of reorganization. Changes included the transcription of oral law to written, and often subtly altered documents, and the true election of kings by the commons became more aptly described as acclamation or acceptance. The longstanding claims by the commons that kings could tax citizens no more than earlier rulers had taxed their ancestors were met by challenges from above in the eleventh century. Flatlanders fought these trends but eventually were forced to accept many changes (Thurston 2001). This was not always the case for the more territorially segregated uplanders.

The Småland pastoralists shared Scandinavia's strongly egalitarian, anti-authoritarian ideology and, unlike more accessible populations, were mostly self-governing, even after incorporation into the body of the state.

Collective action theory postulates that if a state is to impose and collect taxes in an orderly manner, without inviting protest or social upheaval, it must first bureaucratize. The principals—rulers and their immediate proxies—must develop a set of rules, laws, codes, and expectations for how its agents—in Sweden, a force of royal sheriffs, bailiffs, and fief-holders—carried out state imperatives among the populace. In many instances the commons had the ability to properly monitor and rein in agents who were abusers, and to address public concerns (Blanton and Fargher 2009:141). In many parts of Sweden by the fifteenth century, representatives of the various farming districts appeared directly before the king to make a public record of their demands or claims. Yet in Småland, because of its late and more unwilling incorporation into the state, appeals were less direct: they were carried out in local or regional venues as if the state had never enacted itself (Cederholm 2007). In many ways, Småland, between the twelfth and sixteenth centuries, continued to be a separate nation not only in the eyes of the Smålanders but in the perception of the Crown.

When the centralizing agricultural state proclaimed its broad new powers, it came into irreconcilable conflict with tightly knit forest pastoralists who adhered to ancient apparatuses of egalitarianism and who were poorly integrated into any extant protections. In Småland, the system was limited to officially sanctioned tax farming by ruthless royal officials, compounded by the fact that these agents imposed additional taxes and dues on their own, without any central authority, and used brutal tactics to collect them. Since the Smålanders brought their issues only before local courts and not directly to the king, the situation would grow dire. That conflicts arose is not surprising (Blanton and Fargher 2008, 2009). Småland's identity as a set of autonomous "kingdoms" within a larger state outlasted the unification era: Småland long represented the yardstick by which Swedish kings could measure the integration of their state. It often became necessary to make an example of them.

Livelihood and Dwelling in a Time of Subjugation

In addition to examining the building of state urban and military infrastructure, as noted above, a program of landscape archaeology has examined local responses to these processes. Of the 12 "small lands," two have been studied—Tveta and Vista—which border Lake Vättern and Visingsö Island with its royal seat (figure 8.2). Both were coherent political and administrative units in prestate and state times. I have conducted archaeological research in four settlement districts (figure 8.3) that exhibit coherence and continuity, because they represent contexts at "ground zero" to royal authority, and then at ever-increasing distances from central state authorities: the island of Visingsö, where the Swedish monarchs made their full-time permanent home; the nearby and unusually (for the Plateau) agriculturally rich Skärstad Valley; the more distant, smaller Bredestad Valley; and the iron-rich region south of Jönköping town, the Södravätterbygden. These four locales likely had different relationships and obligations to the state, thus differentially reflecting its impacts. All contain sites dating prior to, during, and after state formation and expansion. This work is ongoing, but some preliminary conclusions may be drawn.

Livelihood

The study of complex land-use history and shifting labor organization can aid understanding of Småland's development. It is clear from the juxtaposition of finds at upland archaeological sites that one of the main strategies used by the inhabitants was "occupational pluralism"—the constant combining and shifting of resources, processing technologies, and land/labor rights and strategies. McCann (2000:486) notes that:

> occupational pluralism represents a strategy for family survival in a marginal world of work. It is a response to those situations in which no single activity provides an adequate income to meet family needs. To this end, contributors to the family's economic welfare might follow a number of seasonal activities—some subsistence, some that earn cash—as they try to gain a modicum of well-being.

While McCann goes on to say that in the current day and age, this often fails, leading to the disintegration of families and communities, in Sweden's

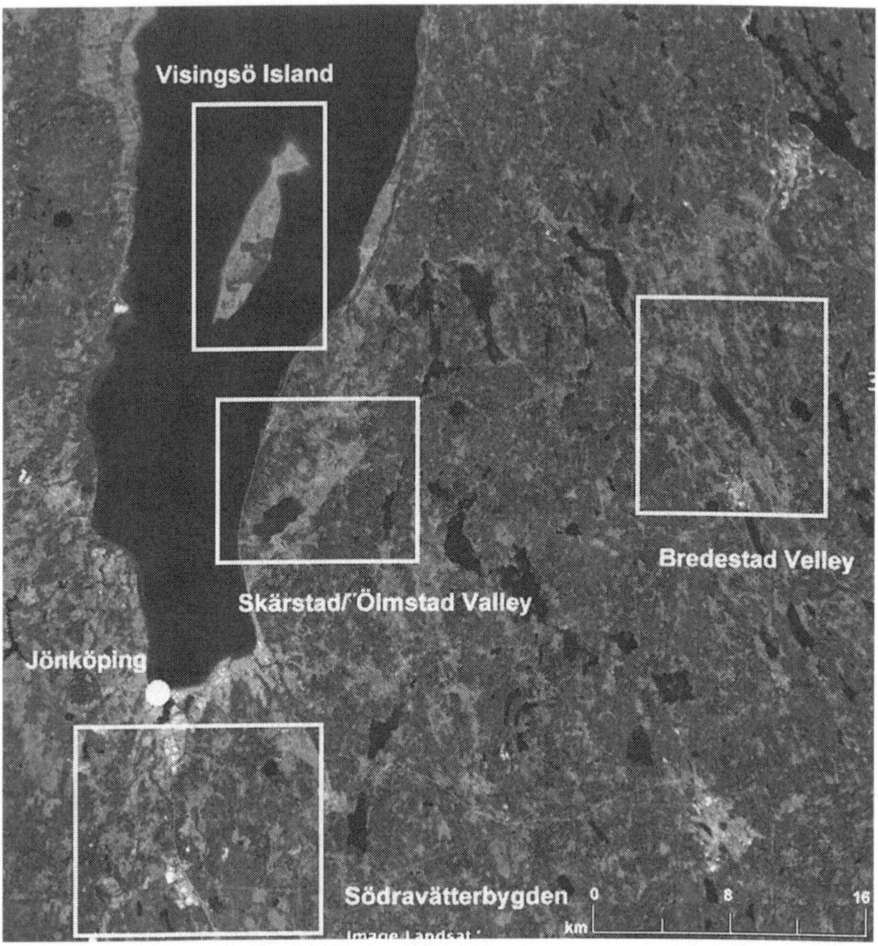

FIGURE 8.3. *Four study blocks with varying proximity to the royal establishment on Visingsö Island.*

uplands it was a centuries-old way of life that was successful, and is still successful, even in current times. On the Plateau, it long comprised livestock raising combined with small-scale cultivation, charcoaling, iron production, woodworking, tar-manufacture, and arboriculture.

There are also proxy data for changes in workforce organization, labor-intensive strategies historically interrelated with proximity and kinship among pastoralists (Fratkin 1989; Grandin et al. 1991). Enormous stone heaps from clearance, accomplished with *svedjeburk*, or slash and burn, thickly dot even tiny clearings, some formed into long, high field boundaries that stretch off into the forest or crest over hilltops. They have been dated by OSL and carbon-14 to waves coinciding with the 1000s–1100s, the 1100s–1200s, and the 1300s–1600s periods (Häggström 2004). Montelius (1953:42) noted that nineteenth-century *svedjebruk* "required so much labour that a single farm could hardly undertake the clearing . . . Felling was a heavy job, but most hands were needed . . . to prevent the fire from spreading . . .Several farms—ten or so—joined together in a common clearing [in addition to] individual clearings." In the 1600s and probably earlier, formal "associations" of 2–4 large households (Vestbö-Franzén 2005) of about 10 persons each (Jansson and Kristensson 2004) cooperatively swiddened, planted, and later herded and grazed the fallow, also cooperatively. Some sites reveal the creation of terraces, not for cropping but simply to create a platform for daily activities on terrain impossible to otherwise build upon.

The pastoral labor pool was increased through household cooperation, "borrowing children," or hiring workers (Sieff 1997) to facilitate herding, milking, and foddering; well, byre, and corral construction; animal droving and marketing; and cultivation. Interherded species require different grazing and watering regimes, thus their own habitual "tenders" (Arhem et al. 1981; Coppock et al. 1986; Cossins and Upton 1988; Fratkin and Smith 1994).

The study of production technologies reveals whether skilled, unskilled, or mixed labor was used. Iron production requires skill plus hard labor; charcoaling and tarring, less skill. Changing economies may point to changing workforces. In the well-studied Bergslagen north of Småland, blast-furnace operations, similar to those of slightly later date in Småland, are found beginning around 1150, where iron and steel were produced in well-documented peasant cooperatives (Florén et al. 2003:78), exemplified at the archaeological sites of Lapphyttan and Vinarhyttan. Metal production at increasingly higher levels was of utmost import to the Crown, and the legal status of the "free miners" was outlined in special legislation from the king, acknowledging and regulating the cooperative nature of organization by granting industrial charters and forbidding a blast furnace from being jointly owned by more than eight households, while individual smiths were responsible for the quality of iron that they themselves decarburized (Gordon and Reynolds 1986; Magnusson 1984, 2002, 2009). Finds at Lapphytten are consistent, revealing exactly eight twelfth-century refining hearths in the immediate vicinity of the smelting operation (Magnusson 1984:61). This traditional collective ownership, which had both communal and individual attributes, continued forward through the medieval era, exemplified in the fourteenth-century Kopparberg charter, indicating that shares of the mine were proportional to ownership of smelters (Rydberg 1979).

Thus, archaeological and paleoecological data tell us that cooperative organization and occupational pluralism continued to be the main strategy for the forest farmers of Småland during their contested unification with the Svear state. The economic impact of the state, direct or indirect, may be seen in the intensification of agropastoralism and forest industries through the series of colonization waves into ever-higher and more difficult terrain. Pollen, animal bones, and settlement remains tell of behavioral changes, but only indirectly of social transformations.

Dwelling

Across much of Scandinavia, recognizable settlement foundation waves are sometimes broadly apparent, to the point where they are almost assumed. In Småland, these waves and phases of organizational and demographic response are somewhat unique, and must be seen in the face of changing conditions. Toponym studies, correlated with thousands of archaeological investigations, are a reliable general indicator of site foundation eras in Scandinavia (Agertz 2000; Brink 1984, 1990). Villages still extant today were recorded by the 1200s–1300s in cadastral registers, as were those abandoned after recording. The place name (Brink 1984, 1990) denotes the foundation of the archaeological "original" settlement, usually within sight of the historic (modern) village locale. These sites, including those abandoned prior to recording with no modern analogues, are found through soil chemical and pedestrian survey.

There were several expansions before the Svear conquest, similar to broad changes across Scandinavia usually interpreted as small foundation waves representing slow population growth in a region without much available farmland for expanding extant villages. In general, these correspond to Roman Iron Age sites (AD 1–400), Migration-era or post-Roman sites (AD 400–550), the Vendel period (AD 550–800), and the Viking Age (AD 800–1050).

While it is important to understand the pan-regional generalities of these settlement waves, it is also important to note that Småland differs significantly. Across Scandinavia, a huge wave of "hamlet foundations" is linked to tax increases around AD 1000–1100 as multiple states emerged, many interpreted as "planned" settlements. These small villages of several farms share the place name suffix *–torp*, sometimes abbreviated to *arp/up*, meaning "new settlement dependent on an older settlement." In Småland it is a distinct but unusually patterned phase; among the four study blocks it occurs significantly only on

Visingsö island, and may correlate with the twelfth-century establishment of the royal residence on Visingsö rather than the late Viking Age. Indications are that few were founded further afield, which may show the era's meager extent of state power or lack of obedience to state will.

Place names unique to the region are plentiful. Highly dispersed one- or two-farm sites with the suffix *-hult* form a large colonization, mostly of the 1200s, coinciding with the establishment of Jönköping. A wave of similar sites ending in *-ryd* appears to have begun largely in the 1300s. *Hult* and *ryd* mean "artificial clearing" and are more apparent and very numerous in the upland valleys and dense forest areas far from the royal island and town; they continued to be founded through the 1500s. The Black Death impacted the region in the later fourteenth century, but a new expansion beginning around 1500, seen especially in the dating of field-clearance cairns, may correlate with the rise of Swedish-Danish warfare and military supply needs, as well as the introduction of the Brahe family's direct local rule in the region.

The difficulty of enforcing tax collection, labor obligations, and new unpopular religious practices (here, the introduction of Lutheranism) in dispersed settlements has been noted by archaeologists, historians, and ethnographers in many global contexts. Examples include Scott's (2009) well-known examination of the evasive strategies of the hidden, dispersed peoples of upland Southeast Asia; the practices of colonial-era Maya, about whom the Spanish complained frequently (Farriss 1985:218); similarly early Spanish-colonial frustrations in Tiwanaku (Bandy and Janusek 2005); and countless others.

Dwelling, livelihood, and change

As frequently inferred by archaeologists, foundation waves of many small, dispersed sites additionally suggest intensification and increased efficiency, a degree of autonomy, egalitarian principles of land tenure, and perhaps attempts at concealment (Drennan 1988; Emerson 1997; Peterson and Drennan 2005), despite the move into higher and higher valleys with ever-worsening conditions for the pastoral agroeconomy. This view is borne out in numerous ethnographic studies (Bentley 1990; Blarel et al. 1992; Hassane 2001; Netting 1989; Tan et al. 2006; Udo 1965).

Each successive wave of these later upland expansions pioneered into higher and higher elevations, from 200 masl, to 250, to 300, to 350 and higher—from valleys with soil to those with almost no cover over bedrock—more hidden but also more difficult to occupy and cultivate (figure 8.4). While pioneers may have been practicing forestry-related commercial activities, they were

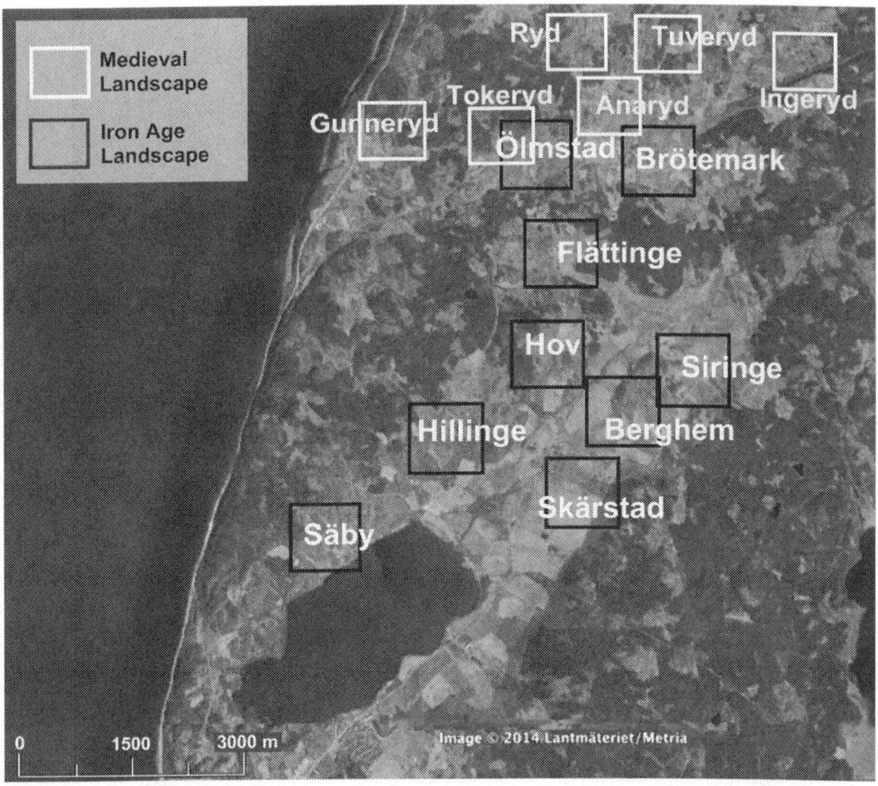

FIGURE 8.4. *Detail of Skärstad-Ölmstad Valley showing changing landscape use from the Iron Age to the Medieval period.*

also carving out their daily subsistence, which would have been much more difficult at high elevations than in the lower-lying upland valleys.

The relative magnitude and distribution of such pioneering suggests moderate to significant intensification. Areas close to the royal residence had planned *torps*, perhaps established by decree for direct royal support, accounting for some of this expansion. However, *hult* and *ryd*, names almost unknown outside Småland, suggest that the major push came from perhaps a local movement from "below." In some areas, such as the Södravätterbygden south of Jönköping town, close to the Taberg iron-ore region, such settlement intensification was surely related to industrial production: wood and charcoal are proxies for the production of iron and tar, and dateable charcoaling features show sharp increases in number and size through time.

"THEY DEFENDED THEIR RIGHTS"

Blanton and Fargher assert that states should effectively offer voice, public goods, and fair tax-collection procedures to taxpayers in order to avoid conflicts, implying—in the neutral tone of theoretical language—that a "rational state" should see the benefits of such arrangements. They do, however, carefully call such a system of government a "sociocultural construction" (Blanton and Fargher 2009:141), which for this case study must be strongly emphasized. The phrase, "They Defended Their Rights," is translated from the work of Cederholm (2007), a contemporary Swedish historian, who makes especial note of the fact that the system of collective government long preceded the state in Sweden, and it was not the rulers who offered it, but the farmers who demanded they adhere to it. Despite the more negotiated conditions elsewhere in Sweden, the Swedish Crown effectively continued to treat the Smålanders as a conquered territory with clearly separate origins, perhaps because the Smålanders themselves in some ways insisted on it.

Blanton and Fargher define resources drawn from outside the state as "external revenues"; it is possible that Småland, conquered yet still in early modern (and recent) times oft-conceptualized as remaining apart, was treated as an external revenue source. Most conflicts between the fifteenth and seventeenth centuries, between rulers and subjects, were over the outrageously high cost of war, and the taxes and other burdens related to it. Since Småland was on the border, it was unduly impacted by violence, burning and laying waste, rape of women and girls, the extraction of resources, and calls for mustering soldiers from local settlements. It is possible that the rest of Sweden benefited from the Crown's treatment of Småland as a colony of sorts rather than a province to which appropriate governance should apply. The Smålanders, through their extreme reluctance to participate in the state and their inward-turned political stance, may have exacerbated this. Given the brutal nature of tax collection, and the lack of recourse enjoyed by other regions through direct lines of communication with the Crown, tax evasion, rule-breaking, and strategies of concealment from state surveillance are predicted by collective action theory as rational responses to an oppressive state.

In an entirely prehistoric context it might be difficult to prove that outcomes were as violent as we know they were, but archaeologists should not fear suggesting such courses. While it is plausible that the expansion waves and their marginal locales simply show extreme pressure from the state to pay higher taxes, the fifteenth and sixteenth centuries also record the most conflicts over tax evasion. A sequence fraught with tension concluded in a dramatic cascade of events that cemented, rather than ended, the place of the corporate tradition and its practice of shared governance and collective action.

After King Gustav Vasa raised taxes sharply, banned cross-border trade with Denmark, and banished rurally cherished Catholicism in 1542 (Cederholm 2007; Hallenberg et al. 2008; Katajala 2004), Niels Dacke, an impoverished local sub-elite who had already paid blood money for killing a sheriff over the heinous practices associated with tax collection, appears in historic texts as a general, riding in command of a farmer army against the state. Dacke's rebellion (*Dackefejden*) spread throughout southern Sweden and along the Danish border, and Vasa's army of foreign mercenaries was massacred by crossbow-armed farmers in the dark forests and steep, rocky terrain, where their military tactics were useless. Vasa, who was himself put in power by an earlier farmer army, signed a peace treaty with Dacke, who raised the ban on trade, lowered taxes, and restored the Catholic church, according to local desires. When Vasa broke the treaty and the commoner's forces were decimated, Dacke was wounded, outlawed, and died on the border of Småland and Denmark in 1543, and his family was executed shortly after.

Despite the ultimate failure of the historic figure, Dacke, the corporate mode of governance fared far better than the rebel leader. After the Dackefejden, Vasa stepped back from the taxation practices that had inspired the Smålanders' rebellion and paid special attention to their complaints. He and his descendants deferred frequently to the farmers who brought issues before the Crown, to ensure the security of the throne (Hallenberg et al. 2008). One might say that while the Smålanders may have been paying more tax than their ancestors, their final expansions and intensifications may have been ones that were of little direct benefit to the state. In other words, later kings of Sweden began providing access to the collective and reciprocal structures found elsewhere in Sweden, which had been modified from Iron Age practices and continued in an Early Modern form. This was what the Smålanders had agitated for, and the reason they initiated violent conflict. While the upland farmers had to accept the less powerful version of "voice," the state also accepted a lower level of taxation and control—a final negotiated reciprocal arrangement that in many ways follows a continuous thread to the present.

CONCLUSIONS: RETHEORIZING A "CEMENTED" RECORD

Kingship, over the course of the study period, was emerging from a heterarchical and weak structure, developed during the more corporate, collectively oriented Iron Age, in which rulers were elected and had more limited power.

After AD 1200, and especially during the 1300s–1400s, there were attempts to create legislation to increase royal power at the expense of the aristocracy and the commons, often using certain tax exemptions for the upper classes, which encouraged nobility to coopt local producers. In the 1500s, modeling their aspirations on the political and economic power of continental kingship, there was a push toward absolutism.

Blanton and Fargher have argued that bureaucratization—the building of organizational structures providing or accommodating voice, distributing public goods, and collecting taxes justly and fairly—is necessary if rulers determine that they must collect, or increase the volume of, revenue. In the absence of these institutions, the theory postulates that people will migrate, evade tax collection, organize demonstrations, or rebel.

The Småland case study offers insight into these processes. The small kingdoms or chiefdoms were already politically complex in the Iron Age, displaying a strongly corporate structure that provided voice in the form of the assembly, public goods in the form of the reciprocal obligations between chiefs/kings and their supporters, and fair and just collection that was apparent in the voluntary nature of tribute given for their support. The social code that permitted warriors to abandon an unpopular warlord and seek out another, and the ability to overthrow or assassinate overbearing leaders, was a leveling mechanism insuring the perpetuation of the system.

As the centralizing Swedish state moved into Småland and attempted to collect taxes, there was no process of bargaining or negotiating with the locals to accommodate their Iron Age traditions and achieve their cooperation by offering them voice or benefit of public goods. Instead, hired ruffians and tax farmers were used to oppress the populace and coerce tax collection. From the cross-cultural perspective of Blanton and Fargher's work, such tactics are weak and lacking in infrastructural power. A ruler may claim absolute powers, but fail to produce results when the commoners are in revolt and his own tax collectors are violently abusing local people and robbing the state by imposing personal taxes that are still collected in the name of the king, thus blackening the reputation of the ruler as trustworthy. It was not until the products of the Plateau became vital to state objectives that rulers realized how much the revenue was needed, and how weak the state's power in the region actually was. The Smålanders on the other hand, as predicted by collective action theory, acted rationally by evading taxes, migrating away from royal surveillance, and eventually fomenting uprisings. The defeated state responded by shifting from coercion to strategies for achieving cooperation, which lie at the roots of modern democracy in Sweden.

These phenomena were not unique to central Sweden but were endemic in all Scandinavian regions where similar Iron Age traditions preceded the Medieval state. In the ninth and tenth centuries, Norwegian chieftains fled to Iceland, which became a republic in AD 930, pointedly having no inherently hereditary leadership, about which the sagas said, "only the law is king." To the south, as Scania was incorporated into Denmark, protests and eventually tax rebellions in the eleventh and twelfth centuries were aimed at overturning changes instituted by the new, broader powers of kings, and in Denmark itself, an eleventh-century tax rebellion resulted in the ruler's death (Thurston 2001). Later rulers met the same fate over perceived unfairness and mistreatment of the populace, banishment of sub-elites who challenge their authority, and especially over taxation. Royal dynasties were pressing for more powers, but those invested in the older system strongly resisted.

The remarkable corpus of theory espoused by Blanton and his coauthors help us to understand the roles of both rulership and subjecthood along the continuum of government as a dynamic, recursive process, the materiality of which can be studied archaeologically, and the parallel social aspects newly retheorized with implications for past and current contexts.

9

The Bakitara (Banyoro) of Uganda and Collective Action Theory

Peter Robertshaw

In recent years Rich Blanton, his colleagues, and his students have championed collective action theory as a theoretical framework for the investigation of the development and workings of state-level societies in which they emphasize the agency of subalterns, political cooperation, and checks on despotic leadership. This is an alternative to neoevolutionary perspectives that focus on the strategies by which emergent elites acquire and legitimize power so that they can exercise authority over their subjects. In a sense, collective action theory is a bottom-up approach to understanding state formation that examines how power is negotiated between rulers and subjects and argues that in many states elite power is constrained by the populace at large. Borrowed from political science (Olson 1965; Levi 1981, 1988), collective action theory (Blanton and Fargher 2008) supersedes the concepts of network and corporate power strategies formulated by Rich and colleagues in their "dual-processual" theory for the evolution of Mesoamerican civilization (Blanton et al. 1996; see also Blanton 1998a). Recently Rich and Lane Fargher have completed an ambitious cross-cultural evaluation of the importance of collective action in the formation of premodern states (Blanton and Fargher 2008). Their results "support the contention that state formation is a process involving rational social action on the part of taxpayers as well as rulers" (Blanton and Fargher 2008:252). The data from which this conclusion is reached are a sample of 30 states from various parts

DOI: 10.5876/9781607325338.c009

of the world and several time periods. As it turned out, the state that scored the lowest on the authors' measures of the importance of collective action in the workings of the state, and hence that could be deemed the most despotic of the 30 states, is the one referred to in the book as that of the Bakitara, who reside in western Uganda (figure 9.1). Thus, I have the dubious distinction of being an archaeologist who studies what was arguably one of the world's most despotic premodern states. This state is more commonly referred to as Bunyoro but Blanton and Fargher's use of the term, Bakitara, is understandable because they derived much of their data from the ethnography of John Roscoe (1923). However, Roscoe himself had little justification for using this term (Sutton 1993:39), whose origin and etymology is unknown but is sometimes applied in the form "Kitara" or "Bunyoro-Kitara" to a broad geographical region encompassing both the nineteenth-century kingdom of Bunyoro and its probable antecedents. If this is not confusing enough, it must be noted that in Bantu languages, such as the one spoken by the people of Bunyoro, noun prefixes are employed. Thus, the people of the country of Bunyoro are the Banyoro (alternatively, Abanyoro; singular: Munyoro), who speak a language called Runyoro. However, when used as an adjective, the noun prefix is normally dropped: for example, Nyoro kingship.

My initial gut reaction to the designation of Bunyoro as the most despotic of premodern states was one of skepticism, perhaps because I did not want to be associated, even intellectually, with a despotic state. However, while one might quibble about minor details of Blanton and Fargher's interpretation and subsequent scoring of the ethnographic data on the Banyoro in compiling their "collective action measures," a point or two here or there would not alter the basic conclusion that nineteenth-century Bunyoro was in some sense a despotic state, this despite the fact that the literal translation of "Banyoro" is "freedmen." I suspect that most present-day Banyoro, at least those of a royalist persuasion, would object to the term "despotic state," viewing the authority of the king and his court in more benign terms, but for the moment I can think of no suitable synonym.

In this chapter, I plan to do several things: first, I will shed light on the workings of the nineteenth-century Nyoro state to determine how such a state was able to function without exercising undue coercion on its populace and indeed with the apparent consent of its taxpayers. I will discuss both the external revenues that fueled the economy in the nineteenth century and the other revenues that seem to have had a deeper history, one that promoted greater negotiation of authority between the king and his people. Second, I will explore the role of ritual activities in the negotiation of the tensions

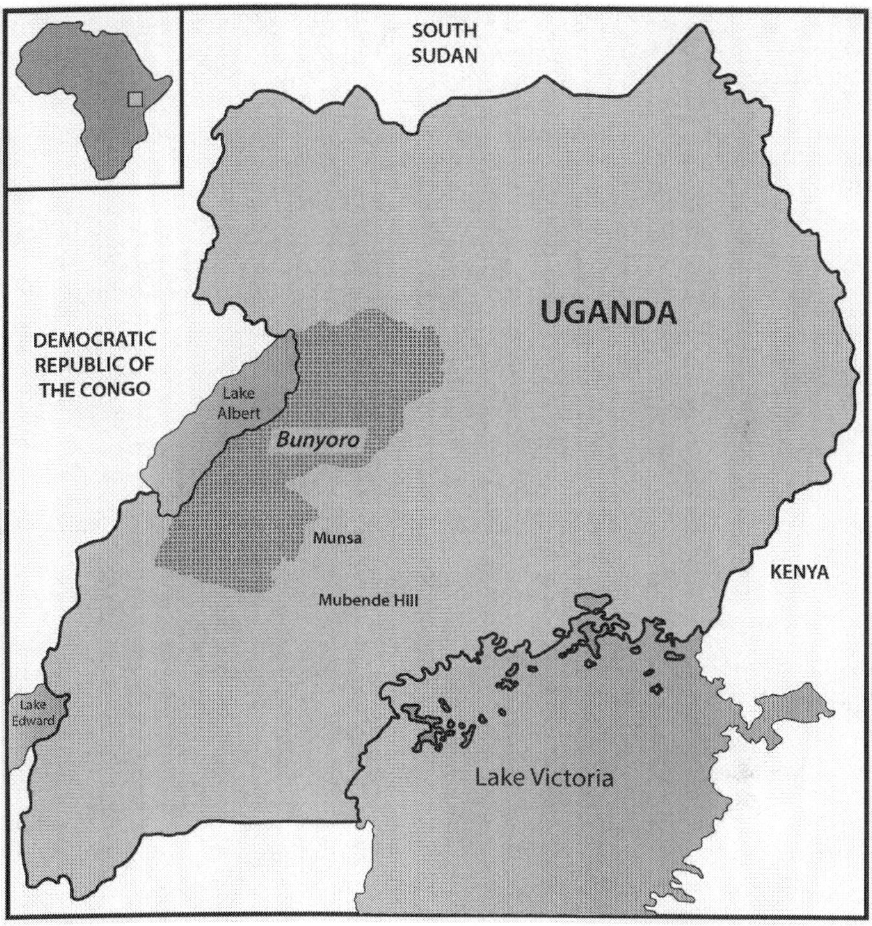

Figure 9.1. *The approximate location of the kingdom of Bunyoro in the nineteenth century and of sites mentioned in the text.*

between state and community. This exploration raises the question of whether the king's subjects might exercise the option of shifting their allegiances elsewhere. Thus, my third topic will be an examination of the topic of mobility, which in turn leads to consideration of how rulers and ruled may have sought to assemble "wealth-in-people" (Guyer 1993). Fourth, with the constraints of archaeological data in mind, I will explain how the materiality of Nyoro kingship was entangled with both power strategies and collective action that negotiated the tension between the state and the people. Fifth, I will revisit the process of state formation in Bunyoro. While the nineteenth-century state was despotic and derived much of its wealth from external revenues, the same cannot necessarily be said of earlier centuries, so a diachronic perspective may shed light on the dynamic balance of power between rulers and ruled.

In previous essays (Robertshaw 1999a, 2003), I applied the dual-processual theory of state formation (Blanton et al. 1996) to Bunyoro; here I reevaluate some of the archaeological evidence that I used to argue for a chronological separation of exclusionary and corporate power strategies, arguing instead that both strategies may have operated simultaneously as the tension between the state and the people was negotiated. Finally, my conclusion will very briefly consider one implication of my discussion that may serve to illustrate the value of applying collective action theory to African prehistory and vice versa.

THE NYORO STATE IN THE NINETEENTH CENTURY

In Blanton and Fargher's cross-cultural survey, low scores on the measurements of collective action correlated with an economic emphasis on external revenues. Such external revenues, they argue, would be favored by rulers because they would then be less economically reliant upon taxes (tribute) garnished from their subjects. As a result of this, rulers would be less beholden to their subjects and could in principle maintain a stronger grip on power, assuming that bureaucratic/infrastructural means existed by which rulers could exercise their authority (Blanton and Fargher 2008:252–254). Roscoe's (1923) ethnography of the Banyoro was the source of most of Blanton and Fargher's data; this ethnography was based on several weeks of interviews at the royal court during the early period of British colonialism and is written in the past tense suggesting that, more often than not, Roscoe's informants were remembering the halcyon days of the late nineteenth century under their great and last independent ruler, Kabarega (Kabalega). The same is broadly true of Beattie's (1971) ethnography of the Nyoro kingdom based on field research in the 1950s.

Taken together these ethnographies reveal a highly despotic and increasingly centralized kingdom in the nineteenth century, during the course of which external revenues derived from exports became ever more important (see also Doyle 2006:51–52), so much so that the trade in ivory and guns was a royal prerogative. Indeed, permission to hunt elephants was granted only to a class of professional hunters controlled by the king (*omukama*) and the populace was forbidden to trade in ivory on pain of death (Uzoigwe 1972:446, 451). All markets, of which there were many throughout the kingdom, belonged to the king, who appointed *abahoza*, a term translated as "political agents/ tax collectors," to oversee them and ensure that revenue flowed to the king; even some wives of the king were abahoza (Uzoigwe 1972:450). Moreover, in Kabarega's reign there were four different markets in the neighborhood of the palace (ibid.). Much wealth was also derived from the export of slaves in the nineteenth century; foreign slave merchants supported Kabarega's bid for power in the war of succession following the preceding king's death (Steinhart 1977:32).

Kabarega is also credited with military reforms that created a standing army with companies (*barusura*) of soldiers who were often under the leadership of foreign mercenaries appointed directly by the king. These companies were stationed throughout the kingdom and indeed expanded the kingdom through conquest, gaining their livelihood by plunder (Steinhart 1977:21–22). Thus, Kabarega instituted a potent source of authority that was mostly independent of the populace and cut across existing class and clan loyalties, further centralizing royal power and expanding his ability to exercise royal authority across the kingdom.

However, while it is clear that external revenues became increasingly significant for Bunyoro's economy over the course of the nineteenth century and that royal control of these revenues, together with the *barusura*, promoted highly centralized authority, such revenues seem at first to have been supplemental to internal revenues. As Beattie (1971:139) remarked, "Everybody should give to the king." Much, if not most, internal revenue derived from the king's personal delegation of authority not only to territorial chiefs but also to people in more lowly political posts. "All subordinate political authority in Bunyoro was held, and was seen to be held, as the direct gift of the king himself, and at his pleasure" (Beattie 1971:147). "The grant of a chiefship by the *Mukama* [king] was essentially the bestowal of rights over a particular territory and the people in it . . . to be given an 'estate' was to be granted political authority over it" (Beattie 1960:37). It is possible that the appointment of chiefs was a practice either established or formalized by Kabarega, as one Nyoro historian has noted that Kabarega changed chiefships from hereditary to appointive (Kihumuro-Apuuli 1994:62).

From their subjects, chiefs received tribute in food, beer, and labor, much of which was passed to the king, who also received ivory, animals, and iron from his chiefs (Beattie 1971:130, 166). Similarly, the king frequently toured his kingdom, exacting food and labor from the areas he visited. Furthermore, all the women in the country belonged, at least in theory, to the king, who did not marry, since the king could not occupy the subordinate status of being somebody's son-in-law; Beattie (1971:143) surmises that kings had "scores, if not hundreds" of royal wives, noting that Kabarega is credited with fathering 140 children.

As with the ethnographic observations on external revenues, it is probable that these observations on internal revenues[1] also derive from the nineteenth century, but it seems reasonable to assume that these internal revenues had deeper historical roots. Such an assumption is perhaps supported by ethnographic data on the ways in which the king was obliged to reciprocate by giving generously to his people. Collective action theory would certainly encourage us to predict that, with less external revenue, the king would need to offer more in return to his taxpayers (Blanton and Fargher 2008:252–254), so the Nyoro ethnographic observations of the king's "generosity" may well reflect an earlier time.

The king was expected to give generous gifts to individuals and to sponsor great public feasts. This was reflected in some of the king's official titles, including *Mwebingwa*, "he to whom people run for help when in need" (Beattie 1971:141). Indeed, the king seems to have been a pivotal figure in organizing famine relief (Doyle 2006:31), while he is also credited with ensuring that traditional medical practitioners were distributed across the kingdom and seconded when necessary to areas of disease outbreaks (Doyle 2006:32). The king's gifts were not all strictly utilitarian; his delegation of power carried with it a delegation or sharing of ritual authority or potency termed *mahano* (Beattie 1971:117–118). Moreover, the king was symbolically identified with his country, requiring him to stay healthy, maintain a state of ritual purity, and perform daily rituals for the good of the country (Beattie 1959, 1971:105–107). Kabarega himself also seems to have used the threat of attack by the neighboring kingdom of Buganda as a means to appeal for national unity and reconciliation (Kihumuro-Apuuli 1994:66).

The fact that the king both received tribute (tax) from his people and provided them with gifts, feasts, famine relief, and ritual support, as well as military protection, draws attention to the negotiation of power and authority between rulers and ruled that is integral to the workings of collective action. It also illustrates the essential tension between the populace's desire for autonomy and their need for state-supported security that the historian Jan Vansina (1990:232) has argued lies at the heart of all politics in equatorial Africa (see also Doyle 2006:15). This tension or duality is recognized by the Banyoro themselves, who have reported that kingship and government stand in opposition to a set of terms expressive of community-based loyalties. Indeed, the Nyoro term for "government," *bulemi*, incorporates both the idea of "ruling" and oppressive "weight," *bulemezi* (Beattie 1971:6–7). Furthermore, the king was regarded as the "ruler or master, not the father of his people" (Beattie 1971:104), the king's lineage having been "chosen long ago to rule us," as recounted in testimony given to Beattie (1971:100). Inequality is pervasive in Bunyoro, as expressed in the proverb, "people are [only] equal in the grave" (Beattie 1971:7). It will come as no surprise to find then that the tension between state and community is evident also in the panoply of rituals associated with the kingship.

Royal Rituals

On the basis of collective action theory it has been predicted that more collective polities would be characterized by rituals associated with rulership that might serve to build trust between rulers and ruled (Blanton and Fargher 2008:203). However, rather than attempting to fathom the role of ritual per se in the negotiation of collective action, Blanton and Fargher (2008:206) attempted to measure the extent to which there was public monitoring of the behavior of rulers, as well as the extent to which rulers controlled "ideological resources" and could use these resources to render themselves in some sense sacred and, therefore, not bound by the same rules as ordinary folk. Presumably the performance of rituals in very public settings would permit public scrutiny of both the ruler's lifestyle and his commitment to the collective (Blanton and Fargher 2008:22, 203; Golden and Scherer 2013:402). The opposite, however, might not be true of more despotic states; certainly one would expect rituals to reinforce royal authority but there would seem to be no a priori reason for assuming that such rituals would be preferentially performed in private rather than in public settings. Private rituals could presumably contribute to the mystique of power, excluding commoners from sources of creative power, but carefully orchestrated public spectacles might equally well serve to reinforce the legitimacy of royal power (see also Fleisher and Wynne-Jones 2010). However, we may also ask how the rituals themselves, not just their setting and performance, served to negotiate power. However, before I embark on an examination of Nyoro rituals of kingship, it should be noted that a distinction between the king as a political and economic agent versus the king as a nexus of state rituals is, if not entirely arbitrary, a distinction that may well be lost on the Banyoro themselves, since instrumental power, power that controls people's actions, and creative power, the power that manipulates and invents forms of meaning (Schoenbrun 1999:139), are interdependent (see Robertshaw 2010).

A review of the literature, primarily the writings of Beattie (1959, 1971: chapter 5), on rituals of Nyoro kingship reveals that the tension between ruler and ruled to which I have already drawn the reader's attention is present too in the panoply of rituals, a conclusion that gives credence to Beattie's assertion that "ritual and ceremonial clustered about the Nyoro kingship because it was the centre of secular power, rather than the other way about" (Beattie 1971:107–108). The king's right to rule was founded upon myth and history but confirmed by the coronation ceremonies performed when a new king claimed the throne—an event frequently preceded by a civil war among rival princes, which in turn served to remind people that peace was only possible with a king firmly in control of the state. The coronation ceremonies took place at a temporary royal enclosure; an initial purification ceremony was reported as being witnessed by "crowds of people" (Roscoe 1923:128), while later when the king sat on his throne, "crowds of the chiefs and better class people pressed forward, wishing him long life and congratulating him" (Roscoe 1923:130). At his accession, the king was also admonished to rule wisely and justly, and made to swear not to frighten his nation and to help without distinction both rich and poor (Beattie 1959:141, 1971:117). Thus, some of the ritual surrounding the king's coronation appears to have been conducted in public and involved an attempt to curtail any tendency toward autocracy. In addition, even those parts of the coronation ceremonies that took place in relative seclusion in the palace were attended by various servants and assistants, who often held honorary titles; perhaps more important, each appears to have represented a particular clan (K.W. 1937). However, other parts of the coronation stressed the king's right to rule and his ownership of the means of violence, notably the handing over to him of artifacts, including a spear, a bow and quiver, a dagger, and a stick. He also struck a hammer on an anvil to signify that he was both the head of all the blacksmiths and he himself like a hammer (Beattie 1959:140; 1971:112; K.W. 1937:296).

Once installed on the throne, many of the rituals of kingship were part and parcel of the king's daily life (K.W. 1937:298–299; Roscoe 1923:91–107); these rituals were required of the king for him to maintain his state of ritual purity, as was appropriate given his symbolic identification with the state. The rituals included food taboos and several daily ceremonies involving cattle and milking. Similarly the king remained in relative seclusion for most of the monthly new moon ceremony and the annual ceremony in which he blessed the country (Roscoe 1923:107–112). These activities all took place within the royal enclosure (*kikali*), a complex of several functionally specific buildings, access to most of which was strictly controlled (Roscoe 1923:73–86). Thus, much royal ritual took place in seclusion, an observation that perhaps accords with the fact that the king was considered to be separate from and superior to everybody else; he was always addressed in the third person and a special vocabulary existed for his person and his activities. The king also possessed considerable royal regalia (K.W. 1935:160; 1936:77, 1937). Therefore, it appears that after the king had been crowned, most royal ritual took place away from the public gaze and could perhaps be interpreted within the framework of exclusionary power strategies (Blanton et al. 1996), despite the fact that some of these activities were commonly considered, at least by Roscoe's informants at the court, to have been undertaken in order to bless the country and its people (Roscoe 1923:93).

Despite the privacy of the regular royal rituals, the king was apparently required to appease the populace after he had ruled the kingdom for nine years, nine being regarded as a very auspicious number by Banyoro. This seems to have been a remarkable ceremony or set of ceremonies, at least for Western sensibilities. Accounts of the ceremony vary (see Beattie 1971:113–114; Nyakatura 1973:205–207), but central to it was an oath of peace sworn by the king in which he pledged, inter alia, not to become angry, not to punish his people, not to kill anyone, not to wage war, and not to be ungenerous. However, what is striking about this ceremony was the massive amount of human sacrifice that either accompanied it or followed it. According to Fisher (1911:130–131), the wife of an early colonial missionary, the sacrifices included 30,000 cattle and 200 princes, the latter killed by being thrown into a large furnace, as well as a royal servant who was sacrificed instead of the king. In Bikunya's account the king and two others climbed into a pit which was then filled up to their necks with human blood from victims sacrificed next to the pit. Once the blood in the pit reached the right level, the king had to climb on top of all the corpses and repeat the oath of peace (Bikunya 1927:52, cited by Beattie 1971:113–114). Nyakatura's account is less gruesome but nevertheless speaks of the execution of large numbers of people, including numerous royal servants (Nyakatura 1973:205). As Beattie (1971:114) remarked, we cannot be sure that these rites were ever performed; they certainly were not witnessed by any of the authors of the accounts and probably not by any of their informants either. However, this does not detract from the fact that such a ceremony was widely believed to have existed and that the oath of peace was accompanied by a requirement, at least according to Bikunya and Nyakatura, that the king relinquish his authority to the senior members of his government.

It is tempting to view this remarkable ceremony as symptomatic of the tension between kingship and populace, with the king being required, albeit at a very auspicious moment of his reign, to publicly recognize his duties to his people, perhaps even to relinquish his authority. An analogous public ceremony, the Orun festival, took place among the Yoruba of West Africa, at which the gods decided whether or not to let the king continue to rule (Blanton and Fargher 2008:208; Trigger 2003:510–511). However, it is the scale of the human (and animal) sacrifice that accompanied the oath of peace that draws attention to the importance of the Nyoro ceremony. The clearly drawn connection between kingship and human sacrifice evident in this ceremony is also present in some other African states, including Yoruba and Buganda (as noted by Blanton and Fargher 2008:208, 213), and spectacularly so in the royal burials at Kerma in the ancient kingdom of Kush (Bonnet 1990, 1992). This can be interpreted as a reminder of the importance of "wealth-in-people" as the sine qua non of success in politics in many parts of Africa (see below), with the destruction of this wealth through sacrifice being the ultimate statement of royal power. One might think that such an awe-inspiring destruction of wealth would serve to reinforce the Nyoro king's authority, not encourage him to give it up. Perhaps the dual nature of the Nyoro ceremony served to stress the power and importance of the kingship and the state, while reminding the king that he himself could be replaced. Be that as it may, our account of Nyoro royal rituals serves to highlight the tensions between the king and the populace. It may also suggest that for long periods of time the balance of power seemed to lie with the state, as expressed in the daily, monthly, and annual rites that were conducted in the confines of the royal enclosure, mostly hidden from public scrutiny. Nevertheless, at long intervals the king's subjects may have had the opportunity either to curb the authority of the king or at least to remind him that his authority could rightfully be challenged.

Mobility and the Composition of Wealth-in-People

Of course, there may have been other avenues by which some of the populace might challenge the king's authority, including allying themselves with princes at the borders of the kingdom who chose, albeit at their peril, to spurn the king and establish independent polities. People may also have chosen to emigrate, but we have very little information on this practice other than Buchanan's study of clan traditions that focused on immigration rather than emigration (Buchanan 1974). However, some idea of how mobility worked in practice *within* Bunyoro is provided by a study of village composition undertaken in 1966 (Charsley 1970). The village under study, located on the periphery of the former Nyoro kingdom, was said to "possess an air of stability and permanence" that was in fact "illusory" because half of the households in the village had moved there since 1960, though the total size of the village had not increased. Mobility was made feasible by an abundance of available agricultural land, presumably a reflection of low population densities, that was neither bought nor sold, as well as by an ethos of good neighborliness (Beattie 1960:61–66). Why did households move? Sometimes there were economic incentives, such as escaping from the crop predations of elephants or moving close to a new arterial road. Sometimes people left an area where they did not get along with their neighbors or felt threatened in some way, perhaps by sorcery or vague spiritual forces (Charsley 1970:17). Where did they move? People mostly relocated to places where they had relatives, agnates or affines, from whom they might expect to receive friendship and support. However, some individuals attracted many more migrants than did others. These individuals shared a higher social standing than most that was derived from one or more sources: the traditional political system, particularly ties to the king; kinship as the head of a large family; and, finally, modern economic and occupational status. Furthermore, these individuals tended to be firmly and historically rooted in their communities (Charsley 1970).

Charsley's study serves to remind us of the concept of wealth-in-people, which was first developed in the context of studies of African politics as a counterargument to the emphasis placed on material wealth as a basis for legitimizing power (Guyer 1993). As Vansina (1990:251) remarked in reference to a broad swathe of societies in equatorial Africa, "Wherever possible, wealth in goods was still converted into followers." People who were able to attract others to their community not only gained access to labor, the reproduction of labor, and the products of that labor, but they could also assemble people with varied knowledge across a broad range of topics, not simply specialized technological expertise. Leaders were able to bring together and mobilize different bodies of knowledge in a process that has been referred to as knowledge "composition" rather than simply "accumulation," a perhaps subtle but important distinction (Guyer and Belinga 1995).

Thinking about wealth-in-people and knowledge composition in the context of both collective action theory and the particular case of Bunyoro encourages the suggestion that knowledge possession combined with ease of mobility could provide individuals with considerable flexibility in terms of where to place their loyalties, a decision that would seem likely to have been made by the individual and his close kin rather than by any collective. Several or many individuals could in theory choose to settle with a leader who might provide an alternative source of authority to that of the state. Such leadership might be based on authority that ultimately derived from possession of some form of creative power. Indeed, in the nineteenth and twentieth centuries and probably in earlier times too, there were shrine sites at various places within the Nyoro kingdom whose guardians seem to have derived their authority from sources of creative power and whose ritual functions commonly invoked the spirits to bless people with fertility. The guardians of individual shrines were often members of particular clans and the authority of these guardians was acknowledged by the Nyoro kings (Robertshaw 2010).

The king and the state also attracted followers and sought to "compose" bodies of knowledge. Indeed, it seems probable that the king and his palace would generally have been the most desirable destination for migrants, provided that they could call upon preexisting connections with kin, affines, or even friends at the capital. The state's competence in composing knowledge and attracting followers is evident from the very numerous specialized offices, duties, and occupations, most with titles, that existed at the court. Moreover, each position was usually occupied by a member of a particular clan, thus perhaps ensuring representation at the capital from a wide range of constituencies. The functions and duties of many of these royal retainers revolved around the performance of ritual acts that served to maintain the king's state of ritual purity (Roscoe 1923). Thus, the state, too, seems to have harnessed creative power to build its wealth-in-people. Moreover, the king, at least perhaps in the nineteenth century, seems to have been sufficiently successful at attracting followers, despite the competition from the communities centered on shrine sites, that he needed to offer relatively little in terms of concessions, such as public goods or shared authority, to his people.

The Materiality of Nyoro Kingship

The attraction of followers and the harnessing of creative power were intimately bound together with a material component. The materiality of Nyoro kingship illustrated once more the tension between the state and local communities. Regalia are frequently mentioned in the ethnographic and historical literature of Bunyoro as being intimately entangled with royal authority. For example, when Rukidi, the first king of the Bito dynasty, to which Kabalega also belonged, came to power, he sent for the regalia left behind by the previous, Cwezi dynasty. The messenger sent by Rukidi carried one of the two drums that were part of this regalia to the new capital, while the other drum was said to have followed of its own accord. Moreover, when Rukidi beat the great drum at his accession, it made a suitably impressive noise, proving that Rukidi was not an imposter (Fisher 1911:120; see also Beattie 1971:53–54). Thus, the behavior of the regalia validated Rukidi's claim to the throne. In similar vein, if the king was too ill to carry out his duties, he would be represented by the royal spear, which had its own name (Roscoe 1923:118). The list of regalia is impressively long (K.W. 1935:160, 1936:77 for what may be only a partial list; see also Nyakatura 1973:181–185, 188–190) and an impression of the pervasiveness of the regalia can be obtained from a visit nowadays to Kabalega's tomb. As mentioned earlier, some of the regalia, such as spears and knives, symbolized the king's coercive authority but the regalia as a whole was by no means under the sole possession of the king; clearly the regalia also spoke to what Blanton and Fargher have termed "principal control" (see also Blanton 1998a; Levi 1988). Numerous individuals had named positions involving regalia, from a "head regalia-man" (*Mujaguzi*) to persons with very specific tasks, such as the man who looked after the king's personal drum. Many of these and other offices were linked to particular clans (Beattie 1971:124). Beattie (1971:125) explicitly recognized that the myriad duties of the specialized establishment of the palace, including the regalia,

> served as a means of integrating the kingship with the Nyoro people as a whole. For it involved a great many different individuals and . . . several different categories of individuals, in a common interest in the palace, and so in the kingship itself. All of Bunyoro's traditional craft specializations were represented, as also were a considerable number of Bunyoro's numerous clans. This last point is of particular importance, for the clan system, as the focus of local rather than nationwide loyalties, may be regarded as having stood, at least in traditional times, in some measure of opposition to the kingship. The vesting of particular palace offices in particular clans went some way to negate this opposition, by integrating the clan system with the palace organization, and so with the kingship.

However, as Lane Fargher (personal communication, 2012) has noted, it is not clear from this description whether clans were corporately organized. Did the clan membership appoint or recommend one of their number for the particular palace office vested in their clan, or was the individual chosen by the king, at whose whim he may then have served? The available literature offers no clear answer to this question, despite the frequency with which the assertion is repeated that particular offices were hereditary within particular clans. What seems more certain is that the shrine sites scattered through the kingdom (see above) were "intimately connected with particular clans" (Buchanan 1974:228) and that the "tensions inherent in this development of priestly power worked as a check to limit the power of the Mukama [king]" (ibid.:227). Moreover, the Nyoro proverb, "The Mukama rules the people, but the clan rules the land" also indicates that, at least prior to Kabarega's usurping of the right to appoint chiefs, the clans mitigated the king's power beyond the capital (see also Uzoigwe 2013:22).

While the regalia served to legitimize royal authority while simultaneously negotiating that authority with the clans at the palace, other items of material culture were part and parcel of both the assumption and delegation of royal authority in other economic and political spheres. In particular, during the coronation ceremonies, a man said to represent foreigners presents the king with an elephant tusk and two copper bracelets (K.W. 1937:293), which may be interpreted as signifying the king's control of the ivory trade and of the import of copper, the wearing of which appears to be associated with high status.[2] Although this connection has not yet been investigated, it is perhaps supported by the observation that members of the king's clan who were clearly related to the king could claim from him a brass ankle-ring (Beattie 1971:99). When the king delegated political authority, he also conferred both its associated spiritual power (*mahano*) and artifacts that signified that authority, notably crowns, spears, and knives (Beattie 1971:102, 118). Finally, when visited by foreign dignitaries, the king gave gifts of salt and iron hoes, the two most valuable commodities produced in his country.

If the regalia and other artifacts of the Nyoro kingdom played a role in negotiating the tension between state and populace, the palace itself seems to have served first and foremost as an agent of exclusionary power strategies. The royal enclosure (*kikali*) was always located in such a way that it was highly visible, in addition to being by far the largest in the country. The six-foot-high fence of elephant grass surrounding the royal enclosure was said to have a circumference of about two miles and enclose perhaps hundreds of buildings. Prominent among these was the court-house (*kaluzika*), which incorporated the throne-room (*hamulyango*), reported as being often 40 yards across and 80 feet high at its apex. However, it was not just the size of the enclosure and buildings that impressed; the whole complex of royal houses and rooms was governed by a plethora of rules and restrictions to access that served to highlight the king's power. Only the king, it appears, could enter every room through every door, though daily rituals even prescribed where the king should be within the palace at set times in order to maintain his ritual purity (Roscoe 1923:73–86; Nyakatura 1973:202–204). Moreover, the plan of the royal enclosure (Roscoe 1923:86–87) shows numerous fences and screens that restricted both the access to and the visibility of many houses and courtyards (see the Kuba kingdom of central Africa for a comparable example [Vansina 1978a:137]). It would be intriguing to discover whether the massive royal enclosure described by Roscoe, which must have entailed the efforts of a large labor force, was only a nineteenth-century phenomenon financed primarily by external revenues or whether it was also typical of palaces in earlier times. It is also worth bearing in mind that every new king built a new capital and that the capital may have been relocated on several occasions during each reign (Roscoe 1923:73–74). As yet none of the capital sites of the kings of the Bito dynasty has been explored archaeologically, though the general locations of several of them are known (see Nyakatura 1973).

BUNYORO BEFORE THE NINETEENTH CENTURY: THE ARCHAEOLOGICAL EVIDENCE

I have argued in this chapter that the despotic nature of the Nyoro state in the nineteenth century, reflected in the scores on the collective action measures devised by Blanton and Fargher (2008), was the product of a historical moment in which the state took full advantage of its position toward the periphery of an economically booming world system to generate substantial external revenues and consolidate royal power and authority, a point on which Blanton and Fargher (2008:46) concur. However, the kingdom already possessed a well-established system of generating revenues that was inevitably entangled within a dynamic negotiation of power and authority between the state and the populace. This negotiation took place in a variety of arenas, of which I have highlighted here the composition of ritual, wealth-in-people, and materiality.

Of course, negotiations over power and authority began long before the Nyoro state was formed. In earlier publications, I proposed that it was a shortage of people, particularly a shortage of female labor, that was the engine of efforts toward political centralization in Bunyoro in the early second millennium AD (Robertshaw 1999a, 1999b, 2003). This proposal rested on the premise that precolonial African kingdoms and chiefdoms exhibited low population densities, in contrast to the high population densities commonly encountered among acephalous societies (Shipton 1984). I argued, following the ideas of Rich Blanton and colleagues (1996), that emerging leaders used exclusionary power strategies to accumulate and compose wealth-in-people. By about the mid-fifteenth century, there had emerged several larger polities wherein corporate power strategies, based on staple finance, became instrumental in the construction of large earthworks at several sites (Robertshaw 2001, 2002). These earthworks appear to have been abandoned around the end of the seventeenth century, with a concomitant shift to the kind of peripatetic Nyoro capitals recorded in later ethnographies; a shift that might reasonably be associated with the establishment of the Bito dynasty (Robertshaw 1999a:60; 2003:163) and that must also be considered within the context of climate change (Robertshaw and Taylor 2000; Robertshaw et al. 2004; Lejju et al. 2005).

This reconstruction can now perhaps be profitably revisited in the light of insights provided by collective action theory. The evidence for exclusionary power strategies, prior to the fifteenth century, comprises the monopoly of prestige goods, notably ornaments of glass beads and metal, perhaps especially copper; the real or symbolic control of iron production; and the harnessing of creative power, as reflected perhaps in the initial occupation of shrine sites such as Mubende Hill (Lanning 1966; Robertshaw 1994:108). On reflection I may have given too much interpretive weight to the relatively rare discoveries of glass beads and metal ornaments, particularly in light of the fact that at Munsa they are mostly associated with female, sometimes juvenile, human skeletons (Robertshaw, Murphy, and Ambrose 2012), which either calls into question their status as prestige goods or prompts us to reconsider the age and gender distribution of elite power. However, the sample of burials, particularly of men, from this period is small and hence injects a note of caution, while the occurrence of these ornaments at several major sites reasserts their importance. When it comes to the shrine sites, however, these could equally be seen as the harnessing of creative power as a form of collective action rather than as an exclusionary elite strategy, since they may not have been under elite control, as indeed I have argued for the use of these sites in later times (Robertshaw 2010). Finally for this period, the question of the control of iron production merits further study, particularly since Iles (2010) has drawn attention to the variation that existed in iron-working technology style across the Kitara region, including the probable introduction, tentatively dated to the fifteenth century, of new technology involving the use of an additional manganese-rich ore in smelting. Iles's results suggest, at least to me, that elite control of iron production could only have been symbolic, as was described ethnographically.

While we need more fieldwork to generate new archaeological data and hence to shed light on the power strategies of the early centuries of the second millennium, it is perhaps easier to infer the existence of tension in the negotiation of power and authority between rulers and populace from consideration of the earthworks of the mid-millennium. Previously, I interpreted the earthworks at Munsa as monuments that "not only expressed group solidarity in material form but also encircled and metaphorically captured the power and legitimacy of earlier elites" (Robertshaw 2003:161).

However, rather than seeing the earthworks solely as the product of corporate power strategies funded by agricultural surpluses collected from taxpayers, as I did previously, they can also be seen as materializing the essential tension between rulers and populace. On the one hand, the very long outer trenches at this and other sites indicate collective action, in a very real sense, likely aimed at keeping elephants out of agricultural fields (Robertshaw 2001); on the other hand, the symbolism of the innermost trench and what lay within its circumference may be reinterpreted in terms of my earlier discussion of the Nyoro palace and could be seen as a place of seclusion, likely the locus of royal rituals, separating rulers from ruled while also drawing attention to itself through its central location on a hill. This perspective aligns well with the oral tradition recorded in the 1950s describing the death of Munsa's ruler, Kateboha, who was killed by his people after they had employed subterfuge to gain access to this inner sanctum. Kateboha is said to have built the earthworks to protect himself, his property, and his daughter from the populace, for whom he was a hard taskmaster, requiring them to cultivate the land and executing any man who slacked off (Lanning 1959; see also Robertshaw 2001:28). The abandonment of the earthworks, probably around the end of the seventeenth century, seems to have ushered in a suite of changes that resulted in the establishment of the Nyoro state, aspects of which were explored earlier in this chapter. Unfortunately there is a dearth of archaeological evidence for this crucial period.

CONCLUSION

In this chapter, I have argued that the highly despotic nature of the Nyoro state in the nineteenth century was a product of a historical moment in which external revenues fueled the economy. The importance of internal revenues in earlier centuries probably provided a stronger basis for the negotiation of power between the king and the populace than existed in the nineteenth century. However, even for the nineteenth century evidence indicates an essential tension between rulers and ruled. While this tension was notably expressed in ritual, it was also both materialized and negotiated in regalia and in the location and architecture of the royal enclosure. My brief examination of the archaeological evidence for earlier centuries suggests that Munsa and other earthworks represent a material expression of the tension between royal authority and collective action.

While the balance of power between state and populace varied over time, in part at least as a result of the relative economic importance of internal and external revenues, the fact that collective action was still employed when state power was at its peak in Bunyoro suggests that the negotiation of authority may have been a dynamic process in all states. Therefore, it seems equally improbable that either total state power or its opposite, a victory of collective action, could have held sway, at least for more than a brief historical moment. If that is indeed the case, then perhaps we should challenge claims of the existence of states without rulers, such as that of the Inland Niger Delta (IND), in which heterarchy has metaphorically vanquished hierarchy (e.g., McIntosh 2005:189). It has become a widely accepted tenet of African archaeology that the IND represents complexity without hierarchy (e.g., LaViolette and Fleisher 2005:333–336). While my analysis of Bunyoro may offer theoretical challenge to the heterarchical IND, recent field research in Mali has provided a model of state-generated landscapes, manifest in later times, that may explain both why hierarchical states could have easily been overlooked and why trading cities like Jenné-Jeno and Dia show little or no evidence of state organization (MacDonald and Camara 2012). Looking beyond Africa, I am struck by the similarities between the Nyoro state under Kabarega's despotic, albeit perhaps beneficent, rule in the nineteenth century and the political structure found in the Mixteca Alta region of Oaxaca, Mexico, during the Postclassic (Fargher, Heredia Espinoza, and Blanton 2011:317ff.). Such comparisons reinforce the fact that the analytical tools engendered by collective action theory and allied concepts have much to offer in our efforts to elucidate past political systems.

NOTES

1. In this chapter I have contrasted "external" revenues, where wealth was derived from the profits of exporting ivory and slaves, with "internal" revenues raised from within the kingdom—for example, in the form of food and labor. However, Blanton and Fargher's (2008) definitions of these terms are subtler and rather different. They consider "internal" revenues to be those "collected by the state from free constituents or taxpayers," while "external" revenues are "sources owned and/or controlled directly by the state," including landed estates and control of "serf-like laborers" (Fargher, Heredia Espinoza, and Blanton 309). The application of these definitions to the Nyoro case is difficult: on the one hand, the existence of chiefly rights over people suggests that the revenues from these estates are "external"; on the other hand, the very term "Banyoro" means "freedmen" and peasants were not serfs in the sense of being bound to the estates on which they lived. Mobility was apparently always an option. Given these contradictions, I have chosen here to use "internal" to describe revenues raised from within the kingdom, while recognizing that others may wish to reclassify them as "external."

2. K.W. offers a different interpretation, noting that the king touches these items, thereby denoting that the king is "the head of all rain-makers."

10

Cognitive Codes and Collective Action at Mari and the Indus

Rita Wright

The revival of cross-cultural and comparative studies is providing new explanatory frameworks for understanding the varied dimensions of power in early states worldwide. Richard Blanton (1998a) and his colleagues (Blanton et al. 1996) have been at the forefront of introducing alternative pathways to complexity that now are accepted as basic to theory-building. In this chapter, I discuss limitations to power in two early states based upon the concepts of a cognitive code and collective action (Blanton and Fargher 2008): the Near Eastern kingdom of Mari (Fleming 2004) and the Indus civilization in South Asia (Wright 2010). The evidence from Mari is based on the royal correspondence of King Zimri-Lim comprising 3,000 letters from a palace archive that were translated and analyzed by Daniel Fleming (2004). The Indus evidence is almost exclusively from material remains (Wright 2010).

INTRODUCTION

The cross-cultural and comparative study of early civilizations has a long history in studies of Old World states. The two most influential contributors are V. Gordon Childe (1934, 1951, 1964), a prehistorian, and Henri Frankfort (1956), a Near Eastern archaeologist and Egyptologist, both of whom focused on the exclusionary nature of early states. Largely based on studies of ancient Egypt and Mesopotamia, Childe considered the means by which rulers extracted food surpluses as

DOI: 10.5876/9781607325338.c010

an overriding factor in their centralized political economy. Frankfort, on the other hand, claimed that Mesopotamian and Egyptian ideologies were the principal bases for the dominance of its leaders.

While current research continues to be influenced by these early works, there has been mounting evidence that their interpretations of the Egyptian and Mesopotamian states need revision. Bruce Trigger's (2003) comparative and cross-cultural studies of seven early states, selected for study based on strong archaeological evidence and written sources either in the form of local texts or ethnohistoric accounts, involved a close analysis of kingship among other factors. His studies demonstrated that the work of kings was often mediated by self-governing associations and councils composed of important lineages and other interest groups. Similar doubts about the centralization of power at the hands of leaders in a broad range of societies in the New and Old Worlds (Nichols and Charlton 1997) complement his research. In addition, Susan and Roderick McIntosh's research at Jenne-jeno in the Middle Niger (AD 400–1100) has provided evidence for a self-organized community in which multiple authorities existed, none of which monopolized power (R. McIntosh 2005). In fact, a range of societies in the African examples provided by Susan McIntosh indicate that kings had relatively weak control (S. K. McIntosh 1999). In another study, Elizabeth Stone and Paul Zimansky used spatial distributions to demonstrate that southern Mesopotamian societies were "populous, entrepreneurial," based on a dispersed pattern of settlement at Mashkan-shapir in which administrative, residential, and production areas were "both independent and connected to the public sector" (Stone and Zimansky 2004; Stone 2007:219).

The McIntoshes, Stone, and Zimansky drew on theoretical models introduced by archaeologists in their attempts to refine understanding of the complexity of leadership and power in early states. Carol Crumley introduced the concept of "heterarchy," which she defined as "the relation of elements to one another when they are unranked or when they possess the potential for being ranked in a number of different ways" (Crumley 1995:3). Even in the most despotic states, local hierarchies and "nodes of social power" can be complementary or potentially conflictive (Brumfiel 1992; Brumfiel and Fox 1994; Janusek 2008:30). Heterarchy and hierarchy, therefore, can coexist (Crumley 1995; Feinman and Marcus 1998), as in the case of mediating bodies like the councils known in Mesopotamia (Jacobsen 1943). The self-organized landscape that the McIntoshes describe for the Middle Niger included settlements with different functions that were not hierarchically but heterarchically ordered in flexible power relations

Blanton's dual-processual theory and subsequent publications have not been widely cited in Old World studies. In this essay, I review the ways in which the concepts of cognitive code and collective action have been employed in explaining the limitations to power in the interpretation of corporate strategies by Daniel Fleming (2004) in the case of the Mari and in my research on the Indus civilization (Wright 2014). The theory posits two forms of power, one that is exclusionary (network), in which leaders monopolize sources of power, and a corporate strategy, in which decisions are shared among groups within the society. According to Blanton, limitations to power are the result of a symbolic cognitive base in which symbols, such as ritual sanctification, limit the exercise of power (Blanton 1998a:152). The Mesopotamian citizens that held offices and acted as mediators in setting limits to exclusionary power are prime examples (Jacobsen 1943).

In their discussions of collective action, Blanton and Fargher (2008) have added new dimensions to the concept of cognitive codes that are of specific relevance to the present chapter. They suggest that cultural codes may be less "influential" in societies in which the dimensions of the world order are more secular (2008:293). This interpretation more closely follows Fleming's research on the Mari and mine on the Indus civilization. As I demonstrate below, among the Mari and the Indus collective action in some segments of society was based on an ideology that was born out of a cognitive code embedded in longstanding social and political orders. This factor, and others to be discussed in the following, has important implications for basic questions regarding premodern and modern political processes and pathways to complexity in the Mari and Indus.

THE KINGDOM OF MARI

The site of Tell Hariri (ancient Mari) is located in present-day Syria near the Middle Euphrates as the river bends west and north (see figure 10.1 for the location of Mari at the bend in the Euphrates). It was settled for the first time in the twenty-eighth century BCE (Fleming 2004, 6:n. 11). When it was a fully developed state, it extended over 100 ha and was rebuilt and destroyed on several occasions. By the mid-third millennium, it was contemporary with several polities, including Ebla, ancient Nagar (Tell Brak in the Khabur Basin) and Akkadian and Ur III states in southern Mesopotamia. The city was abandoned with the conquest of King Hammurabi at around 1761 BCE.

The evidence for systems of rule described here are from texts that fall within a 13-year period during the reign of Zimri-Lim, a "tribal king" and "master of a major city center" at Mari (Fleming 2004:1, 2). What distinguished

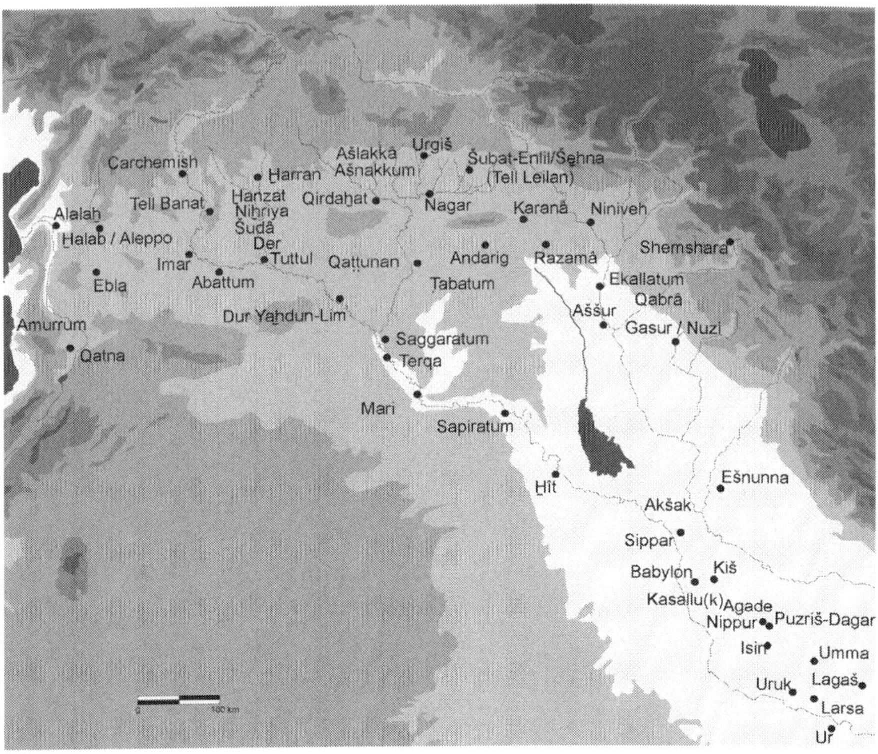

Figure 10.1. *Archaeological sites in northern and southern Mesopotamia. (Drawn by Tom McClellan.)*

Zimri-Lim's rulership was a longstanding system in which decision-making was based on negotiations among tribal groups and collective and exclusionary ruling bodies that acted interdependently (Fleming 2004:19). The sustainability of a longstanding agropastoral component in Mari's political economy was pivotal to the state's exclusionary and corporate structure.

At least by the mid-third millennium, the tell was an enclosed mound comprising palaces and temples that were residential and administrative. Positioned midway between southern Mesopotamia and polities to its north and east, the site's location is considered by its current excavator, Jean-Claude Margueron (cited in Fleming's [2004] text), to be ideal for the control of commercial river traffic and the movement of goods among trading partners in several polities, especially the shipment of wood to southern Mesopotamia. Although its location on the Middle Euphrates offered an invaluable water source for agriculture and pastoralism, the valley was not sufficiently wide to sustain large-scale farming and major populations, and the nearby steppe was too arid for dry farming (Porter 2012). According to textual sources and archaeological evidence, the people at Mari engaged in a mixed economy that was carried out by sedentary and mobile populations that were linked by kinship ties. Mobile pastoralists were dispersed spatially, but their symbiosis with sedentary societies served an ideology in which, although physically apart, they were "conceptually together" (Porter 2012:13).

It is in this sense that Zimri-Lim's rulership among his tribal group, the Sim'alites, differed from his predecessors and the traditional forms of rule carried out in other states in northern and southern Mesopotamian (e.g., at Ebla and by the Akkadian or Ur III kings, and of course Hammurabi) that are often cited as models for exclusionary power (though see my earlier discussion of the research of Stone and Zimansky).

The Mari State: Rethinking Pastoralism, Ethnography, and Prehistory

Before describing the details of Zimri-Lim's reign, it is necessary to review some of the issues that have excluded pastoralism from anthropological views of state leadership. New evidence based on comparative studies has challenged evolutionary models that marginalized pastoralist societies. Evolutionary stage typologies consigned tribes to a position in between bands and chiefdoms (e.g., Fried 1967, 1975; Service 1975). In archaeology, pastoralists often were conceived as one step ahead of hunters and gatherers, thus falling out of evolutionary paradigms when plants were domesticated (but see Hole 1991, 2000; Zagarell 1989).

Agropastoralism was a latecomer in discussions of sedentary peoples since it often was associated with fully formed states. Unlike agriculturalists, pastoralists were thought to be untethered from the landscape and to be organized in egalitarian societies. These views of pastoralists are being turned around based on re-re-evaluations of ethnographic and prehistoric research, offering a perspective from many Old World contexts in which pastoralists are more closely aligned with agriculturalists and mixed economies (Frachetti 2008a, b; Hanks 2010; Linduff and Rubinson 2008; Honeychurch and Amartuvshin 2007; Kohl 2009; Porter 2000, 2012; Rogers 2012; Salzman 1972, 2002; Szuchman 2009) than previously acknowledged. In addition, Anne Porter's evidence for social differences in burial contexts at Tell Banat challenges older views that pastoral organization precluded "differences in wealth that lead to social stratification" and separated them from class-based societies (Porter 2000, 2012).

Finally, these studies demonstrate that pastoralism is not a unitary concept. Daniel Rogers suggests that anthropologists need to alter their perspectives by looking outward from centers (adopting "a decentralized vantage point") and viewing mobility as a central concept (Rogers 2012:9).

These views set the stage for Fleming's interpretation of Mari leadership under Zimri-Lim's reign and its tribal confederation. Using texts that contain almost day-to-day accounts of decision-making, his book predates much of the recent analyses of tribe/state relations, with the exception of the works of Anne Porter. He draws on Blanton and his colleagues, whose researches are cited throughout his interpretation of the Mari texts. His description of the Mari state provides a detailed analysis of the political structure in the millennium that preceded the rise of Zimri-Lim as its king and contributed to state organization during his reign. I provide here a sketch of his findings and synthesis of a wealth of nuanced levels of organization existing during Zimri-Lim's reign.

Table 10.1 outlines the structure of the state and some critical points regarding how it operated.

During the reign of Zimri-Lim political power was bifurcated so that the king was at the head of the core people and the groups dominated as outsiders. Both were mutually dependent sedentary and mobile populations, each of which had a separate social structure with different roles for their leaders. The Sim'alites, the core people, were tribally affiliated with Zimri-Lim. Although they consisted of more than five divisions (perhaps aligned with herding groups), these divisions had lost all political function in his relations with this tribal constituency. The king dealt with a broad-based body politic that functioned in a corporate mode in which he consulted with "elders" who were not attached to specific units. Whether they met at a hamlet, village, city, or specific center is inconsequential, as the main point is that it was a "collective expression" of the population acting as a "body of people" (Fleming 2004:170).

In distinction, the king's rule of the three elements defined as "outsiders" was palace based. Although Yaminite populations extended far beyond the territories dominated by Zimri-Lim, he maintained supremacy over them mainly through their concentrated settlement in the district governed from Terqa. Each of the five tribes of the Yaminites had a king that ruled individual towns but followed a power structure in which there were no domains in which Zimri-Lim's authority could be refused (Fleming 2004:166).

It is this "independence" of mind and collective "ideology" born out of the management of access to grazing as "whole communities" that was at the heart of a cognitive code practiced between the king and the Sim'alites. Although

Table 10.1. Mari: a collective of plural leadership

	Zimri-Lim as King	
	(defined by Mari as his capital city and "Hana" = Sim'alite tribe as his people)	
	*	
	The Bin-u Sim'al (Sim'alites) as core people	
	*	
	Groups dominated as outsiders:	
1. Four districts with governors, including Mari itself.	2. Vassal kingdoms (mainly north of the Khabur)	3. Bin-u Yamina tribe (Yaminites), ruled Terqa district and governor

administrative correspondence regarding sheep makes references to shearing, there are few concerns for breeding and management. The texts also do not include references to grazing rights, restrictions on the use of land, or specifics of their location. The king did not own all land and pastoralists appear to have grazed and moved their flocks without restraint (Fleming 2004:167). It is this ideology of consensus building among the Sim'alites that embodies the corporate code, which Fleming believes provided the mutually beneficial counterpart to the state's exclusionary structure in his relations with the Yaminites. These restrictions on Zimri-Lim's exclusionary power make it a perfect example of the dual-processual model.

Ideology in this case was deeply embedded in the fabric of the social and political order that served as an obstacle to exclusionary power in Zimri-Lim's relations with the Sim'alites. I turn now to a similar cognitive code and collective action in my discussion of the Indus civilization.

THE INDUS CIVILIZATION

The concepts of cognitive code and collective action are of equal interest in understanding structures of power in the Indus civilization. For the Indus, I concentrate on its community infrastructure as a framework from which to observe the nonexclusionary aspects of its political economy (see figure 10.2 for the location of the Indus civilization and sites referred to in this chapter). This example focuses almost exclusively on the civilization's material culture. Unlike Mari, the Harappans did not have lengthy texts that recorded activities, much less the day-to-day activities of the sort in Zimri-Lim's archive. Inscriptions are found on stamp seals and small "tablets" or etchings on pots in which only a few signs are recorded (Meadow and Kenoyer 2000). A

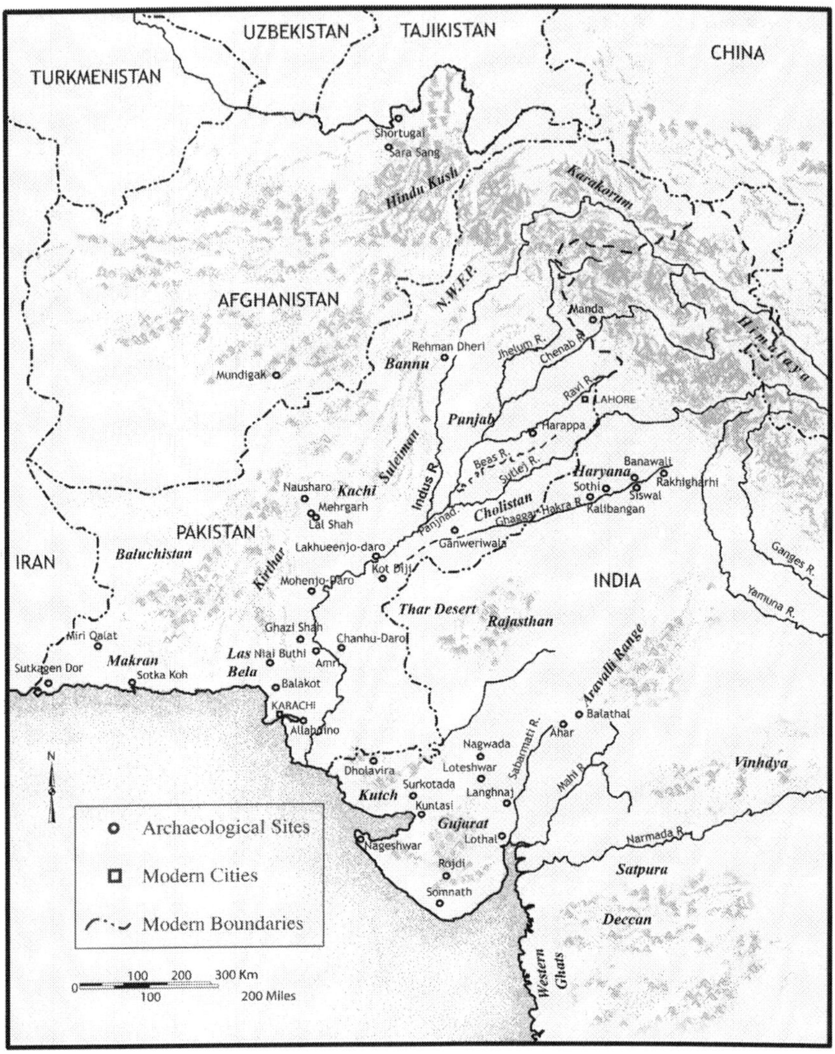

FIGURE 10.2. *Locations of Indus civilization sites.*

signboard with 12 inscriptions was found at the entrance to one of its cities and may indicate fairly widespread literacy, but for the present, we can only speculate, since Indus inscriptions have not been deciphered.

In *Collective Action in the Formation of Pre-Modern States* Blanton and Fargher used macroregional clustering of 30 states to identify "shared aspects of the cultural and social histories of state formation" (2008:3). Of relevance to this discussion is a cognitive code for South Asia that they trace in a five-phase sequence that begins with the Indus civilization and ends with the Mughal Empire. Taking into account the region's broad geographical scale and regional and cultural diversity, they focus on polities "that were based on Hindu and Buddhist political theory and culture" (Blanton and Fargher 2008:60). Using a published interpretation of the Indus civilization, they describe it as "corporately organized," one that lacked "a ruler cult," and that possessed "an egalitarian ethic" and "comparatively limited wealth differentiation" (Blanton and Fargher 2008:62). They rely on a widely cited paper by Daniel Miller (1985a), in which he outlined a long cycling of "egalitarian and more centralized forms" based on Vedic Hinduism and Buddhism and their "distinct theories of rulership," as the cognitive code that was in place in early and subsequent phases in the history of South Asia that ended at around 300 BCE. They are uncertain whether the Indus represented an "initiating phase" of "collective orientation" similar to what is known from the later South Asian states and a second urbanism in South Asia (Blanton and Fargher 2008). I find this interpretation problematic in view of the limited archaeological evidence with which to establish continuity between the Indus civilization and phases one and two (Wright 2010:325, chapter 11). While I am in agreement that the Indus political economy employed a corporate strategy and collective action, the cycling model that links the Indus civilization to Hindu Buddhist ideologies proposed by D. Miller (1985a) cannot be supported by the existing evidence, and I offer a different perspective.

In chapter 12, "Collective Action and Political Evolution," in *Collective Action in the Formation of Pre-Modern States,* Blanton and Fargher (2008) point to selected features (apart from the phased recycling referred to in the preceding paragraph) of the Indus that argue for collective action. They include its planned cities, population numbers that reached 50,000 and "vast communal grain storage facilities" (Wheeler 1968; Blanton and Fargher 2008:291). Although there are other features that argue for collective action, such as the extensive public amenities at Indus centers and at some rural settlements (Wright 2010), my focus here is on collective action among craft producers and merchants.

The Indus Civilization: Geography, Urbanism, and Ecologies

Working toward that end, I provide a brief background to the varied geography of the Indus in order to update important misconceptions about this civilization. The discovery of the urban centers of Harappa and Mohenjo-daro in Pakistan in the 1930s has left a lasting impression that settlements were located solely on the rivers of the Indus valley. The subsequent discovery of large numbers of Indus settlements in northwest India as well as on the Ghaggar-Hakra River in India and Pakistan, and two major centers at Rakhighari and Ganweriwala, have received less attention (figure 10.2). The Ghaggar flowed into Pakistan from India, where its name changes to the Hakra, a river that flowed into the lower Indus and as far as the Arabian Sea. These rivers along with the Indus placed the city of Mohenjo-daro in an ideal location between two hydrographic systems. A fifth center at Dholavira is in Gujarat, not on a river at all and may have been a port city.

Not surprisingly, the ecology is varied when the totality of the civilization is considered, so we should expect significant differences in organization. Studies at Harappa (Weber 2003) have identified an agropastoral, double-cropping system, while in the south in Gujarat, pastoral practices and cultivation of crops differed. There, proximity to the sea and port locations opened to a wider world and intercultural trade with contemporary societies throughout the Greater Near East (Wright 2010:215ff.).

Finally, in many early accounts the civilization was described as homogeneous, but recent research has revealed regional differences in city plans, settlement patterns, and the pace of urbanization (Wright 2010:126ff., 81ff.).

Indus Heterarchy/Hierarchy, Cognitive Models, and Collective Action

V. Gordon Childe's interpretations on the early Indus state were based on intellectual trends of the day and were not so different from what Blanton argues against. As discussed earlier, Childe paid little regard for the economic and social infrastructure of early states, such as Stone and Zimansky's (2004) documentation of spatial arrangements reflective of independent specialists in Mesopotamia. In distinction, Childe assigned control of production and distribution to centralized leaders and overlooked restrictions on exclusionary power in the form of independent specialists and collective action.

In the Indus, recent research on craft production and its specialized technologies and spatial distributions suggests that the variability observed by Stone and Zimansky followed similar patterns (Wright 2010:145ff.). I draw on the production of ceramics and seals, their uses and distribution. During the peak of urbanism, Indus production was intensified (new products, large numbers of producers and output); diversified (products, new skills, and uses), and specialized (divisions of labor and distribution of resources), and objects were produced for internal and external distribution. Artisans worked in large and small-scale workshops. Some were independent of the state while others may have been state controlled. They produced mundane and prestige goods from locally available materials and from distant resources.

Extensive studies of craft producers have been undertaken at Harappa, one of the main Indus centers (Wright 2010:chapters 6 and 7). The production of ceramic vessels and stoneware bangles required related skills in which clay was processed, shaped, and fired but in unique ways. Locally available clays were used for both, but they were refined and standardized differently, according to the desired end product. The production of vessels required substantial trimming, mixing of pigments for paints and slips, and firing in two-chambered kilns (Wright 1989, 2010:152–166), while stoneware bangle production involved several steps in reshaping by grooving or cutting forms shaped on the potter's wheel. In a final step, they were fired in airtight containers to provide optimal control of atmospheric conditions and to attain higher temperatures than in vessel production (Vidale 2000:92). Many bangles bore Indus signs as if produced for individuals, or were possibly inscribed with a name.

The independent, noncentralized organization of vessel production is based on contextual data from the excavations at Harappa. A ceramic workshop in the city was associated with a mudbrick wall believed to be part of a residential building. During the urban period, potters built a large two-chambered kiln replacing a small pit kiln that had been built in a previous period. We interpreted the presence of the two kilns as evidence for an early pre-urban era of small-scale production and the later intensification of production in the urban period (Wright 1991, 2010:187). This time capsule of social change demonstrated continuity among specialist producers in which a craft was handed down over several generations. Although there were changes in the production process over several hundred years of producing ceramic objects, their decisions about timing, technologies, and social arrangements were organized without any apparent control from groups outside of their production units.

In early ethnographies, ceramic production was often described as "a labor intensive and time-consuming" craft in which production was limited (Foster 1995:100) and economic yields were low. Because ceramics were thought to be produced by a single individual, the "potter," their organization had little to offer in shaping society, an assumption partially based on misplaced census reports in which a single entry designated the name of the "potter" (Miller 1985b). In fact, as subsequent research has demonstrated, pottery production more often requires a cooperative network of craft workers who perform various tasks that cannot be accomplished by a single producer and that are essential to the final product. They work together in tightly organized groups in which decision-making requires consensus building and collective action to succeed (Coburn 2011; Kramer 1997; Wright 1991).

The organization of the stoneware bangle industry differed. Situated outside of a residential area within a walled enclosure at the city of Mohenjo-daro, it fits the exclusionary model often associated with specialized forms of ceramic production that are highly standardized and controlled. Perhaps because stoneware bangle production was an innovation associated with the urban period, it has been interpreted as an "industrial craft under administrative control" (Vidale 1989:178). According to a neutron activation analysis of a sample of bangles from several sites, their production appears to have been limited to Mohenjo-daro and Harappa and controlled by a restricted group (Blackman and Vidale 1992).

My second example, the production and use of seals, has a long history in the greater Near East and South Asia. Conceptually, seals and sealings are related to what Dennis Frenez and Maurizio Tosi refer to as a "Transcultural Administrative Sealing System" in which objects were sealed for storage purposes and safe-keeping, a practice known from the sixth to fifth millennia BCE (Frenez and Tosi 2005). By the third millennium, the shapes of seals signified their culture of origin; for example circular seals were from groups in Arabia, while Indus seals were square. The latter were produced from steatite and engraved with large mammals, small stands, and Indus script (Franke-Vogt 1989). At Indus sites, seals are present in great numbers and widely distributed at small and large settlements, in households, public buildings, and workshops.

Based on contextual data—stylistic and technological studies—Indus seals are associated with trade and merchants (Frenez and Tosi 2005). At the small site of Chanhu-daro, a seal workshop was separated from other craft production. Among the production debris associated with the seals, there were net weights, a standard of measurement associated with exchange systems and merchant activities (Mackay 1943). In another study, an analysis of the stylistic elements of seal iconography, Paul Rissman (1988) identified features that were sufficiently distinctive that he was able to identify the work of regional artisans. He referred to them as "schools" of producers who were organized in guild-like structures, possibly within the context of a family group (Rissman 1988), a finding that complements identifiable stylistic variations found in other studies (Vidale 2005). Close analyses of technological features have identified Indus production techniques that vary regionally (Green 2016) and interregionally (Pittman 2013).

Textual evidence for contemporary merchants who traveled abroad and are known from Mesopotamian texts complement these interpretations. In

Mesopotamia, merchant and artisan groups were organized into professional communities, often linked through family ties (Garfinkle 2002), many of which were independent of the state. Merchants are known to have traveled to Mesopotamia, as attested by the words of an Akkadian king (ca. 2350 BCE) who boasted of having seafaring traders from Meluhha (interpreted as the Indus civilization) in his harbor. In a subsequent period (Ur III, ca. 2000 BCE), other texts record a "village of traveling merchants" from Meluhha in the village of Guabba in Mesopotamia (Vermaak 2008). These Meluhhans had become acculturated, as evidenced by their Sumerian names (Parpola et al. 1977:145), suggesting that generations of Meluhhans had traveled to Mesopotamia and became acculturated over a period of several hundred years.

Our best examples of regional variation in Indus seal iconography and technology and traveling merchants unattached to a central authority come from a recent study by Steffen Laursen (2010). His research is based on the discovery of round seals on the Arabian Peninsula in Bahrain. Although round seals are common on the Arabian Peninsula, Laursen believes that the technology of seal production was spread by entrepreneurs from the Indus who transmitted a technological package comprising sealing, writing, and weight technologies. Sometime around 2100 BCE these "breakaway" entrepreneurs established a hybrid form of the technology in Bahrain (Laursen 2010) and by 2050 BCE they were acculturated, much like the traveling merchants in Guabba. Based on the analyses of the sequence order of Indus signs (Parpola 1994), Laursen notes that among the 28 seals he researched, only three have sequences that parallel those found on Indus inscriptions. Others included signs with the double image of an inscription, which is not true to the normal order of inscriptions produced in the Indus. Further, in another group of seals from Failaka, Mesopotamia, and Iran, inscribed twin images and improperly ordered script have also been discovered. Laursen describes the signs as "pseudo-inscriptions" designed to convey a strong message of autonomy (Laursen 2010:36).

These "errors" are typical of altered forms of writing when contact occurs over long periods of time and a language becomes pigeonized. They are clear signs of the development of new forms that are not true to the original languages but that are understood by persons engaged in transactions from different language groups. Possibly the seals belonged to a merchant of Indus origin who lived in Mesopotamia or the Gulf and who either adopted a local language or refashioned his own into a version of the original, signaling a difference from the Indus script that is consistent with the examples of collective action carried out by the merchants who produced and used the seals.

CONCLUSIONS

In a recent work, Blanton looks at the marketplace and argues against earlier antimarket thinking in substantivist debates and the idea that the commercialized West was "unique in human experience" (Blanton 2013:23). His views on marketplaces, which he discusses in great detail, are closely aligned with the interpretations offered here with respect to collective action among pastoralists, craft producers, and merchants at Mari and the Indus. Like marketplaces, pastoralists, craft producers, and merchants carved out social spaces that shaped institutions, social relationships, and domains, establishing ties of sufficient political and economic import that they limited exclusionary rule. The evidence from the Indus and Mari provides a compelling example of the efficacy of the corporate model and offers a structure for understanding collective action. Both in Mari and the Indus they were especially useful tools with which to model strategies among pastoralists, where a strong, materially based ideology, drawn from textual sources, excluded centralization among the Sima'lites. In the Indus, the examples of craft production and merchant activities, when complemented by textual and modern examples of producers and merchants, demonstrated the presence of whole communities (guilds, occupational specializations) and collective action that predates modern states.

Acknowledgments. My thanks to Lane Fargher and Verenice Heredia for inviting me to participate in this volume and in the earlier session at the SAA. I especially wish to thank Tom McClellan for providing me with the Mari map (figure 10.1) and to Dan Fleming, my colleague at NYU, for countless talks about the Mari evidence. I met Richard Blanton for the first time too many years ago to remember when he appeared at the site of Harappa in the Punjab, Pakistan, during our field season. Rich was on a major trek through the Old World, nourishing his interests in comparative studies, stopping at important sites along the way. It was a treat to have him there and to share his observations. Later, I asked him and his colleagues to contribute to a series I edit on Case Studies in Early Societies (Cambridge University Press). The result was *Ancient Oaxaca* (Blanton et al. 1999). Rich's productivity has been an inspiration and his advancement of theory influential in my own efforts to provide a comparative perspective on the Near East and South Asia. This chapter was revised in response to peer reviewers, to whom I am grateful. All errors are mine alone.

11

"We Shape Our Buildings and Afterwards Our Buildings Shape Us"

Interpreting Architectural Evolution in a Sinhalese Village

Deborah Winslow

I take my title from Prime Minister Winston Churchill's address to the House of Commons on October 28, 1943, in the middle of World War II.[1] The House of Commons was meeting then in the Lords Chamber; the Commons Chamber had been destroyed by German incendiary bombs (Parliament 2012). Churchill's brief was to argue for an early rebuilding with two architectural stipulations. First, the new Chamber should retain its traditional oblong shape, eschewing a contemporary fad for semicircular form that, he contended, undermined the party system. A half-circle encourages mingling but the oblong forces Members to signal clearly their party allegiance and renders "crossing over" an act never to be undertaken without serious consideration. Second, the new Chamber should not easily accommodate all members and there should be no reserved desks. Here his case was that the intimacy and crowding of smaller spaces encouraged "episodes and great moments" and imparted a sense that, as he put it, "great matters are being decided, then and there." Churchill's purport was that no less than the future of British Parliamentary democracy would be shaped by how the Members shaped their new Chamber (Hansard 1943).

Churchill premised his case on the conviction that the material spaces people create have the effect thereafter of molding the lives of their creators. In 1994, Richard E. Blanton published an innovative monograph that is grounded in this same estimation. *Houses*

DOI: 10.5876/9781607325338.c011

and Households: A Comparative Study is an unusual volume for an archaeologist but in its grand scope and unconventional methodology it is also classic Blanton. Just consider the numbers. To produce this comparative study of peasant domestic dwellings, Blanton hand coded an incredible 289 variables for 324 houses in 26 communities in China, Taiwan, Japan, Thailand, Java, India, Sri Lanka, Nepal, Iran, Iraq, Syria, Turkey, Jordan, Lebanon, Yemen, Egypt, Mexico, and Guatemala. The result is a unique and valuable database that enables the systematic, quantitative testing of theoretical propositions about houses and households over time and space and at multiple interacting levels of scale. But what most marks this work as a typical Blanton tour de force is its methodology. As in his work on world and regional systems (Blanton 1976), Blanton appropriates theories and techniques from outside of anthropology, retrofits them to new uses, observes the wonders that emerge, and then hastens to share these rich and novel possibilities with the rest of us.

TECHNOLOGIES OF COMMUNICATION

Despite its forays far afield, *Houses and Households* was solidly grounded in archaeological objectives. Blanton's underlying goal was to improve inferences about the evolution of the state and its effects on the living standards of ordinary people such as might be derived from excavations of domestic structures. To that end, he had recourse to both contemporary ethnographic accounts of peasant housing and theories of the built environment, including the work of the Polish architect and city planner, Amos Rapoport. Like Churchill, Rapoport (1969) contended that houses, whether Houses of Parliament or the humble abodes of a peasant village, do far more than merely give shelter. They also impinge on human consciousness, they mold the world they contain, and they affect the larger world in which they themselves are contained. What makes these impingements, moldings, and containments possible, Rapoport and Blanton tell us, is that houses impart information; houses are in fact technologies of communication.

Two architectural communicative modes in particular are critical for Blanton's analysis; he terms these the "canonical" and the "indexical." Canonical communication takes place primarily in the intimate spaces of a house's interior and its immediate surrounds. A house's floor plan, for example, may bring people and activities together, may segregate them, or may simply define them. These interior delineations act to materially represent and so to reproduce the conventional or canonical social order. Indexical communication, on the other hand, occurs when a house is seen from outside. From this vantage point, displays of cultural markers, emblems of achievement, and signs of taste differentiate and place—or *index*—the house's inhabitants within a larger social order. In addition, a house's design invites or discourages the flow of people between inside and outside, determining how easily those who dwell within mingle with those who do not.

But what makes Blanton's work especially useful for social scientists is that he went beyond these insights to devise and adapt a practical, quantitative system for representing these qualitative spatial realities.

BLANTON'S METHODOLOGY

In Blanton's words his spatial analytics comprise "a methodology, grounded in graph theory, that allows me to derive measures of scale, complexity, and integration of houses, in a manner facilitating cross-cultural comparison, and which could be applied to diachronic comparison" (Blanton 1994:23). This methodology, which builds on the ideas of spatial syntax such as those developed by Hillier and his colleagues (Hillier et al. 1976), begins by taking the simple data of a dimensionless floor plan and reducing it to a "planar graph consisting of nodes (or vertices) and edges" (Blanton 1994:26). Blanton's graphical representations are dimensionless, incidentally, because surprisingly few published ethnographies included measurements. Nonetheless, working from the floor plan alone, he derived three useful metrics: scale, integration, and complexity.

- *Scale* is a measure of size as represented by the number of nodes, or "physically defined architectural spaces in the residential compound" (ibid.:52). The example in figure 11.1 has four nodes—three rooms plus the outside. This made it a very small house in Blanton's sample, which had a mean of 10 nodes per house, with an impressive high of 19 for the Chinese houses and a low of 7 for the South Asian and Mesoamerican ones.
- *Integration* refers to the degree of linkage among the nodes or number of possible routes or circuits between rooms. The same number of rooms can be linked in multiple ways. For example, a builder who wants to economize but still have several rooms might end up with one of New Orleans's linear "shotgun" houses. Adding in hallways, staging areas, stairs, and courtyards increases costs but also affords zones of privacy and functional differentiation. Blanton found a simple count of the number of circuits to be the most useful measure of integration. The example house of figure 11.1 has a single circuit because there is only one possible route to get from one room to another.

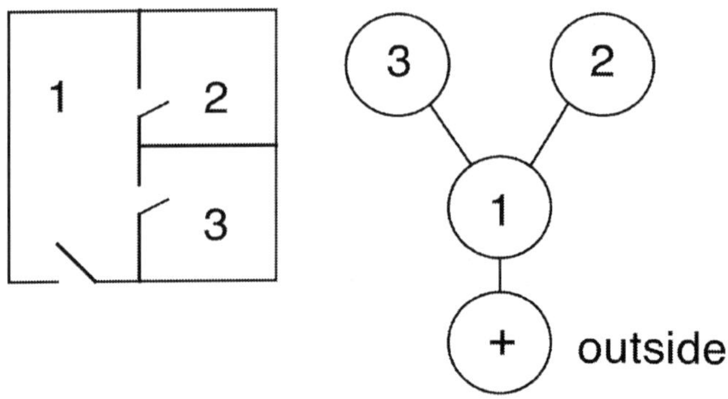

FIGURE 11.1. *Floor plan and its graph. (After Blanton 1994:37.)*

- *Complexity*, the third measure, refers to a house's internal differentiation and levels of accessibility. As Blanton points out, a house can have many rooms and still be quite simple if all rooms are entered off a central courtyard. But if some of the rooms can only be reached by going upstairs and then proceeding off a common landing the layout becomes significantly more complex, even though the number of rooms is unchanged. Blanton attached a numerical value to this variable by counting the number of links separating every pair of rooms or nodes, summing them for each room, and then rank ordering the sums. A lower rank indicates that a room is more easily reached than spaces with higher ranks.

Table 11.1 demonstrates this calculation for the figure 11.1 floor plan. This simple house has only two hierarchical levels: Room 1 is the most accessible space while the other spaces are less accessible, although none more so than another.

Armed with these tools, other data as available, and information about ornamentation and other communicative elements, Blanton compared domestic architecture across six geographic regions. For example, he found that houses in China, Nepal, and parts of the Middle East have more rooms or nodes, more differentiated internal circuits, and more complexity of function and hierarchical levels than do most houses in Mesoamerica and India (Blanton 1994:50–75). The peasant houses of China, Nepal, and parts of the Middle East also had more elaborate canonical communication—that is, internal differentiations that reinforce social distinctions—and greater expression of indexical indicators of wealth differences. In contrast, in Mesoamerica and in some areas of South Asia, canonical communication—practices designed to conserve and reproduce the social order—rested more in the public rituals

Table 11.1. Accessibility ranks of the nodes in figure 11.1 reveal two hierarchical levels.

	Room 1	*Room 2*	*Room 3*	*Outside*	*Sum*	*Rank*
1	x	1	1	1	3	1
2	1	x	2	2	5	2
3	1	2	x	2	5	2
+	1	2	2	x	5	2

of well-integrated communities than in the strong intergenerational control of private spaces (Blanton 1994:192–193). Blanton's analysis chapters are rich in such observations and insightful interpretations. *Houses and Households* illustrates beautifully the power of controlled cross-cultural comparisons even when data are sparse and secondhand.

But what if one's data are more abundant, the fruit of long-term field research and firsthand observation? Then Blanton's analytical tools equip us to discern patterns amidst the inevitable surfeit of ethnographic detail. As it turns out, this is a complementary process: the minutiae of daily life may at times be distracting trees that blind us to the forest, but they are also important guides to the mechanisms that produce the order that Blanton helps us to see.

NEW HOUSES IN WALANGAMA

For me, *Houses and Households* has particular resonance. Since the mid-1970s, I have followed the improving fortunes of a rural Sinhalese community that I call Walangama, "Pottery Village," nestled among the coconut estates that dominate the Kurunegala District of central Sri Lanka. Much of the old-fashioned village I first knew has slowly disappeared over the years, recast by a construction wave that has filled Walangama with modern houses quite different in appearance from those they displaced. Blanton's ideas led me to ask: Does this changed architecture mean also changed socio-architectural effects? Do the houses communicate as differently as they look?

To begin to think about these questions, I began with the changes I was noticing in the social world of the village. The story turned on two kinds of events: on the one hand, the great events of globalization, economic structural adjustments, and social transformation. And on the other, a young Walangama entrepreneur whose pioneering activities effectively inserted the macroeconomic shifts into the local village economy.

First, the big picture: both Walangama's economic prosperity and its architectural transformation were closely linked to changes in their traditional caste occupation, pottery making, abetted—perhaps surprisingly—by economic liberalization (Winslow 1996). After the watershed elections of 1977, Sri Lanka opened its markets, let its currency float, and invested in industrial infrastructure to attract foreign businesses to new Free Trade Zones. The government also undertook two major public investment initiatives: the Accelerated Mahaweli Development Program and a Public Sector Housing Program. The Mahaweli Project comprised a series of hydroelectric dams built to provide electricity for industry and irrigation for Dry Zone farming (Karunatilake 1988). The housing initiatives began with the Hundred Thousand Houses Program (1978–1983), which granted new houses primarily to the urban poor. It was followed by the Million Houses Program (1984–1988), which was targeted more on rural communities (Yap 1994:ch. II). Together, these two programs absorbed nearly half of all public investment in the 1979–1985 period (Athukorala and Jayasuriya 1994:79).

In Walangama, a completely unpredictable but highly productive local-level synergy developed between these two initiatives. It came about because of the far-reaching vision of a young Walangama man called Sumana Mudalali (Sumana-the-Trader). The son of a village coconut trader, Sumana Mudalali had gotten to know the Mahaweli area while making his rounds to supply traditional Walangama pottery to shops and markets. In the early 1980s, he was there just as the effects of increased irrigation were beginning to be felt not only in agriculture but also in dairy farming. Milk is marketed to Sri Lankans primarily as milk powder and as cow's milk yogurt and buffalo milk curd, much of which is sold in disposable clay pots. Crudely made from inferior clay and used only once, these pots were not part of Walangama's traditional repertoire. But when Sumana Mudalali saw the truckloads of curd pots arriving from Matara on Sri Lanka's south coast, he realized that Walangama's much greater proximity would give it a competitive advantage. Sumana Mudalali secured a contract with a dairy farm and then returned to Walangama to acquire the pots he had promised to deliver. But to do so, he first had to recruit potters to make them.

One of Sumana Mudalali's most effective tools for convincing potters to risk the new enterprise proved to be the government's housing construction program. The Million Houses Program did not dole out houses directly but instead used the rural banking system to make available very low-interest loans to income-qualified households. These relatively small loans were to be used for house renovation or new construction with the expectation that the receiving households would fill in the gaps with their own labor, materials, and funds (Yap 1994:ch. II). Although it appears that the majority of Walangama households exceeded the income ceiling, they managed to obtain loans nonetheless. The loans ranged from Rs.1,500 to Rs.7,500, comparable to just a few months of potter household income. But Walangama families reported spending far more, as much as Rs.100,000. They made up the difference with bank loans, savings, and, especially, by taking advances against future pottery production first from Sumana Mudalali and then from other village men who joined him in the booming business of buying and selling pots for Dry Zone dairy farms (Winslow 1996:719). By 1992, this little village of 200 or so households was turning out around 100,000 handmade curd pots a month; by the mid-2000s, with the help of local technological innovations, production was ten times that. Consequently, Walangama incomes have been more than able to compensate for the spiraling inflation and declining government services so typical of economic liberalization, wherever it occurs (Winslow 2003, 2009).

Sumana Mudalali was able to call on real and classificatory kin to begin and later to sustain his new business. Almost all Sri Lankans use a Dravidian kin-term and marriage system, which encodes a preference for cross-cousin marriage. Walangama residents address each other with kin terms almost exclusively and frequently refer to their community as "one family," despite an economy firmly grounded in a household-based mode of production. But the potters also were motivated by the fact that Sumana Mudalali offered them advances both in cash and in bags of cement, loads of lumber, and other scarce building materials. In a 1992 interview, Sumana Mudalali laughingly described the heaps of construction materials that took over his house and yard in those early years. His potter kin repaid him with a steady supply of curd pots to fill ever-larger contracts. Other men (and one or two women) soon followed Sumana Mudalali's entrepreneurial lead and curd pot production and new houses spread in tandem throughout Walangama.

TAKING BLANTON TO SRI LANKA: THE FLEXIBLE AND INCLUSIVE *PIL GĒ*

In order to understand how the new houses might have shaped as well as been shaped by a changed village life, we first need to consider the houses they replaced (figure 11.2). I present below (figure 11.2a) the floor plan and Blanton graph of an old Walangama house. It was built around 1940, although the kitchen, with its woven palm frond walls, would have been replaced many times over the years and latrines appeared in the village only in the 1960s. The

Figure 11.2a. *Walangama pil gē. Built around 1940, this is the oldest extant house in Walangama. Due to a property line dispute between brothers, it is soon to be taken down. Note the swept front yard (midula), the veranda (istōpu), and the veranda room (istōpu cāmara). (Photo: 2013, Deborah Winslow.)*

house is a traditional *pil gē*, a raised mudbrick structure with plastered walls and palm-thatched roof, about 750 square feet (70 m^2) in area. The wide front veranda (*istōpu*), facing the *midula* (swept front yard) and open to passersby, is typical of this type of house.

The *istōpu* serves as an outside foyer, a transitional space between public and private that both invites and channels visitors. Passersby first step off the public path, walk across the *midula* and then, with permission, take a seat on a mat or chair on the *istōpu* to talk, rest, escape the sun, and chew a little areca nut and betel leaf. Typically, the *istōpu* is a primarily male space. Unless they are strangers or past the age at which they might lend a hand with domestic chores, village women are more likely to go around the side of the house to sit near more-private domestic areas at the rear than in the public space at the front. Interestingly, the *istōpu* also has a little isolated room, separate from the main rooms of the house. This room was used for storing rice, giving birth, and for housing overnight guests who were not well enough known to be allowed

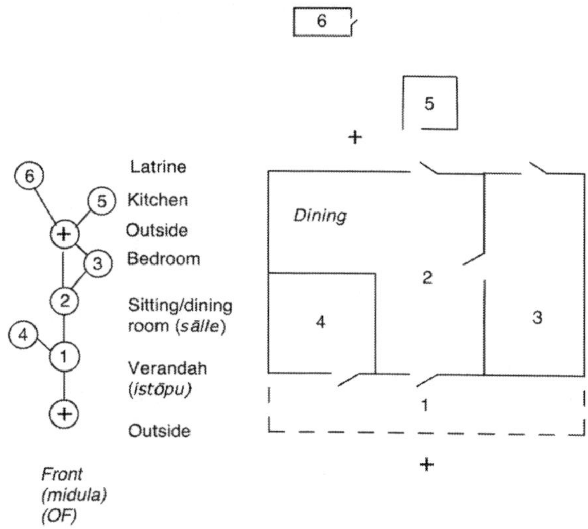

FIGURE 11.2B. *Floor plan of typical Walangama pil gē.*

into the intimacy of the house's interior. It was also sometimes used to isolate girls experiencing the highly polluting status of first menstruation (Winslow 1980:608). Finally, the *istōpu* was where particularly important guests, such as Buddhist monks or marriage negotiation delegates, were greeted formally with bows, foot washing, and betel leaves before being led inside.

Passing through the front door, a visitor enters the *sālle*. Until the advent of television in the 1990s, the *sālle* was not a space where residents of the house sat around casually; it was a reception and sitting room for guests. Within it, but as far from the front door as possible, there was an area set aside for eating at a table, although women and small children more commonly took their meals seated on low stools in the kitchen. Today the *sālle* functions both as a reception area and as a family room.

From the *sālle*, one can access the bedroom and the house's rear exit, which leads to an outdoor space that does not have the public formality of the *midula*. The area right around the back door is used to store water and cooking pots, farming implements, and other household necessities. A few steps further and one finds the kitchen, a simple structure constructed of woven palm fronds whose airy weave allows hearth smoke to escape. Beyond the kitchen, a tiny brick structure houses a water-seal latrine. If you explore a Walangama compound a bit further, you may find a small household vegetable garden, a roofed platform that serves as a pottery workshop, and a shallow pit kiln for firing pots.

TABLE 11.2. Accessibility ranks of an older Walangama house

Space	*1*	*2*	*3*	*4*	*5*	*6*	*O/F*	*Sum*	*Rank*
1	x	1	2	1	3	3	1	11	2
2	1	x	1	2	2	2	2	10	1
3	2	1	x	3	2	2	3	13	3
4	1	2	3	x	4	4	2	16	4
5	3	2	2	4	x	2	4	17	5
6	3	2	2	4	2	x	4	17	5
O/F	1	2	3	2	4	4	x	16	4

Following Blanton, I performed an accessibility rank analysis of this *pil gē*'s rooms: I counted the links between every pair of rooms, summed them, and then rank ordered the sums. Five hierarchical levels emerged. I had anticipated that the *istōpu*, the most public space, would have the lowest rank, but it did not. The best connected, the most central of all the spaces turned out to be the *sālle*, from which one can most easily access all of the delineated spaces. More predictably, the highest ranked and least accessible are the latrine and the kitchen, both tucked into the privacy zone at the rear of the house. In between are the *istōpu* (Rank 2), the bedroom (Rank 3), the non-communicating room off the *istōpu* (Rank 4), and the *midula* (marked "O/F" for Outside Front, also Rank 4) (table 11.2).

Blanton alerted us to the importance of the canonical messages communicated by house architecture, the spatial delineations that express and reinforce particular, conventional categorizations, separations, and alignments. Overall, the *pil gē*'s architectural canon communicates a sense of a family that lives and works together, open to the rest of the village while clearly distinct from it, and with a minimum of internal differentiation for either work or social status.

There is first of all a clear but also graduated separation between public and private; the house welcomes and shelters even casual visitors, but only as far as the *istōpu.* From that point on, one clearly needs a more formal invitation to proceed. While the *istōpu* is always open, the door into the *sālle* is closed and literally barred when the family retires for the night; its reopening the next morning is a public signal that the household is again ready to welcome the world. The front doors of elderly villagers are monitored by concerned neighbors who will go check for illness or even death if a door fails to open at first light. The isolation of the side room directly off the *istōpu* and outside the family threshold is thus quite marked. When the room is occupied by a child enduring her first menstruation, the symbolic separation was poignant, on the same accessibility level as the *midula* and out of the house completely.

The *pil gē* kitchen is not the center of public life as it is in many Western homes. Rather, it is hidden and protected, the least accessible space in the house except for the latrine. Similarly, the dining area is shielded from casual gaze in keeping with the fact that food and its effects are a focus of generalized concern. Villagers dislike being watched while eating; some believe that an envious gaze might attract the attention of demonic forces and illness. More positively, the kitchen is said to be the realm of a goddess, Shriakantha, who comes at night to check it for cleanliness and to eat the bit of rice left at the bottom of the cooking pot just for her.

A *pil gē*'s canon reveals little concern with individuating differentiations. The architecture of the house allows for but does not insist upon separation by either gender or age. The kitchen is a place of women's work but it is not forbidden to men. Men may tend to occupy the public space of the *istōpu* and women the private space of the kitchen, but the separation is far from rigid. Women are frequently found chatting and playing with children at the front of the house while men sit on mats behind, weaving palm frond branches, repairing tools, or simply enjoying a cup of tea in the kitchen.

Other sorts of separations, too, are muted. Over the years, this particular house was lived in by at least five people and sometimes several more, yet there is only a single bedroom. Furthermore, with two doors, one into the *sālle* and the other directly outside, the bedroom is midway in the accessibility rank and not particularly private. It could be claimed for privacy when needed, perhaps because of illness or to accommodate a newly married couple. But Walangama people almost never sleep alone; beds are occupied at night by several women and young children while men and older children sleep on floor mats in whatever space is convenient.

Thus the *pil gē* communicates a message of flexibility and togetherness rather than separation and rupture. There is a concern for privatizing food preparation and consumption without isolating it all together. Only the deep impurities of first menstruation and toilets are true zones of separation. Is the same true of the new houses? To answer this question we turn now to another floor plan and Blanton graph.

CALIFORNIA STYLE IN WALANGAMA

Walangama households undertook their house-building projects in stages. Bags of cement and piles of lumber and hardware were purchased piecemeal

Figure 11.3a. *Walangama California-style house. In Walangama, the style dates from about 1990 but this house was built in 2011. Its distinctive feature is the carport-like structure that replaces the veranda (istōpu) of older houses. However, the carport is not large enough to accommodate a vehicle; instead, it is fitted with chairs and used as an istōpu. (Photo: 2013, Deborah Winslow.)*

as funds became available and traders gave advances. A brick foundation might grace a yard for months or even years before the house itself went up. But over a little more than a decade, bright plastered brick walls, red-tile roofs, and cement floors replaced most of the wattle-and-daub or mud-brick-and-plaster, thatched-roof, and clay-floored structures of earlier times.

Walangama villagers generally favored what some called the "California" plan (figure 11.3a), easily distinguishable by a cement-floored carport-like space sheltering the area around the front door. The carport replaces the *istōpu*; houses have one or the other, never both. But despite its appearance and the fact that it is at ground level, rather than raised like the traditional *istōpu*, the carport normally functions as a space for people, not vehicles. Visitors again cross the *midula* but now linger in the carport where, sheltered by the tile roof overhead, they sit on chairs or mats, or perch on the low cement foundation

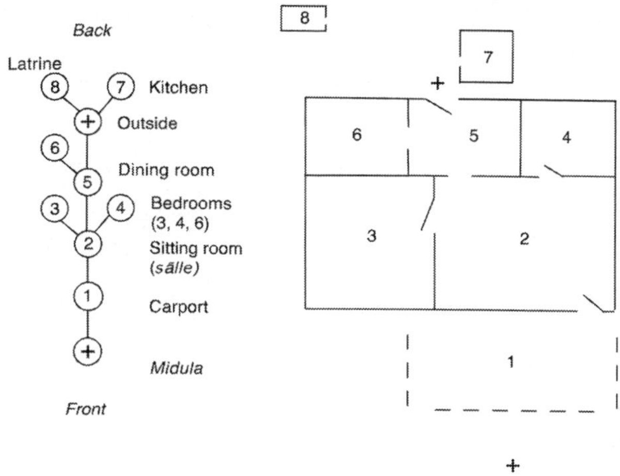

FIGURE 11.3B. *Floor plan of typical California-style house.*

curb that projects into the carport from under the house's front wall. Figure 11.3b displays the floor plan and Blanton graph for a California-style house.

From this floor plan we can see that once the visitor passes the novel carport, he or she is in familiar territory. The first room encountered is again the *sālle*; there are bedrooms off the *sālle*; there is an area reserved for eating at a table that is only slightly more set apart and is again adjoining the back door; and outside the back door are separate structures first for a kitchen and further back for a latrine. By the turn of the century, a few houses had permanent attached kitchens with a raised platform for fire-fueled cooking. They were furnished much like the separate kitchen sheds with floor racks, pots, and spoons hung from the rafters, and mats and low stools on the floor where women work. Even in 2008, while many people told me that they planned eventually to build an attached kitchen, the separate thatched kitchen remained the village norm.

An observant visitor might have noticed that the carport-*istōpu* does not contain the isolation room of the *pil gē*'s *istōpu*. Instead, my own experience is that the small bedroom at the very back of the house (Room 6 in figure 11.3b), which is not reachable directly from the *sālle*, functions in a similar way: it is used for storage, for the sick, and to isolate a newly mature girl (these days, women have their babies in the hospital). Of course, these practices are not discernible from the floor plan alone but depend on the familiarity gained through ethnographic research. However, the floor plan does direct us to the fact that the third bedroom is different than the other two. Once we know that this is the new isolation room, the floor plan also highlights the fact that degree of isolation has been significantly lessened. The polluting girl is no longer kept outside the threshold but remains within the family fold. Interestingly, it is said that there once was a time when girls experiencing their first menstruation would be housed away from the house altogether in a temporary shed built in the back yard; the shed would be burned when the isolation period was over. In this light, the *istōpu* room of the *pil gē* might be seen as an interim step, integrating the girl more with her family during this difficult time. The back room in the California-style house then becomes a further move in the destigmatization of first menstruation, a point that Blanton's spatial analytics helps us to see.

In addition to the back room, the new house has two other bedrooms, allowing potentially for further spatial individualization. However, ethnographic observation reveals rooms and beds are not assigned but continue to be shared. What the additional room seems to provide is the possibility of a space apart. For example, I saw one such room temporarily furnished with a small table, desk lamp, and chair to provide a separate workspace for a child studying for an important exam.

Having more rooms, this house also has more hierarchical levels (table 11.3), seven rather than the five of the old house.

But if we look carefully we can see that the fact of more hierarchical levels suggests a difference that is perhaps more apparent than real. The underlying template, made perfectly clear to us through Blanton's method, is remarkably similar.

- The *sālle* (reception room) is the best-connected space in both houses.
- The dining area is linked directly to the back door in each house, and while it is set apart more in the California house, the separation is not reinforced with a door.
- The *istōpu* or its equivalent, the carport, is next in accessibility
- Next in both houses are bedrooms reachable by one link from the *sālle.*
- Then come spaces that are two links from the *sālle:* the isolation room (4) and the *midula* (O/F) in the old house and the back room (6) and the *midula* (O/F) in the new house.
- Finally, the most isolated rooms in both houses are the kitchen and the latrine.

In sum, then, what Blanton's analysis accomplishes first is to show us objectively that despite the distracting California carport, the fired-brick walls, and

TABLE 11.3. Accessibility ranks of a newer Walangama house

Space	*1*	*2*	*3*	*4*	*5*	*6*	*7*	*8*	*O/F*	*Sum*	*Rank*
1	x	1	2	2	2	3	4	4	1	19	3
2	1	x	1	1	1	2	3	3	2	14	1
3	2	1	x	2	2	3	4	4	3	21	4
4	2	1	2	x	2	3	4	4	3	21	4
5	2	1	2	2	x	1	2	2	3	15	2
6	3	2	3	3	1	x	3	3	4	22	5
7	4	3	4	4	2	3	x	2	5	27	7
8	4	3	4	4	2	3	2	x	5	27	7
O/F	1	2	3	3	3	4	5	5	x	26	6

the red-clay tiled roof, socially significant canonical differences between the old houses and the new are few. Both houses communicate flexibility and togetherness rather than separation and rupture, a concern for privatizing food preparation and consumption, and for keeping latrines hidden. Spatially, the most important difference is the relocation of the isolation room from outside to inside.

This answers the first of my questions: "Does this changed architecture mean also changed socio-architectural effects?" From the perspective of the inside of the house, the answer is no, the kinds of changes that Walangama residents have introduced may make their houses sturdier and give them a changed appearance, but canonically, they have little effect.

But what of the second question: "Do the houses communicate as differently as they look?" To find out, we must consider indexical as well as canonical communication: the meaning of houses from the outside.

WHAT DO THE NEW HOUSES MEAN?

Externally, Walangama's new houses seem remarkably similar. Most are built to the California plan. A few have a frieze of ducks built into the carport trim and some are color washed in pastels rather than brilliant white, but the differences are slight. So much so that the most obvious indexical message appears to be singularly simple. There either is a new, modern house, or there is not; it is a dichotomous variable. A new house tells observers that those who live inside are economically stable enough to find the means to complete such a major project. In a community that prizes economic self-sufficiency and the ability to rise economically (*diyunuwa*), this is not an unimportant message. But Blanton's cross-cultural analysis take us further and permits us to consider why housing is not used more distinctively in Walangama.

One of Blanton's findings was a negative correlation between external housing decoration and a combination variable that he called community integration (Blanton 1994:133–134). In places with higher levels of interhousehold ties and exchanges, where people used a communal cemetery, and where there were communal rituals—the three factors comprising his integration measure—houses were less likely to be decorated expensively and in ways that tried to distinguish one sharply from another. Blanton actually hedged this conclusion, citing limitations of his nonrandom sample. But Walangama bears out these observations: it is a single-caste community with strong interhousehold ties of kinship and marriage; they do indeed use a communal burial ground; and there are significant communal rituals. While household economic autonomy is important there also are strong ethics of communal care and responsibility (Winslow 2009).

Again drawing on his cross-cultural data, Blanton argued that when elaborate external decoration of houses did occur it constituted a kind of information broadcasting in the context of either internal or external social differentiation. Internally, the message was generally one of social boundedness, such as when elites seek to distinguish themselves from commoners. Externally, elaborate house decoration appeared to be an effort to show off wealth and thus exhibit that the residents of the house were worthy of economic and other alliances (Blanton 1994:188–189). After a multiple regression analysis, Blanton concluded that "external decoration of houses is a complex variable that is related both to indexical communication, by way of status anxiety, and to a lesser extent, to social boundedness" (ibid.:147). Again, Walangama fits these cross-culturally derived observations. There is little internal stratification, economic ties with the outside are determined chiefly by caste-based occupation, and those ties are mediated more by trader middlemen than they are by individual households. The persistent simplicity and uniformity of Walangama housing is consistent with Blanton's cross-cultural results and tells us that despite the increase in prosperity, the community has not experienced an increase in internal socioeconomic differentiation (Winslow 2009).

Interestingly, the one apparent departure from Blanton's predictions actually serves to confirm the larger point: the sensitivity of the indexical dimension to social reality. Pottery making is a messy occupation and Walangama yards are workspaces, crowded with drying pots, pit kilns, and shelters for potter's wheels. Potters do not leave the house each day for work in a distant field or town. They work at home, moving between work, childcare, and meals, from wheel to kitchen to kiln to *istōpu*; inevitably some of the water, clay, and ash of their occupation moves with them. Therefore, while Walangama houses are as new and costly as many in neighboring non-potter villages, they do not display pristine paint and tidy yards adorned with decorative plants. I have visited former Walangama residents after they have relocated to new housing colonies. Their houses always stand out as busy sites of industrious labor in contrast to the tidy and quiet domesticity of their new neighbors who leave home each day to work elsewhere.

Overall, the Blanton graphs provided a new understanding of the fluid and integrated nature of Walangama's instantiation of the Dravidian family system. Without them, mere cosmetic differences might have distracted us from important similarities between the old and new houses. On a finer scale, they illuminated subtle changes, such as the shift in the isolation room and the increase in the number of bedrooms. The methods of *Houses and Household* revealed shades of meaning in the phenomenology of Walangama social life that would otherwise remain unnoticed. Blanton himself, speaking, I think, to archaeologists, concluded that "there has been inadequate attention paid to comparative research that would allow us to . . . systematically evaluate our various theoretical frameworks in the broadest possible sense" (Blanton 1994:185). What taking Blanton to Sri Lanka showed me is that it is very much a two-way street: local data and generalizing theory are mutually informing. Without paying attention to particular trees, we would never understand where the forests came from.

"We shape our buildings and afterwards our buildings shape us," Churchill cautioned his wartime audience as they debated the rebuilding of the House of Commons. However, Walangama's architectural tale, illuminated by Blanton's methods and theory, tells us that the relationship between houses and households is not so cut-and-dried. It is instead an ongoing process of mutual shaping and being shaped best understood by combining the twin lenses of generalizing theory and local-level ethnography.

POSTSCRIPT: ETHNOGRAPHY NEVER STANDS STILL

In 2013, after completing this essay, I was again in Walangama. There I found two new architectural developments. Happily, the first confirms the analysis above. California-style carports are being replaced with *istōpus*, bringing form into consonance with ethnographically observed usage. In contrast, the second change may disrupt rather than endorse longstanding social norms. Outwardly, it is but a small deviation: a few families have begun construction

FIGURE 11.4. *Walangama two-story house under construction in 2013. Adding a second story increases the number of hierarchical levels in the house's spatial syntax. (Photo: 2013, Deborah Winslow.)*

of modest two-story houses (figure 11.4). Still, with Blanton's work in mind, one wonders. Does this increase in hierarchical levels matter? Does it perhaps signal a decrease in community integration, even the beginning of a major community transformation? None of the multistory houses had been completed, so I could not yet know if, as was true in the past, cultural practice would mitigate material constraint. But if Blanton has taught us anything, it is that if we want to understand the shaping of society, it behooves us to pay attention to the shaping of buildings.

NOTE

1. This essay draws upon research supported by the National Science Foundation (NSF), the United States-Sri Lanka Fulbright Commission, the American Institute for Sri Lankan Studies, and the University of New Hampshire Center for International Education, all of whose support I gratefully acknowledge. Writing was done while serving at NSF. Opinions, findings, or conclusions expressed here are mine alone and do not necessarily reflect views of the NSF or any of my other supporters.

SECTION 3

Cross-Cultural Studies

12

Network Strategy and War

Peter N. Peregrine and
Carol R. Ember

One of Richard Blanton's most influential papers is the 1996 "A Dual-Processual Theory for the Evolution of Mesoamerican Civilization." There, Blanton and colleagues argue that there are two basic strategies that leaders follow, to varying degrees, to maintain authority: corporate strategy and network (or exclusionary) strategy. Leaders following a corporate strategy attempt to build a power base by developing and promoting activities that reinforce the corporate bonds tying members of the polity together. A common corporate strategy is, for example, to mobilize goods from across a polity for large public rituals or construction projects that bring members of the polity together in corporate-affirming activities. Leaders following a network strategy attempt to build a power base by controlling access to networks of exchange and alliance both within and outside the polity. Thus a network strategy is one in which leaders attempt to monopolize sources of power, while a corporate strategy is one in which leaders attempt to share power across different groups and sectors of a polity.

The paper has been influential because it provides a way of understanding variation in the nature of Mesoamerican polities, some of which appear to have obvious, self-aggrandizing leaders, and others that appear "faceless" despite clear evidence of centralized leadership. The value of this perspective has appealed to many scholars, and it has been widely applied to other prehistoric polities (e.g., Mills 2000; Butler and

DOI: 10.5876/9781607325338.c012

Welch 2006). Despite its influence, the "dual-processual" paper had two flaws. First, it did not adequately clarify that the "dual processes" of political strategy formed a continuum rather than two types of political strategy, so that other scholars often took them as reflecting cultural "types" rather than as expressions of variation in political strategy (e.g., Yoffee 2005:177–179). Second, the paper did not explain how variation along this continuum changed over time.

In recent work the authors of the dual-processual paper have attempted to address these flaws by clarifying the idea that the "dual processes" of corporate and network strategies form a continuum from one more inclusive of citizen participation in the polity's political action and decision-making (corporate strategy) to one more exclusive of such participation (network strategy) (Blanton and Fargher 2008; Peregrine 2001, 2008). The first author of this chapter has also attempted to demonstrate variation on the corporate-network continuum over time; to develop an extended multivariate approach to political strategy; and to argue that balancing stability and the "cost" of implementing political strategies are one reason that variation occurs (Peregrine 1998, 2003, 2009, 2012). In this chapter, we suggest that variation on the corporate-network continuum might be related to socialization for mistrust, unpredictable natural disasters, and external warfare (see also Earle, chapter 12, this volume). And we explore the mechanisms that might underlie these connections.

Ember and Ember (1992a) found that both socialization for mistrust of others and unpredictable natural disasters are independent predictors of warfare frequency. Peregrine wondered whether xenophobia might be more common in network-oriented polities, as leaders might encourage xenophobia to prevent individuals from seeking contact with external polities. While socialization for mistrust (a variable coded by Barry and colleagues [1976]) is not the same as xenophobia, one might presume that if children are told to mistrust others in the community, they might develop more generalized mistrust. Using the Embers' data set, and a measure of the corporate-network continuum of political strategy developed by Bramm (2001), Peregrine (2009) found evidence to support the idea that network-oriented polities socialized for mistrust more frequently than did corporate-oriented societies, and had more frequent external war. By "external warfare" the Embers meant warfare taking place outside of the culture or linguistic group (Ember and Ember 1992a, 1992b). This means that one chiefdom, for example, fighting another chiefdom within the same linguistic or cultural group would not be considered an example of external warfare, but one chiefdom fighting another speaking a different language or being part of a different cultural group would be considered external warfare.

Following upon that work, Peregrine also wondered whether, given Ember and Ember's (1992a: 256) argument that "fear of future economic problems (rather than current problems) is the major motive for going to war," the relationship between warfare and the "fear of future economic problems," as measured by the presence of unpredictable natural disasters, might be associated with network-oriented polities because of their leaders' emphases on the control of external resources. This chapter provides evidence that network political strategies are related to unpredictable natural disasters and warfare, but in a more complicated way than initially expected.

DATA AND METHODS

The research presented here begins with data coded by Carol R. and Melvin Ember for their study of the conditions favoring warfare. The Embers employed the Standard Cross-Cultural Sample of 186 societies, which provided them not only with a large and relatively well-documented set of cases, but also allowed them to use variables coded by other scholars for the same sample (Ember and Ember 1992a, 1992b). The Embers coded 43 variables concerning the type and intensity of warfare, pacification (pacified societies were not used in their study), outcomes of warfare, individual and social aggression, resource problems, and sex ratio. The coding method employed two naïve coders. If the coders disagreed in their coding of a particular case, they developed a "resolved" score for that case. The Embers found that simply using resolved scores yielded weaker results than using resolved scores where coders more closely agreed in the first place, and so they dropped all cases where reliability was not that strong (reliability scores greater than 6). We follow a similar procedure here except that we only use the societies in *eHRAF World Cultures* (http://ehrafworldcultures.yale.edu).

To ensure our data selection and statistical procedures matched the Embers', we first replicated their findings to make sure that the results were still significant in our subsample. The Embers employed a carefully selected subset of their coded cases in their analyses (dropping island societies, pacified societies, and cases with poorer reliability), using only 30 cases in their final analyses. Our procedures were slightly different here in that we did not dichotomize the natural disasters variable because we wanted to maximize variation. They also used a different statistical package (Systat) than we did (SPSS). We found that our data and statistical package replicated the Ember's results satisfactorily (see table 12.2a).[1]

Once we replicated the Embers' results, the first author coded five ordinal variables, each focused on a separate facet of corporate/network strategy. These five variables were then summed to create an interval scale of corporate/network strategy, with more corporate-oriented polities having lower scale scores and more network-oriented polities having higher scale scores. The codebook is presented as Appendix 12.A, and the raw data as Appendix 12.B. While the first author did the coding by himself, he followed the basic strategy identified by the Embers. In their original study they found that cases where the information did not allow for a clear score (i.e., where the coders disagreed) added "noise" to the data, and they dropped them from their analysis. Following that idea, the first author coded only cases where the information implied an obvious score. Because of this selectivity, and because the first author only coded those cases with primary source material available in *eHRAF World Cultures*, only 11 cases are included in the analyses that follow.[2]

RESULTS

First we examine whether unpredictable natural disasters predicts more network-oriented polities. The bivariate Pearson r is .544, $p < .04$, one-tailed, which is consistent with the hypothesis that network-oriented polities are more common where there are conditions of unpredictable natural disasters.[3] Figure 12.1 shows a box plot of the relationship. The second hypothesis is that there would be more xenophobia in network-oriented polities. As measured indirectly by socialization for trust, the hypothesis is not supported, and for reasons we cannot explain ($r = .365$, $p < .27$, two-tailed). We know from the Embers' study that unpredictable natural disasters are related to more warfare, as is socialization for mistrust (both of them presumed causes of warfare). Recall that Peregrine (2009) found that network-oriented polities had more external warfare.

How might these variables relate to each other? It seems plausible to us that network-oriented leadership may be adopted in situations of crisis. Researchers examining the psychology of survival have found that network-style, authoritarian leadership is common in the early stages of a crisis, and their presence often leads to a group successfully overcoming that crisis. As explained by Leach (1994:140–141), "The initial leader will usually be authoritarian, he will be decisive and will lead by example . . . an authoritarian, military style of organization is not only acceptable but may even be welcomed in the initial stages of a disaster." A network-oriented leader, who controls the polity unilaterally or with a small cadre of peers, is well-positioned to respond quickly and decisively to a crisis. In polities with an ongoing threat of unpredictable natural disasters, such a leader might be desired by the members of

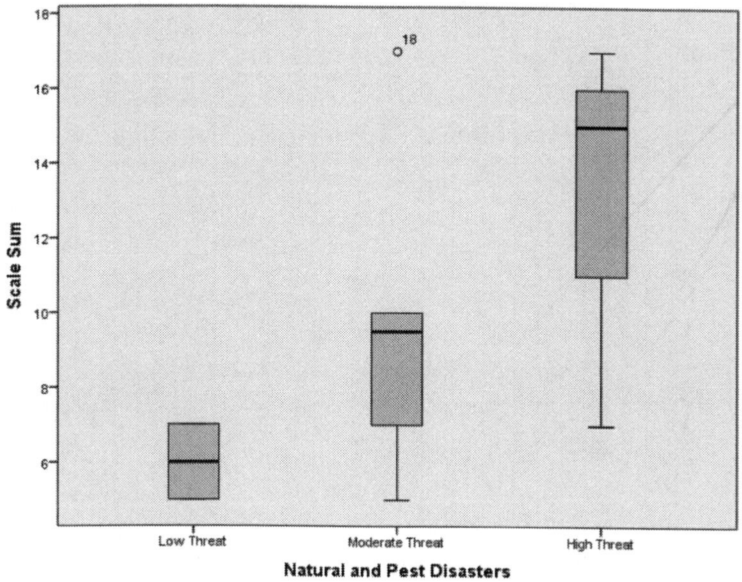

FIGURE 12.1. *Relationship between network strategy and natural disasters.*

the polity, and might be able to maintain power much more readily than a leader employing a corporate strategy. Being attacked by others might also increase the "crisis mode" of a society. So warfare, particularly external warfare, might increase the likelihood of network-oriented societies.

Table 12.1 shows a multiple regression analysis with network-oriented societies as the dependent variable and unpredictable natural disasters and external warfare as independent variables.

The results are not statistically significant, although the beta values suggest that natural and pest disasters is the stronger predictor of network strategy in the model. But we have implied above that a network-oriented polity might also deliberately undertake more warfare, particularly external warfare, to defend external resources. If this is so, a network-oriented polity not only might become more likely with warfare (as we discussed above), but it also might increase warfare, particularly external warfare. (This would be a feedback loop.) To test this idea, we add the network variable to the Embers' original model that has warfare as the dependent variable (Ember and Ember 1992b). (The Embers' model with the societies in *eHRAF World Cultures* is shown in table 12.2a for comparison purposes.)

TABLE 12.1. Predictors of network strategy

	Beta	*Significance*
Constant		.181
Natural and Pest Disasters	.427	.412
External Warfare Frequency	.182	.722
$n = 9$		
$R = .579$		
$R^2 = .336$		**.239**

TABLE 12.2A. Predictors of overall warfare (Embers' model)

	Beta	*Significance*
Constant		.000
Socialization for Trust	−.407	.022
Natural and Pest Disasters	.312	.074
$n = 29$		
$R = .522$		
$R^2 = .272$		**.016**

TABLE 12.2B. Predictors of overall warfare using Ember and Ember predictors

	Beta	*Significance*
Constant		.043
Socialization for Trust	-.563	.049
Natural and Pest Disasters	.435	.123
Corporate-Network Scale	.411	.167
$n = 11$		
$R = .823$		
$R^2 = .677$		**.038**

The result shown in table 12.2b suggests that this expectation may be correct. Not only do all the independent variables have high beta values, but the overall model predicting warfare is better when we add network-oriented strategy (the multiple R is .82 in table 12.2b compared to the multiple R of .52 in table 12.2a). If we make the dependent variable external warfare, the

Table 12.3a. Predictors of external warfare

	Beta	*Significance*
Constant		.041
Socialization for Trust	-.285	.077
Natural and Pest Disasters	.350	.146
n = 25		
R = .472		
R^2 = .223		**.063**

Table 12.3b. Predictors of external warfare

	Beta	*Significance*
Constant		.010
Socialization for Trust	-.707	.003
Natural and Pest Disasters	.694	.045
Corporate-Network Scale	.637	.009
n = 10		
R = .946		
R^2 = .895		**.002**

effect of network polities is even stronger (compare table 12.3a and table 12.3b), the increase in *R* is even greater, and all the independent variables are statistically significant.[4]

How can we make sense of this? We suggest the explanation lies in the unique conditions of polities in which leaders employ a network strategy. Recall that leaders employing a network strategy not only attempt to limit the political participation of individuals, but also to restrict access to resources and knowledge from outside the polity. Thus natural and pest disasters, which result in disruptions to external communication or resources, would have a much stronger impact on network-oriented polities than on corporate-oriented ones. To better control those external resources, network-strategizing leaders might go to war with the external group. Of course, such war could take place within a single group, as in the case of peer-chiefdoms given above, but it might more often occur with external groups, particularly in states.

At the beginning of this section we suggested that a crisis is likely to push toward more authoritarian leadership. But there is a catch. Network strategy is not simply authoritarian, but is based on controlling access to power and resources. The authoritarian nature of network strategy might be beneficial in the initial stages of a crisis, but what about later? Here, according to Leach (1994:140), more corporate-style leadership will be preferred: "The later type of leader will be one . . . who will work with the rest and will organize and minimize differences amongst the group"; in other words, a corporate-style leader. Thus, while a network strategy may work in a crisis mode, it may fail to be accepted when situations improve. We argue that the reason that network strategy improves the predictive power of the Embers' model for external war is because it is in situations where political leaders both restrict access to political authority and use connections to other polities to maintain and legitimate their own authority that they might go to war in the face of unpredictable natural disasters in order to assure themselves access to those resources upon which their authority is based. Furthermore, leaders in network polities are in a strong position to initiate a war—they tightly control the population and are leaders who function well in a time of crisis. External war may appear an easy solution to an immediate or feared resource problem.

CONCLUSION

One of the limitations of the dual-processual paper (Blanton et al. 1996) was that variation on the corporate-network continuum of political strategies was not well explained. We suggested one possible explanation here—that network strategies may be adaptations to situations of resource unpredictability. Because of small sample sizes, we only have suggestive evidence to support this relationship. But we have better evidence for a more complex relationship with a feedback loop such that network-oriented polities themselves may create more warfare, particularly external warfare. We suggest that, because adding variation in the corporate-network continuum improves the predictive power for the Ember's model of war, network orientation may itself be a factor behind increased frequencies of war. We have argued that, in conditions of resource uncertainty, leaders employing a network strategy might find going to war a good, and easily implemented, solution to a potential resource shortage.

We realize that our explanation for the increased predictive power that accompanies network strategy in our regression model for the frequency of war is not entirely satisfying,[5] but we hope the reader might overlook this to accept a broader point: the corporate-network continuum of political strategy that Blanton developed to explain variation in the archaeological signature of

Mesoamerican polities has explanatory power far beyond the narrow contexts within which it was developed. The idea that political strategy is actively promoted by leaders, and that those strategies impact polities in predictable ways, has had profound impact. It represents the best of anthropology's insights, in that it both allows us to understand others and to better understand ourselves and the world in which we live (a world in which there are still corporate-oriented and network-oriented polities, acting in predictable ways). It also represents the best of Blanton's work—strongly theoretical, but also practical, allowing for explanation of variation in the past and the present. Whether focused on cities, political strategy, or collective action, Blanton has demonstrated that anthropology has the ability to develop concepts of great explanatory power, and we hope our brief contribution to this volume has expanded the explanatory power of Blanton's corporate-network continuum, if only in a small way.

NOTES

1. If natural disasters were dichotomized as *no* or *rare* versus *more*, the natural disasters variable would have a higher beta than socialization for mistrust.

2. Relying only on cases in *eHRAF World Cultures* meant that we could not code all the cases in the Embers' sample. We hope to do so in the future. Thus this should be taken as a preliminary study.

3. Natural and pest disasters were recoded into a three-category variable from a four-category one for this correlation and the regression shown in table 12.1. The full variable was employed in all other analyses.

4. We must point out that the sample size upon which these analyses were performed is very small, and we must be cautious in interpreting them. However, we feel that the results are so strong, and the improvement in the regression model so great, that they cannot be ignored, despite the potential problems of such a small sample.

5. Blanton would refer to our explanation as an example of "rinky-dink theory," replete with "ho-hum hypotheses."

APPENDIX 12.A

Codebook

Column 1: Society Name

Column 2: Standard Cross-Cultural Sample Number

Column 3: Differentiation among leaders and followers

1. none
2. leaders have some privileges and/or access to resources others do not
3. leaders have extensive privileges and access to resources others do not, including special housing and sumptuary goods
4. leaders have exclusive privileges and exclusive access to special housing, resources, and sumptuary goods

Column 4: Leader identification

1. none
2. leaders are identified by treatment or appearance
3. leaders are identified by recognized symbols of power or special behaviors
4. individual aggrandizement and/or cult of leaders

Column 5: Sharing of authority

1. leaders share power extensively with others
2. leaders share power with a large cadre of other leaders
3. leaders share power with a few other leaders
4. leaders exercise exclusive power

Column 6: Emphasis of authority

1. emphasis placed on group solidarity and group survival
2. emphasis shared between group and leader, with greatest importance given to group survival
3. emphasis shared between group and leader, with greatest importance given to leader survival
4. emphasis placed on leaders as the embodiment of the group

Column 7: External contacts

1. few or unimportant
2. external contacts are part of leaders' authority, but not exclusive
3. external contacts are key to leaders' authority, but not exclusive
4. external contacts are exclusively controlled by leaders

Column 8: Sum of scale items (columns 3–7)

Column 9: Overall warfare frequency (from Ember and Ember 1992b)

Column 10: External warfare frequency (from Ember and Ember 1992b)

Column 11: Socialization for trust (from Ember and Ember 1992b)

Column 12: Natural and pest disasters (from Ember and Ember 1992b)

APPENDIX 12.B

DATA

Column											
1	*2*	*3*	*4*	*5*	*6*	*7*	*8*	*9*	*10*	*11*	*12*
Nama Hottentot	1	1	1	1	1	1	5	18	—	6	2
Hausa	26	4	3	3	3	2	15	16	15	7	4
Masai	34	2	2	1	1	1	7	18	17	5	4
Somali	36	2	2	2	2	2	10	17	16	5	2
Riffians	42	2	2	2	2	1	9	18	17	4	2
Lapps	52	1	1	1	1	1	5	1	1	7	1
Abkhaz	55	2	1	1	1	2	7	18	17	2	3
Vietnamese	73	4	3	3	4	3	17	18	17	8	4
Koreans	116	4	4	3	3	3	17	18	17	6	2
Klamath	138	2	2	2	2	2	10	16	15	5	2
Siriono	173	1	1	1	2	2	7	1	1	7	1

13

Framing the Rise and Variability of Past Complex Societies

Gary M. Feinman and Linda M. Nicholas

> The most important issue confronting the social sciences is the extent to which human behavior is shaped by factors that operate cross-culturally as opposed to factors that are unique to particular cultures. (Trigger 2003:3)

Cooperation, inequality, leadership, and governance have long been recognized as fundamental characteristics of the human career. Understanding the ways in which humans forge identities, form socioeconomic networks, and establish institutions remain critical issues for the human sciences today (Pennisi 2005). And yet, as Trigger (2003:3) remarked, we are still searching for the appropriate balance between general and specific factors (or processes and history) when it comes to building our frames to examine the range and nature of human behavior. In this chapter we conceptualize the issues surrounding the development and diversity of large-scale human cooperative formations from a perspective that embraces both the critical role for comparison and the multiple pathways that human historical processes take (e.g., Little 2000).

The causes and consequences related to the formation of hierarchically organized societies have been seen for centuries as key hinge points in the human career (Childe 1950; Haas 1982). As the first episodes of these transitions to chiefdoms and states occurred before the advent of writing, a focus on the rise of hierarchically organized societies has long been a core comparative

DOI: 10.5876/9781607325338.c013

issue for archaeologists (e.g., Adams 1966; Wright 1977a). Many years ago, Henry Wright outlined the challenge of this topic (Wright 1977b:215): "The construction of research strategies for the investigation of state origins is complicated by two principal factors: the type of explanations required and the samples available to test them." Since that time, we believe we have made more progress on building that empirical record than on finding an appropriate conceptual frame (although there remain untold lifetimes more to do on the former front as well).

In fact, the last overarching archaeological framework designed explicitly to tackle this suite of questions (Flannery 1972) has in large part stood (fostering highly productive results) for decades (Brumfiel 1992:551). Yet not surprisingly, over many years certain conceptual issues (particularly in regard to considerations of diversity, agency, and scale) have arisen even among those committed to explicitly comparative approaches (e.g., Blanton et al. 1996; Blanton and Fargher 2008; Brumfiel 1992; Chapman 2003; Earle 1997; Feinman 2012).

From the outset, we stress that we do not present definitive answers here, nor is our goal to offer an in-depth review or critique of the broad corpus of literature focused on the rise of hierarchically organized societies. Rather, our intent is to guide and redirect our framing theory or conceptualizations of this question toward comparatively based explanations that aim to understand patterns of diversity rather than uniformity. To date, most of the ways that archaeologists have looked for generality have not been adequately explanatory, widely influential in other disciplinary arenas, or able to achieve broad consensus even within our own field. The recent dearth of strong theoretical explanations for the development of cultural complexity (a point noted recently by Jeremy Sabloff [2012:xvii]) encourages those of us interested in more general and comparative approaches to consider alternative ways to frame this issue.

The critical link between frameworks and scientific advances when we are tackling big issues is illustrated through a reading of Ernst Mayr's (1985:404–408) provocative history of biological thought. Mayr recounts that despite the general accumulation of pertinent biological data, Lamarck's efforts to understand the complex history of life did not prompt a breakthrough because he basically asked the wrong questions. Charles Lyell, in contrast, posed the right questions, but he came up with the wrong answers. Charles Darwin, building on their works, framed the right questions and arrived at answers that we now know were more fully explanatory than those of his predecessors, thereby advancing disciplinary thought. To put it another way: "Discovery requires an aggressive and critical engagement with the status quo" (Gerring 2012:28). Or, as "a brief review of the history of archaeology makes clear . . . advances have come less by resolving major questions than by superseding them with better questions" (Drennan 1987:320).

PRIOR FRAMEWORKS FOR ARCHAEOLOGICAL GENERALIZATION ON STATE ORIGINS

No one would deny that every historical case of archaic state emergence is in certain respects unique (Wright 2009:122) and important to unravel in detail, both to address specific historical questions and to provide the living and future descendants of those cultural traditions with information concerning their pasts and histories. Yet for centuries, scholars also have looked for more general processes or conditions that help explain this suite of evolutionary transitions, the differences as well as the parallels (cf., Pauketat 2001).

Some of the first comparative efforts advanced the notion that early civilizations arose in a specific environment (river valleys) or in response to particular prime movers (Wittfogel's [1957] irrigation management) (e.g., Steward 1955). Such attempts neither adequately accounted for the full suite of empirical cases nor drew convincing causal chains to explain the relevant societal transitions or the specific timing of the focal historical developments (Flannery 1972). Perhaps, over the years, the most cited and discussed of such prime movers has been Carneiro's (1970) circumscription model, which advanced population pressure, geographic boundedness or circumscription, and offensive/conquest warfare as a lineal set of triggers that led to state development. Over the years, both conceptual limitations and empirical shortcomings of this model have been repeatedly outlined (Blanton et al. 1979; Flannery 1972; Roscoe 2000). In turn, these critiques have prompted efforts by the author (Carneiro 1987, 1988) to broaden the original causal scenario to accommodate problematic cases. And yet, despite this tinkering, population pressure remains elusive, the envisioned impermeable or circumscribed region remains historically more mythic than real (Ofek 2001; Wolf 1982), and the proposed empirical link between offensive warfare and the emergence of states has not been shown to be universal.

The so-far futile search for a single, specific prime mover or set of universal conditions that invariably provoked early state development prompted a reorientation toward a series of approaches that more explicitly compared historical sequences in an effort to define more general properties. For example, refining earlier studies by Carneiro (1962, 1968), a recent cross-cultural study (Peregrine et al. 2007:77) uses Guttman scaling on 20 cases from the

Collection of Archaeology in the Human Relations Area Files to outline a broad sequence of cultural change. Basically, their scaling confirms what decades of archaeological analyses have already evidenced, that sedentism, domestication, and increasing population densities preceded the formation of states (cf. Jacobs 1969; Taylor 2012). Yet even this broad-brush scaling does not conform with various historical specificities (region to region), such as precedence of sedentary life before agriculture in the Levant, the emergence of villages and inequality among sedentary foragers in several global areas, and the much later use of metal (compared to this scaled sequence) in Mesoamerica.

Another tack has been to look for general, cross-cultural parallels in sequences leading up to state development. For example, Yoffee (2005:44) has argued that when states arise they tend to be small, ensconced in networks of city-states or peer polities. Out of these networks, a large, more dominant polity can eventually emerge. In sharp contrast, Marcus (1998) has proposed that early states tend to be large polities that oversaw expansive territories, and that only later did these entities sometimes break down into smaller states. Yet neither of these sequences has turned out to be universal when we consider the suite of cases of complex society development. In fact, in the two regions where we have investigated, we see the latter pattern in the Valley of Oaxaca, where one center, first San José Mogote and later Monte Albán, seems to have dominated the region basically from the outset of sedentary villages (Blanton et al. 1999; Kowalewski et al. 1989). Whereas in coastal Shandong (China), several roughly coequal and large centers arose and were rather evenly spaced across the landscape, more in line with the former model (Feinman et al. 2010; Underhill et al. 2008).

COMPARATIVE DATA: DIVERSITY RULES

To date, archaeologists have for the most part looked for uniformities in the underlying or initial conditions, commonalities in the processes leading to, or the basic properties of, early complex societies. But these efforts have all been confounded by seemingly exceptional or nonconforming cases. In a sense, this realization should no longer be surprising. Frankly put, the more data that scholars collect relevant to the emergence of complex societies across the globe, the more variation becomes apparent, and, significantly, that variation is not simply associated with elements that might be considered local, historical, idiosyncratic, or purely culture bound (Earle 1997; Smith 2006:13; Trigger 2003:660–661).

Several recent studies, for example, have compared multiple archaeological sequences of demographic change, village formation, and population nucleation prior to the emergence of states. In one, Bandy (2008) examined 36 cases (in two cases, large villages as he defined them never emerged), finding that there was significant variation in the time lag between the first sedentary communities in a region and the advent of large villages in that area (> 300 people and 3 ha in size) (figure 13.1a). In some cases, there was little or no lag, while in others as many as 4,000 years elapsed. In certain cases one dominant village emerged, while in others, multiple large communities arose (figure 13.1b). Bandy also isolated eight of his cases where states ultimately developed. For each of these eight cases, the initial time lag from the earliest sedentary villages to large villages was 1,000 years or less (figure 13.2a) (a point that we return to), although the lag from the presence of sedentary villages until the rise of states was more variable (figure 13.2b), as was the timing from large village formation to state development (figure 13.2c).

In an amplification of this analysis, Peterson and Drennan (2012) compared 11 archaeological sequences, some overlapping with Bandy's and others not. They examined the process from the first sedentary agricultural villages in each region (figure 13.3) through the advent of more nucleated centers, again finding considerable variation in both the timing of change and the size and number of centers that were founded. They also made rough estimates of the tax rates per citizen in each context and were surprised when the highest tax rates were seemingly being paid in three of those cases in which the central settlements were comparatively small and where more hierarchical formations did not directly develop (more on this as well).

Fostering this point, Fletcher (2012) has reminded us that the cities associated with early states were not always compact; sometimes they were dispersed. They also vary markedly in size (figure 13.4) and seemingly in their basic systems of governance (e.g., Fargher, Heredia Espinoza, and Blanton 2011; Feinman 2001), with some stressing the focal primacy of individual rulers and others exhibiting less personal glorification and greater indications of power-sharing. Michael Smith (2009:28) has noted this diversity in regard to early cities, but the same is at least as valid for archaic states. What we need are comparative approaches that can accommodate and explain elements of diversity in the evolution of human institutions.

BUILDING THEORY TO ACCOUNT FOR HUMAN ORGANIZATIONAL DIVERSITY

Clearly, our focus on the diversity of early states and the processes leading up to them is not meant to discourage comparison or the building of theory on

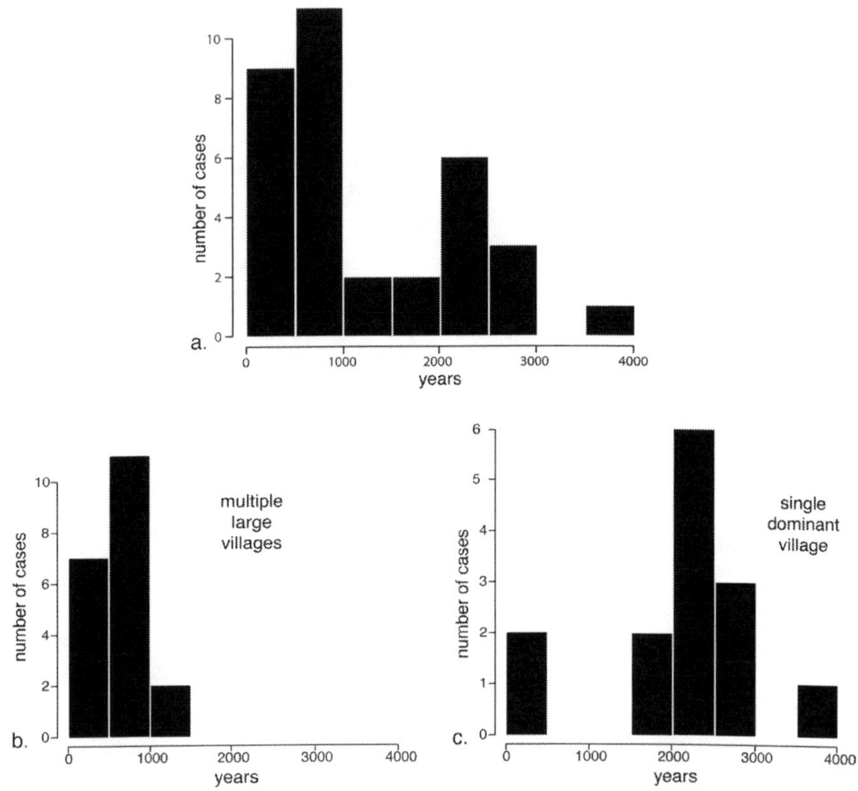

Figure 13.1. *Time from the first settled agricultural villages in a region to the presence of large villages: (a) 34 cases where large villages formed; (b) cases where multiple large villages formed versus those where a single village was dominant. (Adapted from Bandy 2008:figures 2 and 3.)*

a more general level. Rather, it serves to underpin alternative ways to conceptualize this complex issue. Since Childe (1950), it has not proven productive to search for the unilineal pathway, the uniform proximate cause, or the universal set of unvarying properties that can account for early complex societies and their diversity. But fortunately, there are conceptual roadmaps for building and using cross-culturally applicable constructs to account for human behavioral variation (Ellen 2010:399; see also Goldstone 1998; Kiser and Hechter 1991).

Here, we outline four behavioral tenets (table 13.1) that, based on a broad reading of literature in the social and behavioral sciences, we see as at the heart of future framework construction.

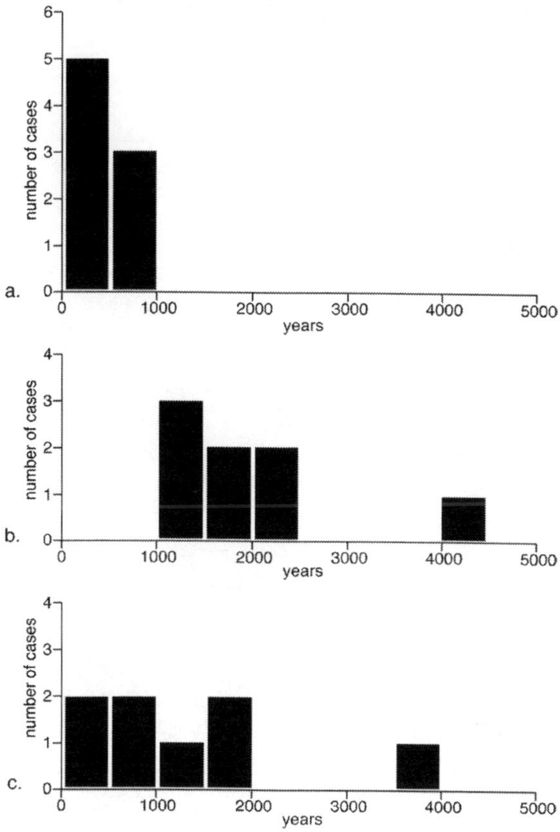

FIGURE 13.2. *Village formation and state development in eight world areas where states ultimately developed: (a) time from the first settled agricultural villages to the advent of large villages; (b) time from the first settled agricultural villages to primary state formation; and (c) time from the advent of large villages to primary state formation. (Adapted from Bandy 2008:336–337, figures 5 and 6.)*

Oddly, none of the five principal paradigms (table 13.2) that have been most influential in anthropological archaeology over the last 50 to 60 years comfortably conforms to all of these tenets, nor, in our opinion, have these extant paradigms found just the right balance between accounting for generalities and specifics.

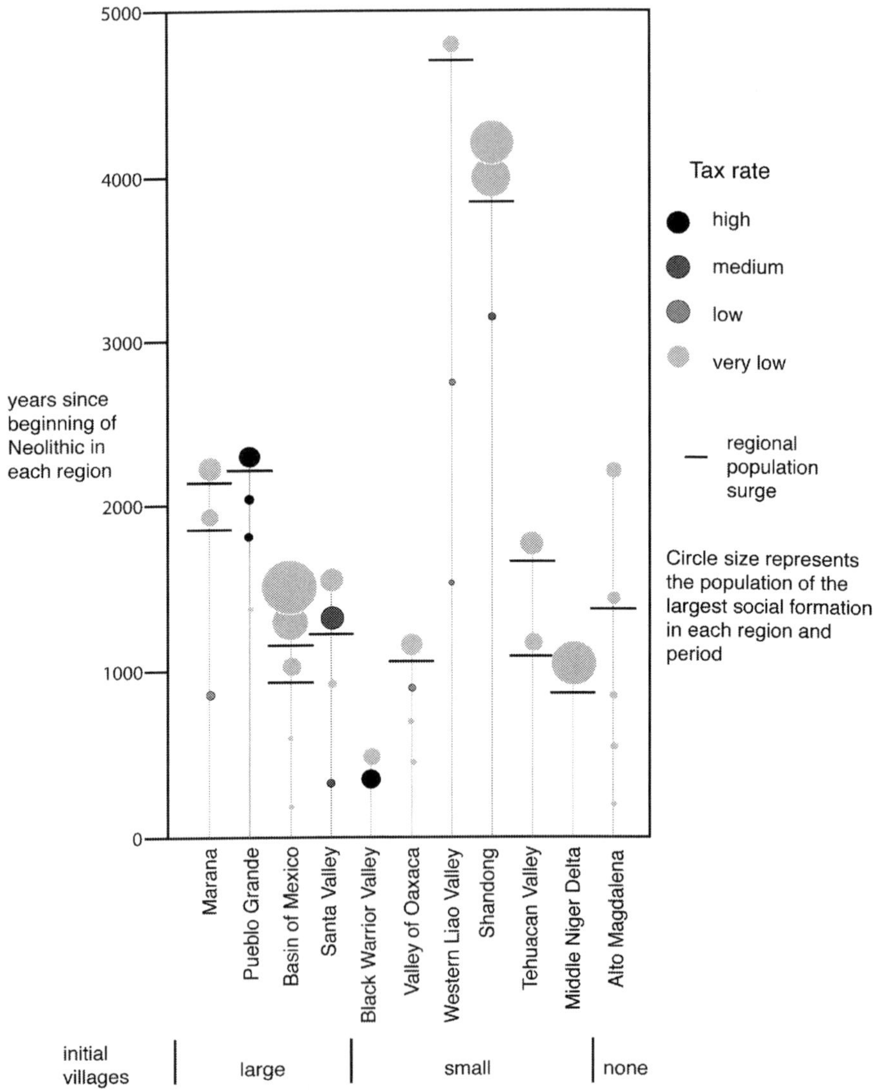

Figure 13.3. *Sequence from the presence of the first sedentary agricultural villages (the beginning of the Neolithic) to the advent of more nucleated centers for 11 archaeological cases. (Adapted from Peterson and Drennan 2012:figure 6.14.)*

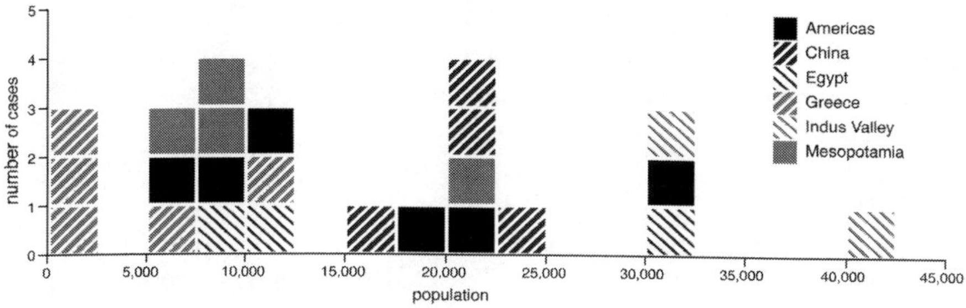

FIGURE 13.4. *Size of the largest cities in early states. (Data drawn from Blanton et al. 1993; Chandler 1987; Culbert et al. 1990; Kowalewski et al. 1989; MacSweeney 2004; Modelski 1997, 1999; Morris 2010; Stanish 2010; Underhill et al. 2008; Yoffee 2005.)*

TABLE 13.1. Human behavior: basic principles

1. The evolutionary legacy of our species has tendencies toward both dominance hierarchies and high degrees of sociality with great potentials to cooperate (e.g., Boehm 1993; Dubreuil 2010).
2. Agency is universal but also constrained by structure and resources (e.g., Schelling 2006; Sewell 2005).
3. Human groupings may be open and permeable to varying degrees, but they rarely are entirely closed for lengthy periods (e.g., M. L. Smith 2005; Wolf 1982).
4. Multiscalar perspectives are essential for understanding human groups, as humans generally participate simultaneously in networks of varying scales (e.g., Parkinson and Galaty 2009; Turchin 2003).

TABLE 13.2. Theoretical frames on the preindustrial past

Theoretical frame	*Agency*	*Scalar focus*	*Boundedness*
Culture history	Elite?	Culture	Closed
Cultural evolutionary systems	Elite	Society	Closed
Marxism/Marx-influenced	Elite	Society, Class (rarely)	Potentially open
Sociology (narrow Darwinian)	All	Individual, Kin	Not adequately considered
Postprocessual	Elite (situational for commoners)	Society	Mostly closed

The first three frameworks tend to give inadequate consideration to the "whys" and "hows" of human groups, often simply presuming their existence, continuity, and closure (e.g., Schortman and Urban 1992:12). Also, these three conceptual frameworks tend to afford little, if any, agency to the nonelite, or most people, in preindustrial contexts (see Brumfiel 1992; Wright 2009:122).

In contrast, recent postprocessual perspectives do speak of commoner resistance, yet it is largely conceptualized as an ad hoc phenomenon—for example, at times of already known societal breakdown or collapse (e.g., Joyce et al. 2001). Through this generally circumstantial or situational consideration of agency, the opportunity to understand the shift or diversity in the rules and practices of cooperation is largely lost. Strict sociobiological approaches do ascribe agency more broadly, particularly for small groups, but they are reliant on narrow definitions of self-interest and have not been able to account convincingly for large social formations and their diversity (Sterelny 2013). On their own, these approaches also fail to account adequately for the role of history and its contingencies or path-dependent aspects (Goldstone 1998:836). Clearly, we need analytical frames (Blanton et al. 1993; Little 2008:242; Parkinson and Galaty 2009) that not only take account of the multiple, scalable, and interlinked networks in which humans (past and present) participate but that also recognize that human groups and networks rarely have been entirely isolated or closed (e.g., Adams and Kasakoff 1975; M. L. Smith 2005).

Although it would be premature to try to outline an encompassing new frame here, there are, in our view, appropriate constructs in the broad transdisciplinary literature on collective action and cooperation that can serve as a foundation. Although much of this theorizing has roots outside anthropology and archaeology, such frames are compatible with the four tenets outlined earlier. These constructs are intended as an intermediate theoretical frame, in the sense of the sociologists Hedström and Swedberg (1996:281): "attention is called to an intermediary level of analysis in-between pure description and story-telling on the one hand, and universal social laws on the other." But also as defined by Blanton and Fargher (2009:135): "a productive middle path between a 'homo economicus' perspective of methodological individualism on the one hand, and normative determinism, on the other."

Recent philosophy of the social sciences also has pointed in this direction. For example, Mjøset (2009) opines that the social sciences require theories that can account for variation and allow for the contingency of history (see also Kiser and Pfaff 2010:573), calling for the decoupling of unilineality and uniformity from explanatory power. Similarly, Little (2000:89) advocates "explanations that . . . highlight both the structural factors that govern change and the multiple pathways that change can take." Such approaches offer a means to account for "outcomes on the basis of both social context and individual action" so that they are inherently multiscalar (Hechter and Kanazawa 1997:208).

Decades ago in an effort to compare diachronically three ancient Mesoamerican civilizations, the senior author and colleagues (Blanton et al. 1993) advanced a conceptual frame that bears certain parallels to these approaches through a focus on parameters such as societal scale, integration, complexity, and boundedness and the variable relations between them. Frankly, this approach gained little traction because we did not fully spell out the expected theoretical links between these parameters (but see Kowalewski et al. 1983), nor did we give sufficient consideration to human agency, instead conceptualizing more at the societal or systemic scale. Fortunately, in the interim, advances associated with human collective action, cooperation, and social networks have provided a basis for constructing new frames that are specifically designed to account for the diversity of complex societies while also being underpinned by the four tenets outlined earlier.

COOPERATION AND COLLECTIVE ACTION

The empirical relationship between group size and organization long has been recognized. As Dubreuil (2008:203) has observed: "many things are debated about the evolution of human societies, but no bands of 50 individuals have ever created a bureaucracy and no society comprising millions of individuals has remained perfectly egalitarian." While we explicitly do not wish to imply that human cooperation in small groups is "natural," ever-present, or easy, there is an empirical basis to suggest that consistent face-to-face interaction does seem to facilitate it (e.g., Apicella et al. 2012; Fowler and Christakis 2010). Yet in repeated synchronic, cross-cultural studies, intensely interactive groups or communities over several hundred people (Dunbar 2011:table 1) generally have some kind of suprahousehold integrative institutions, while sedentary populations numbering more than 2,000–3,000 are almost always hierarchically organized (Feinman 1995, 1998, 2011) (table 13.3).

Various streams of research tie these new forms of cooperation either to mechanisms associated with the sanctioning of free riders (Baldassarri and Grossman 2011; Dubreuil 2010:166–170; O'Gorman et al. 2009; West et al. 2011) and/or human cognitive constraints (Dunbar 1993, 1998; Johnson 1982; Kosse 1994); the two are not mutually exclusive (Hooper et al. 2010).

And yet, despite the strength of the correlation between group size and hierarchical complexity (Johnson 1982:figure 21.1; see also Bodley 2003; Feinman

TABLE 13.3. Organizational thresholds of human groups

Source	*150–200*	*Source*	*2,000–3,000*
Hill and Dunbar (2003)	150	Carneiro (1967)	2,000
Forge (1972)	150	Forge (1972)	2,000
Adler and Wilshusen (1990)	~200	Sinha (1978)	2,000–3,000
		Kosse (1990)	2,000–3,000
		Johnson (1982)	2,400
		Bernard and Killworth (1973)	2,460
		Brown and Podolefsky (1976)	2,500
		Lekson (1985)	2,500

2011), there is considerable variation when one zooms down to more constrained group size ranges. No magic thresholds have been identified; most importantly, the kinds of social adjustments made by groups with increasing size are by no means uniform (Adler 1989; Adler and Wilshusen 1990; Feinman and Neitzel 1984).

For example, in Bandy's comparative data, the eight regions where states ultimately emerged saw the rise of large villages within a millennium (see figure 13.2a), but the subsequent time lag to the rise of states was much more variable (see figure 13.2b). Also, there were 12 other regions in which large villages arose just as quickly but states did not ultimately develop in the examined time horizon. So one set of factors (including available resources and geographic conditions) may have pushed or pulled humans into large communities, but once there, the specific cooperative adjustments that the occupants of these large communities made played a role in whether and how rapidly (or not) those groups established new forms of more hierarchical arrangements or states. So path dependence and the organizational/institutional choices taken also matter.

In seven of the eight cases where states ultimately arose, multiple large communities were established within 1,000 years (the Valley of Oaxaca is the sole exception in that sample with a primate center). And yet, there also were 11 cases or regions where multiple large communities or peer polities were founded within a millennium, but states did not develop. Competition matters and conceivably even may be necessary, but it alone is not sufficient.

Cooperative arrangements entail social contracts, but the natures of human social contracts are variable. This observation is not new, and for years social

Table 13.4. Variation in modes of finance and leadership

Autocratic	*Collective*	*Reference*
Finance-based big-man	Production-based big-man	Strathern (1969)
Individualizing chiefdom	Group-oriented chiefdom	Renfrew (1974)
Wealth finance	Staple finance	D'Altroy and Earle (1985)
Predatory rule	Quasi-voluntary compliance	Levi (1988)
Exclusionary/network	Corporate	Blanton et al. (1996)
Extractive	Inclusive	Acemoglu and Robinson (2012)

scientists working from different approaches and disciplines have recognized that in human groups, even at comparable levels of vertical complexity, power is worn and wielded in distinct manners and that these differences show a strong relationship to the ways that the funds or bases of power are procured (table 13.4). How rulers amass their resources affects the nature of the social contract and leadership. In other words, by definition, leadership is relational, albeit in different ways (Ahlquist and Levi 2011:5).

Each of the studies listed in table 13.4 contrasts different modes of leadership that parallel in similar ways the distinct manners in which funds of power were amassed. Building on rational choice approaches, the political scientist Margaret Levi (1988) outlined a provocative model that endeavors to account for these patterns theoretically, while also offering a means for addressing the micro-macro problem (e.g., Schelling 2006) that has plagued most archaeological approaches to state formation. Levi's model was devised specifically to understand the development of states, albeit in more contemporary times. Nevertheless, recently, this model has been extended, expanded, and tested through application to a large comparative sample of preindustrial states (Blanton and Fargher 2008; see also Fargher, Heredia Espinoza, and Blanton 2011; Kiser and Cai 2003; Kiser and Linton 2002).

The architects of this model propose that more representative or collective forms of leadership will be found where those with power depend more directly on the local populace for their economic underpinnings, whereas exclusionary/autocratic rule is more apt to occur where leaders rely less on their immediate populace and acquire their funds of power from external sources, such as the monitoring of exchange routes, war booty, or the control of spot resources (figure 13.5). In the latter cases, leaders exact less from their immediate populace and so are freer to afford diminished representation and fewer public goods. In large human cooperative arrangements and institutions,

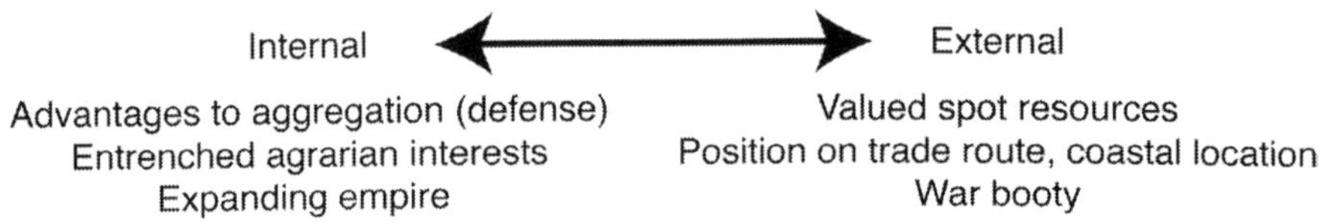

FIGURE 13.5. *Internal and external sources of funds. (Adapted from Blanton and Fargher 2008:254–255.)*

the more rulers depend directly on their immediate sustaining population for their resource support, the more agency and voice that populace is likely to be able to assert and the more public goods are apt to be distributed (Blanton and Fargher 2008:figure 10.2) (figure 13.6).

Revisiting the Peterson-Drennan sample (see figure 13.3), we see that the cases with the highest tax rates were calculated for relatively small polities in which the leaders were neither particularly powerful nor ostentatious/self-aggrandizing. Yet they also note that the higher tax costs were linked to the building of public works and common goods, such as defensive features and irrigation systems (Peterson and Drennan 2012:123–124). Consequently, in accord with the collective action framework, the high-tax case (Pueblo Grande, Hohokam) has both an astonishingly large irrigation system and rather collective forms of leadership/cooperation (see also Ross 2004).

Obviously, this model cannot provide anything close to a complete roadmap to understand this aspect of the diversity of complex societies or early states, since human institutions, such as states, are complicated, and history does matter. Nevertheless, it does present a conceptual and testable basis to understand certain key aspects of the variability in human cooperative institutions. And it links that relationship to the quantities and the means through which resources are acquired by governing authorities as well as the works and goods that are delivered. Here we stress that while such variables as primary productivity, available technologies, and, particularly, population size factor into the kinds and amount of resources these authorities could potentially accumulate, critical too are the specific ways in which (and from where) these acquisitions occurred. In other words, the nature of cooperative arrangements, institutions, and their configurations do matter, and in ways that are potentially explicable.

ADDRESSING EMPIRICAL QUESTIONS AND CONUNDRUMS

One measure of a conceptual frame is whether it can help untangle puzzles and conundrums that are left unexplained by extant paradigms. For example,

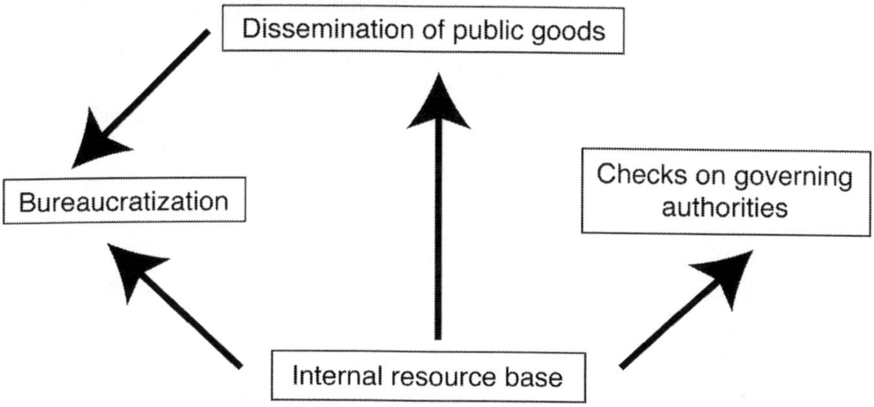

FIGURE 13.6. *Model of collective action. (Adapted from Blanton and Fargher 2008:figure 10.2.)*

if we briefly compare two prehispanic Mesoamerican state societies during the Classic period (ca. AD 200–900), Teotihuacan and the polities of the Classic Maya, we face a number of seeming paradoxes (see Feinman 2001; Feinman and Nicholas 2011 for fuller treatments of this topic). The urban center of Teotihuacan was much larger, with more monumental buildings, and was laid out impressively along a grid system. In contrast, even the largest Classic Maya cities were smaller, with less monumental structures, lower densities, and seemingly less structured urban layouts. And yet, Maya rulers wore their power more ostentatiously, they were buried with highly valued, elaborately crafted goods in special contexts, and they often lived in palaces that most archaeologists can agree as to their function. They also had a highly developed writing system that in large part relates the life histories and real and imagined exploits of named rulers.

In stark contrast, we know the names of no rulers at more monumental Teotihuacan. Their writing system was much more basic, leaving few evidences. Murals at the site depict important figures, but they usually are in groups, wearing masks, and present no details concerning personal history or genealogy (Cowgill 1997:152; Manzanilla 1999:111). Tellingly, if a possible Teotihuacano leader is named, it is through Maya epigraphic studies (Martin and Grube 2008), although there is debate whether this mention represents an individual, office, or military order. Archaeologists cannot agree if the rulers of the site lived in a palace (three completely different locales have been proposed) (Cowgill 1983; Flannery 1998; Sanders and Evans 2006), and researchers keep tunneling into the largest structures to see if they can find a rich ruler's tomb but so far have come up empty.

From a unilinear scale-complexity perspective, one might think that the larger (greater urban scale and density), more monumental polity should have more aggrandizing rulers and more highly developed communication technologies, but that clearly was not the case. Alternatively, some archaeologists have traditionally ascribed these differences to culture or ethnicity; the Maya were this way and Teotihuacanos were another. Of course, that point is valid to a degree, but the later, Postclassic Maya had a different system of rule and cooperation (e.g., Chase and Chase 2006), they used writing differently, and we have not found the degree of aggrandizement evident for the earlier Classic Maya elite.

Clearly, agrarian tribute made up part of the fund of power in both regions, but Classic Maya rulers seem to have relied more than their central Mexican counterparts on external revenues from the exchange of crafted prestige goods and interpersonal networks/elite alliances, and possibly the direct control of spot resources (reservoirs) that they centrally constructed to store water at several major centers (Lucero et al. 2011). In contrast, reliant on internal revenues, the nonelite population at Teotihuacan may have had greater voice and been the recipients of more public works (wide thoroughfares, grid plan layout, large public spaces). Power was more broadly shared, and differentials of access and wealth were flatter. Classic Maya rulers were more flamboyant, and the disparities of wealth appear to have been more fully expressed.

As a working hypothesis, it seems plausible that most first-generation states, at least immediately following their foundations, were underpinned heavily by internal resources, namely the agrarian production of the local population, and hence, more similar in this respect to Teotihuacan. Once established, early state rulers had incentives to provision public goods, such as the construction of defensive features, the building of roads, and the coordination of disaster relief. Flight and various forms of resistance were consistent options for commoners (Van Vugt et al. 2003), and in-migration likely led to potential growth in extractable revenues. Governing authorities may have had few incentives to rule flamboyantly, despotically, or to pursue offensive wars (Kiser and Linton 2002), although defense and the sanction of free riders were often central to their duties as leaders (e.g., Feinman and Neitzel 1984:table 2.6). Of course, as the revenue streams sustaining these polities and the wider worlds in which they were situated always shifted, so could extant social contracts.

Alternatively, as argued for the lowland Maya and also the Mycenaean polities of ancient Greece (Parkinson and Galaty 2007), there are other contexts

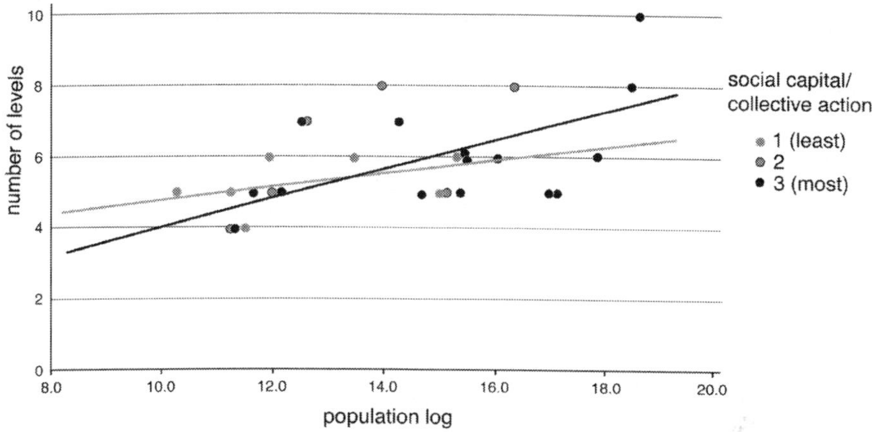

FIGURE 13.7. *Relationship between population size, hierarchical complexity, and social capital/collective action. (Drawn from data in Blanton and Fargher 2008:appendix 3.)*

in which emergent states relied more heavily on external resources. Frequently, these states arose in peer polity networks, in which governing authorities monitored and derived a sizable portion of their wealth from trade and foreign networks as well as the spoils of war. In such instances, power was wielded in comparatively more authoritarian and dynastic ways. Early Egypt, where trade networks going north and south yielded key resources, also may conform to this pattern (Bard 2000; Köhler 2010:40). Fuller assessment of this hypothesis requires better understanding of the economic underpinnings and the revenue-generation practices of early states.

Another conundrum is the vexing variance in the previously discussed relationship between population size and hierarchical complexity. At smaller scales, relatively more collectively organized and tightly interdependent (Curry and Dunbar 2011; Johnson 1982) social groups seem to be able to forestall the formation of new tiers of hierarchical organization, as degrees of cooperation are high. But at larger scales, more collectively organized groups often require greater investments in infrastructure and bureaucracy to maintain and facilitate revenue generation and connectivity over space. Hence these more representative polities often are associated with larger populations, as noted at Teotihuacan and in a sample of historical cases studied by Blanton and Fargher (2008:appendix 3) (figure 13.7). Thus some of the variation in the size-complexity relationship may reflect variance in cooperative social contracts or modes of intergroup connectivity (figure 13.8).

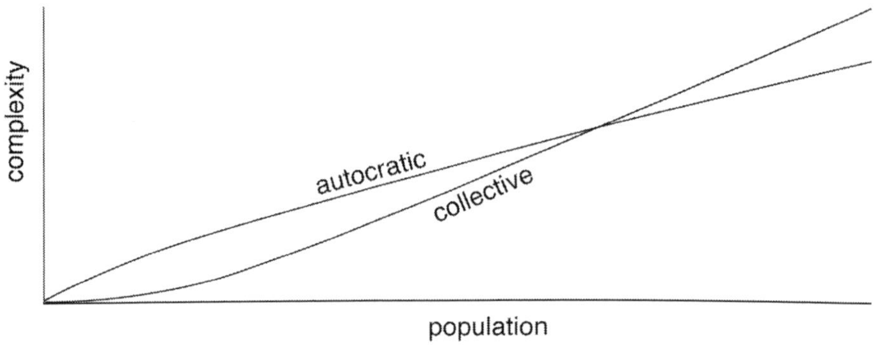

FIGURE 13.8. *Relationship between population size and increasing complexity for collective and autocratic organization.*

CONCLUDING THOUGHTS

Roughly 12,000 years ago, humans embarked on new social arrangements, but these groupings took variable forms and the historical processes that followed were hardly uniform. Understanding the diversity of these social formations and the alternative historical paths that people took in different regions are fundamental elements to understanding why, when, and where complex societies emerged in certain settings but not others.

Through its explicit focus on diversity and its ability to account for change, the approach outlined here is suited to make greater sense of the complexities of early complex society development. That it also provides a "microfoundation for macro-historical and comparative phenomena" (Levi 1988:8) adds to its utility, as the micro-macro problem (e.g., Schelling 2006) rarely has even been addressed by comparative approaches in archaeology.

Working from this frame serves to facilitate the building of theoretical bridges and networks of communication to other disciplines, including other theoretical approaches focused on cooperation (e.g., Kohler et al. 2012) and social networks (e.g., Curry and Dunbar 2011). If the theoretical frames employed to understand the emergence of complex societies were made more broadly compatible with approaches employed to explain the diversity of states over time and space, we suspect that analytical benefits would be realized. At present, our analytical sample and theoretical frames are often constrained by longstanding but poorly justified disciplinary divisions, which utilize distinct theoretical constructs (Gerring 2012:3–4).

Until recently, surprisingly few systematic analyses or conceptualizations have crossed the chasms between classical and anthropological archaeology, prehistory and history, or the recent and the deeper past. If there are solid reasons to isolate certain cases in time and space from broader analytical samples, and at times indeed there may be, then such focused investigations should be based on the precise questions being asked rather than on disciplinary inertia or ossified theoretical tenets and constructs.

As an example, since all larger human cooperative groups, across time and space, were embedded in even wider social networks, we are not convinced that it makes analytical sense to isolate at the outset primary or pristine states from all others (A. T. Smith 2003:83). The corpus of primary states is internally variable itself, and perhaps these examples ought be examined in broader comparative contexts to help define and understand that diversity (Parkinson and Galaty 2007:113).

A recent commentary in the *New York Times* (Gutting 2012) questioned the utility of the social sciences in regard to contemporary political practice and policy because, in the view of the author, research in these fields generally is not constructed to yield predictive results. Certainly, such perspectives are not our overriding concern, nor should we measure our advances simply by this yardstick. And yet, it is clearly unfortunate that our societal institutions and relevant publics often find it difficult to derive practical insights from our discussions drawn from the archives of human histories. In part, the obstacles reflect the manner in which we have presented our findings. If we may hazard an opinion, there may be benefits and insights if the collective "we" made more concerted efforts to frame our concepts and investigations in ways that could potentially facilitate and encourage reasoned practical considerations of the lessons derived from comparative examinations of the past. The frameworks outlined should better position us for just that, while neither ignoring path dependence nor the potential diversity of human relations and institutions.

Acknowledgments. An earlier version of this paper was presented at the Santa Fe Institute. We are grateful to Institute Director Jeremy Sabloff for that rewarding opportunity. Richard Blanton, David Carballo, Mark Golitko, and Steadman Upham offered astute and helpful comments on a draft of the paper. We also thank Verenice Heredia and Lane Fargher for organizing this volume and for inviting us to participate. Finally, we wish to express our great appreciation to Richard Blanton. Over the years, Rich has been an invaluable teacher, mentor, colleague, and inspiration for us, and we are thrilled to have this opportunity to celebrate both the impressive creativity and wide-ranging impact of his research.

14

Pathways to Power

Corporate and Network Strategies, Staple and Wealth Finance, and Primary and Secondary States

Tim Earle

Rich Blanton is a leader of the second generation of processual archaeologists coming out of University of Michigan in the 1970s. Building on the approaches of Kent Flannery, Jeffrey Parsons, and Henry Wright, we believed that social evolution involved systemic change in a broad range of social, political, and economic variables. In the 1980s, just as this cohort of young scholars was coming into its own, a group of "postprocessualists" heavily critiqued the first-generation ideas of Louis Binford (Earle and Preucel 1987). Their harshest criticisms, perhaps, were directed at its unilinear, staged-based formulation of social change. Among processualists, Rich showed leadership in his response, demonstrating fundamental ways by which to formulate a vital processualism. His insight was to focus on process (not typology), which allowed scholars to conceptualize the dynamic and variable structural arrangements of complex human societies (Blanton et al. 1996). For many archaeologists, his work has been inspirational.

Rich's career and mine have parallel intellectual histories. In 1969, I arrived at the University of Michigan, and he was one of the first graduate students whom I got to know. He had recently returned from fieldwork in the Basin of Mexico, and I was fascinated to see how his systematic survey of a large region could yield data to investigate sociopolitical organization. His work was my first lesson on the interdependency between good theory and good data. Following in the footsteps

DOI: 10.5876/9781607325338.c014

of Jeffrey Parsons, with whom Rich had worked in Mexico and whose Basin of Mexico data I had analyzed for my master's paper (Earle 1976), Rich and I independently organized large-scale regional projects in Oaxaca, Mexico, and the Mantaro, Peru. Rich took on a new region that could be compared to Parsons's surveys in the Basin of Mexico, and I took over Parsons's Mantaro Valley project (D'Altroy and Hastorf 2001). The two of us focused on systematic large-scale regional research giving scope to investigate intertwined social, political, and economic processes in complex societies.

From this shared beginning, our careers have had at least three parallels. First was our mutual involvement with the Society for Economic Anthropology. From its beginning in 1981, the SEA has provided a forum for both of us to direct our economy-oriented archaeological research toward a general anthropological audience. Each year, the Society meets to consider a particular economic theme, and Rich's participation in its first annual meeting and landmark publication helped place archaeology firmly within the subdiscipline of economic anthropology (Blanton 1983b). I followed his lead, to become regularly involved in the SEA. Over the years, archaeological participation in the SEA has included many of our collaborators and graduate students, helping to maintain economic anthropology as a core subdiscipline that integrates archaeology, history, cultural, and practice anthropology. Archaeologists in the SEA helped keep alive a strongly materialist understanding on long-term social history, to which Rich and I have been dedicated.

Second has been our parallel commitment to comparative research. Anthropology has a grand tradition of comparison, searching across cultures for common patterns. Rich's book, *Ancient Mesoamerica* (Blanton et al. 1981), compared the archaeological trajectories of three regions, looking especially at broad-scale economic relationships. The success and interest of that volume provided a model for my comparative work on three world regions (Earle 1997) and then on three European regions during the Bronze Age (Earle and Kristiansen 2010). In collaboration with Lane Fargher, Rich conducted a major cross-cultural study of state societies, published as *Collective Action in the Formation of the Pre-Modern State* (Blanton and Fargher 2008; see also Blanton and Fargher 2009; Fargher and Blanton 2007).

Third is our shared focus on understanding the processes of social evolution. Writing independently with coauthors, our seminal articles "A Dual-Processual Theory for the Evolution of Mesoamerican Civilization" (Blanton et al. 1996) and "Ideology, materialization, and power strategies" (DeMarrais et al. 1996) were chosen by *Current Anthropology* (CA) to be published together. Here, Rich developed the distinction between corporate and network strategies as a means by which to organize societies; these alternatives were orthogonal to gradations of complexity, suggesting that social evolution had alternative forms of structuring. His conception of alternative organizing strategies is arguably one of the most influential research directions of the past 20 years. Political organizations must be understood in terms of interpersonal processes that mix and match strategies of power and cooperation. CA's decision to publish our articles together recognized the processual approach that we shared. Subsequently, the corporate-network distinction became central to my work (Earle 2002).

What is to be made of our parallel intellectual histories? Perhaps most important is to illustrate how scholarship develops. Scholars with a common academic background often develop along parallel intellectual lines. The graduate students from Michigan in the 1970s defined shared goals and scholarship focusing on sociopolitical and economic processes to create a distinctive social archaeology. Our similarities were reinforced by repeated contact and interactions at meetings, by our reading of each other's papers, and by a common desire to solve research questions considering the sociopolitical organization of complex societies. The peer interactions in scholarship are evident; a generation, including Rich, Liz Brumfiel, Dick Drennan, and I, among others, followed rather different lines of evidence and interests, but our career paths have been intertwined.

To illustrate how Rich and I have interacted, I focus on our shared theoretical interest in political economy as the means by which emerging elites (or "principals" in Rich's terminology) mobilized resources to finance their operation. My original conception of staple versus wealth finance formed part of his distinction between corporate and network strategies. Here, I argue for the importance of several additional factors that together created the complex pathways for social change. Key factors in the political economy of premodern societies were property rights over "bottlenecks," and warrior complexes that served to control resource mobilization. Thus, the specific relationships among finance strategies, property rights, and warrior complexes affected the overall political-economic continuum from exclusionary to corporate strategies, and this complex overlap of variables appears to have been present in all complex societies from chiefdoms, through primary/secondary states, to modern states. As Lane Fargher commented on a draft of this chapter, where staple finance is associated with commoner claims to rights over landscapes ("internal revenues" in Blanton and Fargher 2008), a more corporate political economy emerges; in contrast, where elites claim personal ownership of productive lands ("external revenues" in Blanton and Fargher 2008), they can mobilize staple finance to fund an exclusionary political economy. The same argument can be made for wealth finance and claims over bottlenecks. Where high-end goods are funneled into markets and taxed like regular goods, the state does not control the bottleneck (internal revenue) and a rather corporate political economy should emerge. Conversely, where the state monopolizes the production or distribution bottleneck (external revenue), we could expect the construction of an exclusionary political economy. Chiefs or kings can deploy warrior specialists to gain or maintain control over bottlenecks. Conversely, state militias or armies in corporate states could be tasked with defense and policing and not with dispossessing commoners for the benefit of a ruling class. Here, I illustrate how political processes created multiple pathways to complexity that resulted in continuous variability across corporate, network, and mixed political economies, showing the need to focus on the processes of social evolution and not evolutionary typologies.

THE POLITICAL ECONOMY IN PREHISTORY

"It's the economy, stupid" is a famous political statement. As a materialist, I believe that economic organization and resulting opportunities for productivity and control gird the development of alternative political institutions. Working with Terry D'Altroy in the 1980s, we formulated a distinction between staple and wealth finance as alternative means by which chiefdoms and states could mobilize surplus. With the publication of Blanton's distinction between corporate and network strategies, I could better conceptualize how the nature of finance was so critically linked to contrasting political strategies.

Staple and Wealth Finance

In my research in Hawai'i and then Peru, I saw that to understand complex societies required an understanding of how resources were mobilized to support ruling institutions (D'Altroy and Earle 1985; Earle and D'Altroy 1989). Although complex societies develop for a plethora of reasons, an absolute necessity was the mobilization of resources to support new institutions of governance and rule. D'Altroy and I defined two alternative means of finance (staple and wealth), and, using the Inca Empire, we showed how these strategies were combined. Our original definitions follow:

> *Staple finance* . . . involves obligatory payments in kind to the state of subsistence goods such as grains, livestock, and clothing. The staples form accounting units

(bushel of wheat or a head of sheep) that have established values. Staples are collected by the state as a share of commoner produce, as a specified levy, or as produce from land worked with corvée labor.

Wealth finance involves the manufacture and procurement of special products (valuable, primitive money, and currency . . .) that are used as a means of payment. These wealth items often have established values with respect to other goods of a similar nature but vary in their convertibility into staples. (D'Altroy and Earle 1985:188)

The mobilization and distribution of staples versus wealth as the currency for finance were quite different. Staples were collected as rent from land owned by elites or their institutions. Because of weight, however, staples could not be moved great distances, and so they were used primarily locally. In contrast, wealth was mobilized by controlling the production and/or exchange of specialty items such as bodily decorations, personal kits (mirrors and razors), weapons, and currency. These items could be produced locally, but their value depended on rarity often tied to foreign origins. To understand how strategies of finance worked, the development and maintenance of these and alternative systems of tribute and taxation all rested on the existence of *bottlenecks*, points in the production and distribution chains that could be controlled. Such bottlenecks were highly variable, but not infinitely so, and they derived from a number of critical conditions (Earle and Spriggs 2015).

For staple finance, the most important bottleneck was land ownership. As originally conceived by David Ricardo (1821), land rent resulted from intensification, which pushed farming into suboptimal lands and required capital improvements such as drainage, terracing, and irrigation on prime lands. For both the use of poorer land and of improved land, the productivity of land in a region would have become sharply differentiated by quality, and the better lands could demand a rent because of their higher than average productivity. Intensification also created a physical marking in the landscape that allowed for the easy division of property. Staple finance was the most basic strategy of mobilization in agrarian chiefdoms and states, and land improved by irrigation is the clearest example of such engineered landscapes creating the foundation for staple finance (Earle and Doyle 2008).

For wealth finance, potential bottlenecks were diverse, but each was subject to different problems of control. The most direct was control over wealth production by specialists attached to the ruling segment (Earle 1987). This bottleneck required a particularly high level of skill so that only a few could produce the wealth items. The manufacture of bronze swords in northern Europe offers a good example for such a controlled production system (DeMarrais et al. 1996; Earle 2004). Alternatively, wealth could be controlled during long-distance trading. If goods moved by water, for example, elites could own the transport vessels. Or chiefs could operate as stationary bandits, positioned strategically at choke points on high-volume routes, such as major rivers. Friedman and Rowlands (1977; Kristiansen 1998) describe the importance of prestige-goods exchange as the basis for social differentiation throughout Europe in the Bronze Age. Exchanges of foreign wealth structured relations among chiefs, whose status was materialized by these special objects.

To analyze the political economy as a means for institutional finance, the researcher must recognize the bottlenecks, how they were created, and how they might be circumvented to disrupt monopoly. The obvious question is: how were contrasting strategies of finance linked to different social formations?

Corporate versus Network Strategies and Political Economy in Pacific Prehistory

Blanton and his colleagues (Blanton et al. 1996) conceptualize corporate and network strategies as alternative political strategies to organize societies, and these structural differences link to various sources of revenues (local vs. external) in the political economy. Political strategies could be combined, changed, and used by alternative political segments to challenge or support each other's search for power. Their original definitions follow:

> In corporate strategies, power is shared across different groups and sectors in society in such a way as to inhibit exclusionary [network] strategies. (Blanton et al 1996:2)

> In network strategies, preeminence is an outcome of the development and maintenance of individual-centered exchange relations primarily outside one's local group. (Blanton et al 1996:4)

I expand on Blanton's recognition of how political differences were linked to political economy by looking at the history of chiefdoms in the Pacific. Chiefdoms are middle-ranged societies that organize populations in the thousands or tens of thousands. Most important is the ability of chiefs to exercise central power. The Pacific illustrates vividly how chiefs came to power in different ways as they crafted their political strategies of expansion and consolidation (Earle and Spriggs 2015). The foundation for Oceanic political strategies appears to have rested originally on prestige-good exchange networks as tied to long-distance feats of valor, but, following colonization, most island societies became focused on corporate strategies and staple mobilization. With the expansion of the Chinese empire, trade in luxury products resulted in the development of "secondary" chiefdoms in the Philippines and elsewhere in the extreme western Pacific. The changing opportunities for control in the political economies of the Pacific can be seen as highly variable and linked to contrasting political strategies involving warrior might and religious legitimation.

In the Bismarck Archipelago through the main Solomon Islands off New Guinea, a regional culture, called Lapita, emerged sometime after 1400 BP (Kirch 1997). Occupying small islands and the coastal fringe of larger islands, Lapita maintained a distinctive maritime orientation. They were involved in fishing, long-distance exploration, trade (obsidian, axes, ceramics) (Kirch 1991), and perhaps raiding (common historically in maritime societies). In conjunction with linguistic evidence, the archaeological record documents many shell valuables and a highly decorated pottery tradition, suggestive of small-scale social hierarchies (Green 2002).

Friedman (1981) argued that Lapita was structured by prestige-goods exchange, similar to that described for European prehistory. Exemplifying social ranking described in the region's ethnographies, status in Lapita culture may have been linked to external ceremonial exchanges of valuables (Kirch 1991). But what were the potential bottlenecks in the prestige-good system that allowed for social differentiation? Where island groups were closely spaced, control over trade and interaction would have been problematic. For more isolated island groups, like the Trobriand Islanders, however, the bottleneck was probably ownership of seagoing craft adequate for successful navigation across broader expanses of open water (Brunton 1975; Earle 2002; Hayden 1983). For example, labor crews and specialists supported by a chief's resources built the elaborate boats used in Kula voyages, and chiefs were their owners (Malinowski 1922). Also, the risks of distant expeditions were seen as requiring elaborate ritual magic for success, and this magic too was the property of the chief. These chiefs were able to dominate trading expeditions, and virtually all valuables went through their hands. In highly elaborate ceremonial displays, chiefs exchanged shell valuables with elite partners as a means to build prestige. We can imagine that such distant voyaging created a small-scale hierarchical structure based both on the special knowledge of ritual and navigation and on the ownership of boats. However, I propose that control over canoes was a rather weak bottleneck, open to many comers, and, thus, limited the degree to which elites could centralize power.

As part of the broadly networked Lapita society, voyagers explored and colonized deep into the Pacific, reaching the islands of Vanuatu, New Caledonia, Fiji, Tonga, and Samoa (Irwin 1992). Exploration was purposeful, part of a strategy to find new resources and establish new venues for political action. Following colonization in the deep Pacific, however, economic opportunities changed to include many isolated islands with relatively greater agricultural potential (Kirch 2000). Partially because of the distance between the major island groups, a network-based trading economy became of much less importance, except perhaps in some Micronesia island groups and in the Tonga archipelago. Alternatively, the opportunities for a highly intensive, staple-based political economy based on corporate strategies became the core of the political economies on the larger islands groups, replacing the earlier network strategies on which Lapita ranking was initially based.

Social evolution of Hawaiian chiefdoms provides a vivid example of one of the pathways for political development. By the time of Western contact, the chiefdoms of several Hawaiian Islands had formed state-like political structures with territorial divisions for taxation and divine rule to legitimize strong central authority (Kirch 2010; Hommon 2013; Earle 2012). Irrigation agriculture was a critical bottleneck in this case (Earle 1980). The intensification of the landscape—especially with irrigation, but also with fishponds; banana, breadfruit, and coconut groves; and dryland fields—created an engineered landscape of walls and spaces that were divided up among chiefs and farmers embedded in an overlapping system of property rights (Earle and Doyle 2008). Commoners received land on an irrigation system or on other agricultural complexes in return for working on plots set aside for the chiefs. The surplus generated supported the chiefs, their attached specialists (including land managers, artisans, warriors, and priests), and labor crews for constructing productive facilities like irrigation systems and fishponds, monuments and roads, and work at major ceremonies (Earle 1978). Such an intensified agrarian base supported the classic corporate structure, providing a particularly clear example of how a property system based on an engineered landscape allowed for staple mobilization. At the same time, because the irrigation systems were built and operated by commoner farmers, one would expect that the commoner could claim moral obligations from their chiefs. Here the role of warriors conquering the land and thus asserting exclusive ownership for the governing paramount chief was essential.

Kirch (1994, 2010) argues that the largest and most complex chiefdoms developed into states on Maui and the Big Island of Hawai'i, where irrigated agriculture was of only secondary importance to dryland farming. These less-productive lands created a more dispersed and lower-intensity agricultural base that was more susceptible to failure (Ladefoged et al. 2009). Initially, it might seem that such a lower-intensity agricultural base would be less likely than an irrigation-based economy to form the footing for state polities. The particular outcome, however, appears to have been linked to expansionist warfare so often identified as the crucible for state formation (e.g., Carneiro 1970). In such situations, warfare, rather than the development of a permanently productive infrastructure, served as the dominant means to increase surplus extraction. Surplus generated from local staple production was used by chiefs to support specialized warriors, who defended the chief, asserted rights to surplus from lands, and expanded by conquest the regions from which surplus could be mobilized. Where a less-intensive productive base existed, warriors thus served to maximize surplus extraction through conquest and in the process broke the ownership of local communities. High-ranking warrior chiefs received local communities as fiefs in return for their loyalty to the paramount. Lower-ranked warriors became local land managers (*konohiki*) and received subsistence land grants, sometimes on newly developed irrigation systems (Earle 1978). State-like polities were made by warrior conquest as a strategy to maximize surplus extraction when agricultural intensification was costly or inefficient. Basically, warrior might provided a means to abrogate the corporate contract in cases where the dispersion of productive resources made it hard for chiefs to control farmers, who produced the surplus for finance.

For a staple finance strategy to work in less-intensive agricultural economies, warriors created an extensive polity based on physical force with a rather "exclusionary" or extractive character that seems to have reversed the typical corporate identity of staple-based polities. Thus, as suggested in conversation with Fargher, territorial conquest converted an agricultural system closely associated with a long history of commoner development into an external revenue source (using Blanton and Fargher's [2008] terms). Following conquest, paramount chiefs were able to treat agricultural lands as personal property. As illustrated by the Hawaiian chiefdoms, the elaboration of an ideology of divine rule may then have been necessary to institutionalize the larger political order. Thus, primary state formation involved a particular hybrid strategy, mixing corporate and exclusionary policies stabilized by an elaborate and theatrical ideology. Following Kolb's (1994) analysis of religious monuments, an early construction of community *heiau* (religious shrine) was probably part of developing a largely corporate strategy for mobilization, but the progressive association of some of these monuments with grand ceremonies involving human sacrifice to the god of war effectively severed this dependent bond between chiefs and commoner farmers.

In the extreme western Pacific, a more typical network political strategy appears to have emerged. As a China-centric trade sphere expanded after 1000 BP, the Philippine Islands witnessed the elaboration of chiefdoms (Junker 1999). Rather than control over land and its staple productions, chiefs depended on, "control over individuals and groups established through the construction and maintenance of alliance networks, which fluctuated in size and composition over time according to the chief's ability to pull individuals and groups into his sphere of influence" (Junker 1990). A local development in ranking propelled local chiefs to search out opportunities to funnel specialty items into the expanding Chinese market. The goal was to obtain valued imports (including Chinese ceramics and metal) that provided the media to expand an already existing prestige-goods exchange network. Although these imports were derived externally, the elaboration of the region's chiefdoms was an internal matter, involving local chiefs controlling the collection of forest products for export and the distribution of Chinese imports. Although most would call the developments of social ranking in the Philippines a response to Chinese contact (thus providing an example of "secondary" chiefdom formation), Junker (1990) makes clear that the Chinese traders provided only expanded opportunities for trade in an existing prestige-goods system. Raiding also was a critical part of the political process, illustrating highly dynamic warrior strategies geared toward directing the movement of prestige goods. The driving forces in Filipino political development were regional, as chiefs actively sought dominance of prestige-goods exchange with a clear network strategy. The existence of this exchange, not the contact with China specifically, drove social evolution.

Oceanic prehistory and societies illustrate a highly variable mix between corporate and network strategies. Chiefdoms arose by developing political economies to generate the surplus in staples or wealth to finance their operations. Chiefdoms were transformed based on available means and opportunities for bottlenecks in resource production and distribution. This variation was further complicated by the active role of warriors in political strategies to control the political economy. My sense is that the conjunction between political economic and warrior strategies might help lay bare the processes energizing emerging institutions along alternative pathways. Although not considered in detail by Blanton's original formation, warfare is perhaps a key element for understanding how different systems of political economy were linked to particular power strategies (see Peregrine and Ember, chapter 12, this volume).

THE EVOLUTION OF PRIMARY AND SECONDARY STATES

Like Blanton, I believe that social evolution should be a central topic of our discipline, and modern archaeology offers the only systematic opportunity to describe the long-term trajectories of change in social evolution. We both reject evolutionary typologies as anything more than heuristic devices and rather seek to investigate comparatively long-term archaeological sequences to discover how common processes resulted in diverse historical developments. When we were in graduate school together, a major focus of research was the attempt to explain the origin of states.

Definitions of states vary, but several traits seem critical. States represent large-scale polities (populations of 100,000s and more) that have diverse ethnic, historic, and economic settings (Johnson and Earle 2000). States and empires were social mosaics spread over large areas, often of thousands of square kilometers, and an ideology of divine rule provided an essential religious superstructure, especially in states dependent on the mobilization of agricultural surplus generated by local commoners (Kirch 2010). The specific characteristics of states were, I believe, less important than the organizational challenges and diverse solutions concerning how to organize a large polity. To effectively control and manage such populations and territories required the development of comprehensive institutions that potentially included a mixture of state religions, bureaucracies, military and police force, judicial bodies, administered economies of irrigation and trade, road systems, and the like. Any premodern state combined some assortment of these institutions, although the particular mix varied for historical and processual reasons. Each state institution had the common job of coordinating large-scale polities, and to accomplish this objective each institution was associated with extensive facilities, personnel, equipment, and special ceremonies. The physical reality of such institutions required a means to finance their operations. In this regard, archaic states were really almost super chiefdoms, which must have elaborated means to finance their expanded reach. States were not simply imagined; they were physically manifest by their new institutions supported through political economies.

A political economy approach should help us understand the emergence and development of states. In their masterful cross-cultural study of states, Blanton and Fargher (2008) propose an extension of the original corporate-network distinction as a means to understand traditional states. They suggest that, when the primary revenue sources for a state were locally derived from its subjects, the state had to provide acceptable services (the common good) to keep their subject populations from resisting tax demands. Alternatively, an external source of revenue for states lessened the dependency of the state on its subjects and allowed for a more self-serving leadership. Their argument derives from a political science approach that considers how states were articulated with the world economy. In the terms originally used by Blanton and myself, in states emphasizing a corporate strategy, commoners control agricultural production and so maintained more of a collective action ethos. However states emphasized more of an exclusionary strategy when rulers claimed these same agricultural systems through right of conquest. Alternatively, the dependency on wealth finance was exclusionary, being based on the control over high-end wealth objects, but a shift toward taxing traders and export producers could create what amounts to a local or corporate revenue source.

Using these insights within a political economy approach, I suggest that the typological distinction between primary and secondary states must be reconsidered. The distinction has long been used rather arbitrarily to suggest that studies of the origin of states must focus on examples of primary states that developed independently. In contrast, secondary states have been of little interest, because their development is seen as an outcome of acculturation (but see Price 1978).

The typological distinction should be rethought in terms of processes of state formation based on how emerging state institutions were financed either internally or externally. Institutions that constitute states "are not exportable" (Price 1978:182). States should not be considered as a trait that can diffuse; they developed both in isolation and equally in association with other states. All chiefly leaders wish to be kings, and the origin of states depended on whether would-be sovereigns could develop systems of finance to support state institutions and to exclude rights of the common horde.

The formation of states was made possible by an ability to finance institutions, based either internally on agricultural intensification (staple finance) or externally on networks of trade and raiding (wealth finance). As ideal types, I believe that primary states were dependent more on local resource mobilization and secondary states were dependent more on external resource procurement. In fact, these strategies represented a graded mix that shifted opportunistically according to operational conditions. Similar to the patterns described for the Oceanic chiefdoms, any state was a fluctuating mixture of power strategies.

Primary States

Primary states emerged where no states had existed before and in isolation from all other states. They are, therefore, considered to be an invention of a new political form, for which researchers can investigate causes of their origin. The normal list of primary states includes Mesopotamia, Egypt, India, China, Mesoamerica, and coastal Peru, and a strong argument can be made to include the Hawaiian Islands as another primary state (Hommon 2013; Kirch 2010). Primary state origins must be studied archaeologically, because they all formed either prior to general writing or in regions isolated from societies with writing. From an archaeological perspective, the characteristics of states and the probable causes for state origins were distinctive to each situation, and a general consensus argues that state origins resulted from interacting variables (Wright 1977). Although I agree with this statement, I would add that primary states all required the sustained and substantial mobilization of resources to support their broadly reaching institutions of power and integration. Although states can be financed in diverse ways, primary states appear to have relied first and foremost on corporate strategies, involving the mobilization of local staples. Relatively high population and agricultural intensification created an engineered landscape (Earle 1978). Perhaps most important were situations like Mesopotamia, Peru, and the Hawaiian Islands, where irrigation agriculture served as a primary basis of intensification; the contrast in both productivity and stability between the wet and the dry effectively caged farmers, creating the bottleneck allowing the mobilization of staples and/or corvée labor as an obligatory rent (Childe 1951; Mann 1986).

Primary states appear to have relied heavily on corporate strategies with staple finance; however, these states additionally developed a military to extend power and maximize surplus extraction. Along coastal Peru, the highly productive and large-scale irrigation agriculture formed the basis for condensed state-like societies that included the Early Intermediate–period Moche and Lima states. But these societies operated much like chiefdoms, without strong political centralization. I have long felt that the city-states of the Middle East or Mesoamerica were variants of chiefdoms, with some incipient institutional characteristics of states made possible by staple finance from their highly productive, irrigation base. Castillo (2007) calls Moche an example of an "opportunistic state," for which central power across a broad region was largely ceremonial with little administration superstructure. In the highlands of Peru, in contrast, irrigation agriculture was initially of much lesser scale and productivity, and this limited staple-based political economy may have created a need to expand the polities' financial base through conquest. Jennings (2011) has argued that the Wari polity developed an expansive empire without first creating a regional state. Conquest makes states, and the logic may rest on the construction of a political economy to suck in new revenue sources across vast areas claimed as the property of the conquering monarch. The local agrarian city-states, with their potential collective-action structure, may have been superseded by imperial conquest that distanced the ruling institutions from their base.

Can you have primary state formation based on trade rather than on staple finance and conquest warfare? Theoretically yes, but practically it does not appear to have happened often. I am particularly intrigued by ongoing work in Micronesia and Tonga that may help resolve our understanding of this issue. I feel, however, that what we are dealing with is a definitional problem: emergent states based on external revenue sources may automatically be classified as "secondary" states. Chiefdoms may emerge based on peer-polity interaction involving control over regional prestige-goods exchanges (Friedman and Rowlands 1977); however, the volume of trade would probably be inadequate to support the needed volume to support a state. Conversely, as primary states expand into empires, they start to use wealth in increasing volume, creating trade systems emanating from their core.

Secondary States

Secondary states emerged within the broader context of existing states, especially when chiefdom-scale societies became linked into external flows of goods that served as a source for wealth finance. The foundational article is "Secondary State Formation: An Explanatory model" (Price 1978). Price envisioned various forms of secondary states.[1] Most important was the expansion of primary (staple-based) states through conquest and intimidation, transforming local chiefdoms or city-states into revenue-producing polities that exported surplus to the dominant imperial state. Her argument seems reasonable and agrees well with the militaristic role of state expansion from a core staple base. Basically, the conquered local polity would have been required to redirect its production, often staple based, toward supporting tributary demands from the external power.

Expanding imperial states alternatively may have targeted areas for conquest where special raw materials or wealth objects were available, and tribute could be extracted and moved across the great distances involved in empires with relatively little transport cost (D'Altroy and Earle 1985). In Mesoamerica, the expanding Classic-period state of Teotihuacan may have targeted regions for conquest with particular resources such as cacao and jade (Braswell 2003; Price 1978:172). The Aztec state subsequently relied on tribute extraction of wealth objects, such as textiles and cacao, from conquered city-states (Berdan et al.

1996). By conquering northwestern Argentina, the Inca Empire targeted a region rich in copper that was extracted for the expanding wealth-finance system of the empire (Earle 1994). These are examples of conquest. State structures were imposed from the outside as a means of direct imperial tribute extraction.

A more important process in "secondary state" formation, however, involved a balanced interaction between central states and periphery developments. States arise on the edge of core agrarian states in situations where direct control and intimidation by the central agrarian states were impractical. Local lords created political economies based on the control of wealth flows toward the core states. The elites in staple-based states, I would argue, increasingly wanted to incorporate wealth finance for a more flexible political economy (D'Altroy and Earle 1985). The desire for wealth as a store of value and as a display of elite distinction that distanced them from the ruled created a heavy demand for both special raw materials and prestige goods from the outside. As status in the core became based on foreign goods, the corporate dependency of the ruling elites on a local populace lessened, resulting in a more class-stratified society, what Blanton refers to as an "exclusionary strategy."

As a result, networks of exchange in high-end prestige goods and materials such as metals would have radiated out into the periphery. Some attempt might well be made by core states to control production and trade through imperial conquest, but I believe that such attempts would be both costly and risky. More common would have been attempts to establish long-distance trading relationships in wealth objects (Stein 1999). This high-end trade then offered the peripheral societies the opportunity to control production and transport in those wealth objects, and thus allowed the elites in the periphery an external source of wealth only weakly tied to underlying staple intensification. As illustrated by the region to the north of Mesopotamia, the result was an aura of trade-based chiefdoms and states arching above the agrarian states in the plain. What Renfrew and Cherry (1986) call "peer polity interaction" would appear to exemplify the emergence of complexity based on network relationships of wealth exchange that offered foreign resource sources for local political economies. Where the movement was extensive and controllable through bottlenecks in production or distribution of the wealth, secondary state formation could be based on control over these networks in a classic exclusionary strategy.

Here the critical article is "Secondary States in Perspective: An Integrated Approach to State Formation in the Prehistoric Aegean" (Parkinson and Galaty 2007). The authors look closely at the Minoan and Mycenaean states that emerged in the Aegean during the Late Bronze Age. Renfrew (1972) had argued that Bronze Age Aegean complexity was autochthonous, resulting from local agrarian intensification and the development of regional networks of wealth exchange. He stressed the primary (independent) development in the Aegean, countering a longstanding belief that European development was simply secondary, a result of contacts with an eastern "cradle of civilization." Parkinson and Galaty (2007:table 2) argue too that the development of Protopalatial Minoan states resulted from local corporate strategies, but that they still represented secondary state formation because of "interaction with Mesopotamia." I don't like the vagueness of "interaction," whether used by Renfrew or Parkinson and Galaty. State institutions were not exportable; instead conditions in the political economy provided the material support for new institutional forms that might be modeled on existing forms seen elsewhere. The emergence of the state-like structures in the Aegean during the Bronze Age was probably based on Minoan and Mycenaean elites being able to control key bottlenecks in the networks of trade in metals and other wealth goods, and then they would have used foreign ideologies to legitimize institutionalization (Earle 2011).

I question whether the distinction between primary and secondary states has real analytical value. Should we reconceive of emerging states as based on common processes that mix and match strategies of local and distant resource mobilizations in the political economy? Thus a "secondary" (wealth-financed) state may well be connected to a "primary" (staple-based) state, and that connection may result in "borrowings" of both symbols and structures of power for state institution building. Processes of finance were the drivers here, not the borrowing of ideas. As Renfrew stressed, it is unclear whether the Aegean linkage to Mesopotamia and Egypt was determining; rather it may simply have been the culmination of control over high-end exchanges that had been developing in Europe throughout the Bronze Age and that had as a part the connection to the imperial societies to the east. The later Iron Age states of the Greeks and Phoenicians were trade-based societies spread across the Mediterranean, competing militarily and commercially to develop and control trade in high-end items that included pottery, glass, metal, wine, and olive oil. Certainly these trading empires were linked into the agrarian states that bordered the Mediterranean, but the bulk of trade and wealth generated was most probably outside of those relationships. The trading empires of the Mediterranean were entrepreneurial and decentralized. The trade-based source of state revenues derived from external sources, but only through local taxation of a class of wealthy traders and estate owners, thus creating effectively a corporate-like strategy.

To maintain and expand state institutions, elites mobilized resources opportunistically by staple and wealth finance. According to logistical concerns for power at a distance, all states became involved in wealth trade as a highly mobile and thus easily centralized means of finance. The ripple effect was to create a world of interacting states more or less reliant on administered trade in high-end goods. These systems can be described as peer-polity interaction, but they appear to me to represent interlinked political economies that created rapidly expanding political spheres of action.

CONCLUSIONS

Rich Blanton is a leading member of a cohort of processual archaeologists focused on complex societies and using a political economy approach to understand organizational differences and change. He and his coauthors provide the foundational distinction between corporate and network strategies for sociopolitical structuring that helped set the agenda for a generation of research. His conception is processual, not typological.

To be able to understand the variation that their distinction represents, I emphasize the necessary role of resource mobilization. Different sources of revenue supported chiefly and state institutions derived from structured political economies, and the ability to channel resource flows from these economies into public coffers stems from controlling specific bottlenecks in the systems. When seeking to explain the emergence of complex institutions, a corporate strategy relied on bottlenecks linked to land tenure. The most dramatic case was chiefly or state ownership of irrigation systems, although any condition of intensive land use linked to a dominant warrior elite created the ability to control local production. A network strategy, in contrast, relied on bottlenecks linked to control over the production and distribution of prestige goods. External flows of goods could be controlled through attached specialization, ownership of transport systems (boats, for example), or channeled routes of high-volume transport. To a greater or lesser extent, warriors might have played a critical role, especially as the scope of political action spread across larger distances through conquest and trade. The importance of a warrior elite probably was responsible for a more exclusionary basis to rule and may have been compensated by increased elaboration of ideologies of divinity. A critical step in the development of trade states appears to have been the abandonment of direct control over trade and rather the imposition of direct taxes on traders and producers that made the sources of revenue essentially local (Blanton and Fargher 2008). But that is another story.

Understanding the contrasting pathways to power must be based on understanding mixed strategies of control over the political economy. This orientation helps to redirect our attention away from typological distinction and toward the comparison of prehistoric sequences and the processes that they represent. Such a research strategy changes the nature of archaeological research, deemphasizing concerns with origins and diffusion. Institutions should not be seen as ideas or technologies that are invented, transferred, and inherited. Rather institutions are complicated organizations based on dynamic and changing political, social, religious, and economic relationships. To understand such organizations, we must focus on the strategies of power and finance that cause the formulation, spread, collapse, and reconfiguration of structured institutions. The longstanding distinction between primary and secondary states can be reconfigured as the contrasting alternative political strategies involved in increasing or decreasing centralization of power through institutions related to economy, force, and belief. Stratified and centralized societies may prove to be quite similar in process, but differentiated by the historical opportunities and challenges that characterize a changing political economy.

NOTE

1. Secondary states included those that were subsequent to earlier states (Price 1978; Parkinson and Galaty 2007). Here the distinctive characteristics involved preexisting institutions of power and finance. But institutions cannot simply be inherited; they must be sustained materially. To understand "subsequent states," as I like to call them, the same analytical approach should apply: we must understand the processes of state governance that rested on institutions of power that required control over bottlenecks in the political economy.

15

Corporate Power Strategies, Collective Action, and Control of Principals

A Cross-Cultural Perspective

Lane F. Fargher

In two pathbreaking publications, Blanton and his colleagues (Blanton 1998a; Blanton et al. 1996) introduced the concept of corporate power strategies as an additional pathway to complexity. In the original *Current Anthropology* article, they somewhat vaguely defined corporate power strategies as the sharing of power "across different groups and sectors of society in such a way as to inhibit exclusionary strategies" (Blanton et al. 1996:2). The definition and identification of corporate power strategies in complex societies had to be vague to some degree precisely because research and theory building in neoevolutionism and much postmodern theory focused nearly exclusively on elites and their despotic strategies (Flannery 1972; Fried 1967; Gilman 1991; Kristiansen 1991; Patterson 1991; Shanks and Tilley 1982; Sanders et al. 1976; Service 1975). In his 1998 book chapter, Blanton set out to some degree to remedy this theoretical deficiency. He defined five characteristics of corporate power (assembly government, corporate regulation of sources of power, "reflexive communication," ritual sanctification of corporate cognitive code, and "semiautonomy" of lower-order subsystems). Building on this work, we (Blanton and Fargher 2008) evaluated the utility of collective action theory for explaining variation in premodern states using a worldwide cross-cultural sample of 30 states. The theory used in this study was largely taken from economics and political science (Levi 1988; Lichbach 1995, 1996; Olson 1965). While Blanton's work on corporate political strategies

DOI: 10.5876/9781607325338.c015

informed the development of our coding scheme, we did not explicitly code for or consider the role of corporate political strategies in collective action. Consequently, we lack a clear understanding of the role of corporate power strategies in the construction of collective polities. For example, does corporate power correlate with collectivity or do they vary independently? Can collectivity arise in the absence of power-sharing?

As such, in this chapter, I set out to better define the relationship between corporate political strategies and collective action. First, I review the philosophical roots of power-sharing in both Europe and China. I selected these two cases because they are the two most important politico-philosophical traditions in human history, as well as being highly sophisticated and antithetical, according to European thinkers. Accordingly, they offer an excellent basis for examining the different ways that state builders confront problems associated with power-sharing. I then use this information to build on and modify Blanton's (1998a) classification of corporate strategies to develop a concise measure of corporate power. Next, I consider the relationship between corporate strategies and collective action using our worldwide sample of premodern states (Blanton and Fargher 2008). Finally, I test my conclusions regarding corporate strategies using two diachronic cases: the rise and decline of the Sui-Tang dynasty and Terminal Formative/Early Classic Valley of Oaxaca. I selected these two empirical cases because they offer a historical and an archaeological perspective on a similar problem faced by state builders (i.e., control of regional or local elites). The results of this study indicate that power-sharing can take various forms (assembly government or bureaucratization) and that it is an effective strategy for control over regional elites.

POWER-SHARING IN TWO CIVILIZATIONAL TRADITIONS

Europe

With the transition from feudalism to the "modern" nation-state, European (and Euro-American) scholars devoted much energy to reworking ideological and political theory to allow for wider participation by citizens in government, to curb the power of the aristocracy and monarchs, and to develop workable models for power-sharing in functioning governments. Assemblies, suffrage, and social contracts were at the center of much of this theorizing (e.g., Hobbes 1651; Jefferson 1999; Locke 2003; Rousseau 1764). The general premise shared by these Enlightenment thinkers emphasized the existence of inherent rights among men (and women, as recognized in modern thinking) living in a state of nature. Thus, the civil state forms when a community of free men gives up certain rights to the body politic in exchange for the protection of other inherent rights. Because all men in the body politic are equal in terms of rights, the limits and powers of the civil state must be set by the members of the body politic through "majority will" (democracy) and cannot be dictated by a monarch or privileged aristocracy. According to Jefferson (1999) and Rousseau (1764), educated men with access to information are able to govern rationally, and it is through reason that they are able to establish order and protect the rights of the members of the body politic. In practice, this theory was translated into men voting on policy and selecting representatives for legislative assemblies.

What emerges from these works is an ideological focus on the equality of men and the importance of merit in the selection of governing officials, especially principals. Thus, these thinkers believed that education, enlightenment, and reason were the capacities that individuals had to acquire in order to govern properly. Moreover, they concluded that legitimate government or the civil state emerges because rational men govern together (shared power) as members of the body politic. Thus, their theory emphasized voice and participatory government as a remedy for monarchy and other forms of totalitarian government that did not respect the inherent rights of free men.

China

Classical Confucian political philosophy shares a number of fundamental concepts regarding power-sharing with its European counterpart. Confucius and Mencius maintained that all men (and women, as recognized in modern thinking) are born with an inherent capacity to do good regardless of social status, but this capacity has to be cultivated to reach maturity (summarized in Chan 2002:3). Through education in ethics, compassion, ritual, social etiquette, and history, a man could become virtuous and live rightly or righteously (rationally) (Rainey 2010:31, 34, 42–43, 45). Thus, Confucius transformed the concept of "gentlemen," which had been applied exclusively to the nobility, to mean a man who lived virtuously (Rainey 2010:42). Those who lived virtuously regardless of social origin could be considered "noble," whereas, those who did not could be considered low (Ivanhoe 2009:21–22, 33, 62, 74–75; Rainey 2010:42).

Mencius went on to espouse the view that, in order to be considered true kings, rulers had to be virtuous and take virtuous men into their service (Ivanhoe 2009:passim; see also Rainey 2010:49). Rulers who did not live virtuously were just lowborn individuals occupying the throne and, thus, could legitimately be overthrown by the people (Ivanhoe 2009:22). Virtuous men, on the other hand, should serve the king and advise him in policy formation. According to Confucius, a good minister constantly admonishes the ruler and points out his errors (Chan 2002:4; Rainey 2010:49). Conversely, the king should work with his ministers, consider their advice, and act wisely upon it. Thus, a respectful relationship arises between the king and his ministers and good government results from virtuous conduct (Ivanhoe 2009:34; Rainey 2010:52).

Moreover, the concept of a social contract is well developed in this tradition. Confucius and Mencius argued that the role of the government was to protect the people and respect their rights (Ivanhoe 2009:3–7, 51–56; Rainey 2010:48, 54). A good king should not interfere with the farmer, the craftsmen, or the merchant. He should create an environment where everyone can flourish, both biologically and socially. Just as the farmer trades grains with the craftsmen for goods, farmers and craftsmen provide their goods as taxes to the king and his ministers in exchange for the service of good government. Yet, as gentlemen, kings and ministers are admonished to collect in tax only what they need to pay the costs of governing and to justly compensate themselves for their services.

The Philosophical Roots of Power-Sharing

From an ideological perspective, the philosophical roots of power-sharing in Europe and China are highly analogous. Both traditions emphasized that an inherent equality existed among all men regardless of social status (Chan 2002:3; Hobbes 1651). For European thinkers, that quality was based in "natural law," whereas Confucius and Mencius identified that quality as the capacity to do good or live virtuously. Both traditions emphasized the idea that education made men eligible to govern (Chan 2002:3; Ivanhoe 2009; Jefferson 1999; Rainey 2010:31). Thus, both traditions emphasized merit over adscription in the selection of governors. Finally, both traditions emphasized that educated men should govern together (share power). For Jefferson (1999) and Rousseau (1764), ruling together took the form of assembly government, whereas the Chinese tradition emphasized a reciprocal relationship between scholar-officials (literati) and a virtuous king (Rainey 2010:42, 49). Thus, both traditions rejected hereditary rule by oppressive and selfish individuals and called for the overthrow of such regimes.

CORPORATE POWER STRATEGIES, COLLECTIVE ACTION, AND CONTROL OF PRINCIPALS

Bureaucratization is consistent with these philosophies (e.g., Levy 2009; Weber 1947:329–334, 392, 404–406; 1978:948), and to some degree both Europeans and the Chinese used it to materialize power-sharing. Bureaucratization serves to fragment power between many individuals (agents and/or principals), limiting the degree that any one individual can monopolize it. Fragmentation is achieved through hierarchization and specialization of duties/responsibilities (e.g., division of powers), the rationalization of bureaucratic routines, and the open and competitive selection of officials based upon merit (Weber 1947:329–341).

Combing the insights from the European and Chinese traditions with Blanton's (1998a) discernments, I propose that corporate power is a complex construct consisting of three variables: (1) the number (or proportion relative to a polity's population) of agents sharing powering within a single administrative territory or division; (2) bureaucratization in the selection and monitoring of agents; and (3) corporate cognitive code development (cf. Blanton 1998a:147). Each variable (which I discuss below) can be conceptualized as ranging from low to high, and, thus, corporate power can be characterized as a continuum. Thus, I propose that states high in corporate power will have large numbers of agents sharing power within single administrative divisions, a high degree of bureaucratization, and well-developed corporate cognitive codes. Conversely, low levels of corporate power can co-occur with network strategies, but higher levels are expected in collective states or where state builders seek to control the agency of principals.

Number (or Proportion) of Individuals Sharing Power

Power-sharing can be achieved by dividing administrative tasks/policy making horizontally (e.g., dividing policing, tax collecting, judicial, and public goods administration into separate hierarchies) (cf. Weber's rational legal authority [1947:328–341]), so that any single administrative or governing agent has limited powers that apply only within a specified political field. Further subdivision of powers along a vertical axis can be implemented to develop clear structural hierarchies that facilitate the capacity of the state to respond to petitions, appeals, and complaints from citizens or officials (cf. Blanton and Fargher 2008:167; Weber 1947:331; 1978:957). Weber referred to such vertical structures as "precise appeal hierarchies" (Weber 1978:957). Another strategy available to state builders is to assign administrative tasks/policy-making to committees or councils to achieve power-sharing (cf. Weber's [1947:392–404] collegiality and Blanton's [1998a:154–155] Assembly or Commonwealth Government). Finally, these strategies, or combinations thereof, can be applied at any hierarchical level from local administrations to principals. The degree to which these strategies are implemented across all hierarchical levels of the state will result in more power-sharing agents and higher degrees of corporate power.

Ideally, this variable should be calculated based on a raw count of the number of officials occupying public offices or administrative positions. However, such information is rarely available for premodern and ancient states. Furthermore, the raw count should be standardized based on the size of a polity's population to facilitate comparison. Ultimately, highly accurate population estimates necessary for such calculations are difficult to ascertain or obtain. Thus, alternatively, coding, as in this case, can be based on qualitative descriptions and the construction of an ordinal variable (see below).

Bureaucratization

Previously, we (Blanton and Fargher 2008:166) developed a specific scale variable to measure bureaucratization "to assess the degree to which an administrative apparatus is consistent with the predicted requirements of collective action." Here, I take a narrower view of bureaucratization (cf. Blanton and Fargher 2008:table 8-1). Thus, bureaucratization in this context is defined as the measure to which recruitment is open and competitive, the degree to which the actions of officials are monitored and evaluated/punished, and the degree to which officials are salaried (cf. Blanton 1998a:159–163; Blanton and Fargher 2008:168–169; Lichbach 1996:167; Weber 1947:312, 335–336; 1978:948, 963–964).

Political agents selected from diverse geographic, ethnic, and socioeconomic backgrounds are expected to be less likely to favor narrow sectorial, class, or family interests (Lichbach 1996:167; Weber 1947:333, 335). Once incorporated into the administrative hierarchy, they are monitored to ensure that they do not use their positions for personal gain or overstep the limits of their posts (e.g., corruption, malfeasance) (Weber 1947:334). Finally, the degree to which officials are paid a salary as opposed to owning or controlling their incomes (e.g., prebends or benefices) provides the state with more flexibility to control and, if necessary, dismiss corrupt officials (Weber 1947:333–336). Again, bureaucratization can be applied at any level in the administrative hierarchy from local officials to principals.

Corporate Cognitive Codes

Here, I define corporate cognitive codes as legal or moral codes that detail behaviors, actions, and ideologies consistent with corporate power strategies (cf. Blanton 1998a:159–163). Although, corporate cognitive codes may take a variety of forms, I identify three as the most important.

1. Ideological constructs that emphasize egalitarianism in the capacity to govern effectively regardless of geographic or socioeconomic origin. Such cognitive codes justify the open and competitive selection of political agents based on merit and serve to reduce the dominance of hereditary nobility (Weber 1947:331, 335).
2. Ideologies that emphasize a shared corporate identity that cross-cuts geographic, tribal, or ethnic divisions. Such ideologies serve to create inclusive identities (e.g., a corporate body politic) that reduce the potential for factions, ethnic divisions, and class distinctions to be used in the development of networks of power.
3. Codes that emphasize proper conduct on the part of governing officials and the equality of citizens in judicial and administrative proceedings (cf. "rationalization of bureaucratic routines" [Weber 1947:328–332, 340, 392]). Built around the previously described ideologies, these codes serve to ensure that political agents comply with power-sharing goals and limit the degree to which certain individuals or sectors of society receive preferential treatment by the state. Where egalitarian legal and administrative codes have been implemented to rationalize bureaucratic functioning, principals cannot act despotically when dealing with citizens because they are enmeshed in structures that require officials to follow codified procedures (Weber 1947:333; cf. Mann 1986:185). Thus, well-developed corporate structures divide and fragment power, making them antithetical to despotism.

A Test of the Corporate Construct

Using the data amassed by the collective action project (Blanton and Fargher 2008), I assembled the quantitative data for my corporate power variable by summing the scores for four of the variables previously developed for bureaucratization (feasibility of commoner appeals and complaints, detection and punishment of malfeasance, officeholder recruitment, and degree of salaried officials) (see Blanton and Fargher 2008:table 8-2; cf. Blanton 1998a:147) and a new variable (degree of horizontal power-sharing) (table 15.1).

TABLE 15.1. Corporate power variables

Variable	*Coding Construct*	*Code in Blanton and Fargher (2008)*
Number of Individuals Sharing Power	Precise Appeal Hierarchy	Feasibility of Commoner appeals and complaints
Number of Individuals Sharing Power	Degree of Horizontal Power-Sharing	
Bureaucratization	Office Holder Recruitment	Office Holder Recruitment
Bureaucratization	Monitoring of Officials	Detection and Punishment of Malfeasance
Bureaucratization	Degree of Salaried Officials	Degree of Salaried Officials

Data were insufficient to code for corporate cognitive code development. The simple variables in my construct for corporate strategies have strong internal correlation (Cronbach's $\alpha = .904$) and thus my construct is empirically valid.

The "degree of horizontal power-sharing" variable was developed from the cultural summaries for each case in the collective action data set. Specifically, I coded the degree to which state builders divided powers among functionally distinct departments or placed power in the hands of supervisory boards or assembles consisting of multiple individuals (table 15.2). I assigned a low score in cases where there was no functional division of powers and individual agents were the exclusive decision-makers in their administrative territories or divisions; an intermediate score when functional specialization or collegiality was present in part of the administrative hierarchy but not across the entire structure; and a high score when functional specialization or collegiality was present throughout the administrative hierarchy.

I then evaluated the relationship among the corporate power construct and the public goods, resource emphasis, and modes of control of principals constructs (the reader is referred to Blanton and Fargher 2008:tables 6–3, 7–2, 9–2 for descriptions because space limitations do not allow for their detailed reproduction in this context). Corporate power strategies correlate strongly with public goods ($r_s = .707, p < .001, n = 30$) and modes of control of principals ($r_s = .754, p < .001, n = 30$), as well as to a lesser degree with resource emphasis ($r_s = .573, p < .001, n = 30$) (figure 15.1). Alternatively, treating resource emphasis as a dichotomous variable (external vs. mixed) verifies the strength of the last correlation. States with an external revenue emphasis score significantly lower on average in corporate power (mean = 7.79, SD = 2.43) as compared with states with an internal or mixed revenue emphasis (mean = 11.1, SD

Table 15.2. Degree of horizontal power-sharing codes

Case	*Degree of Horizontal Power Sharing Code*	*Sources*
Tio	1	Vansina 1973
Bali	1	Geertz 1980; Schulte Nordholt 1991
Aceh	1	Hurgronje 1906; Reid 1975
Perak	1	Gullick 1958
Bakitara	1	Beattie 1971; Roscoe 1923
Pudukkottai	1	Dirks 1987
Nupe	1	Nadel 1942
Japan	2	Bolitho 1991; Hall 1991; Perez 2002; Tatsuya 1991
England	1.5	Cam 1950; Haven Putnam 1950; Holmes 1962; Morris 1940; Morris and Strayer 1947; Waugh 1991; Wilkinson 1940
Bagirmi	1.5	Reyna 1990
Tibet	2	Carrasco Pizana 1959
Thailand	2.5	Rabibhadana 1969
Yoruba	1.5	Law 1977; Lloyd 1971
Vijayanagara	2	Saletore 1934
Inca	2	D'Altroy 2002; Julien 1982; Malpass 1996; Murra 1980
Kuba	2	Vansina 1978a, b
Buganda	1.5	Roscoe 1965; Southwold 1961; Wrigley 1996
Java	2	Moertono 1981
Egypt	2.5	Brier and Hobbs 1999; David 1998; James 1984; Kemp 1989; Kitchen 1982; Montet 1964, 1981; Murnane 1998
Asante	2.5	McCaskie 1995; Rattray 1929; Wilks 1975
Lozi	2	Gluckman 1941, 1943, 1961
Swahili Lamu	2	Horton and Middleton 2000; Prins 1971; Ylvisaker 1979
Ottoman	3	Inalcik 1994; Lybyer 1966
Aztec	3	Davies 1987; van Zantwijk 1985
Burma	3	Koenig 1990
Mughal	3	Habib 1963; Hasan 1936; Sarkar 1963

continued on next page

TABLE 15.2—*continued*

Case	*Degree of Horizontal Power Sharing Code*	*Sources*
Rome	3	Birley 2000; Burton 1996; Eck 2000a, b, c; Galsterer 2000; Griffin 2000; Hopkins 1980; Levick 1996
Venice	3	Lane 1973; Norwich 1982
Ming China	3	Huang 1998; Hucker 1998
Athens	3	Gulick 1973; Hansen 1999

= 2.38) (Kruskal–Wallis Test $s = 143$, $z = -3.07$, $p < .01$; Kolmogorov–Smirnov Asymptotic Test KS = .258, KSa = 1.42, $p < .05$).

Discussion

State builders interested in achieving collective action (e.g., the delivery of public goods in response to taxpayer compliance with internal revenue demands) or limiting the power of principals to act arbitrarily can implement corporate power strategies. In his classic study, van Zantwijk (1985) argued that the *altepetl* (native state of the Basin of Mexico) divided responsibilities for various functions among constituent *calpolli* (neighborhood-scale political subdivisions of the state) in such a way that the various calpolli had to work together in order for the altepetl to function. As such, no single calpolli could dominate the altepetl and make the remaining calpolli subservient.

I think van Zantwijk's insightful description of the altepetl provides a useful analogy for understanding corporate power strategies. They divide power in such a way that political agents are forced to work together, in a reciprocal fashion, in order for the state to function (cf. Weber 1947:393). Thus, state builders employing a corporate strategy create an institutional structure that distributes power among numerous individuals and makes it impossible for any individual political agent, despite his or her position in the hierarchy, to effectively monopolize power because any and all actions must abide by the functional rules of the structure. Consequently, the most powerful agents in the structure (principals) have command over (or direct) the infrastructural power of the state (e.g., Mann 1986), but relinquish personal power in the process.

Conversely, principals wishing to maintain individual power must develop personal relationships with their agents, which can only be achieved by breaking down corporate strategies or operating outside the official structure

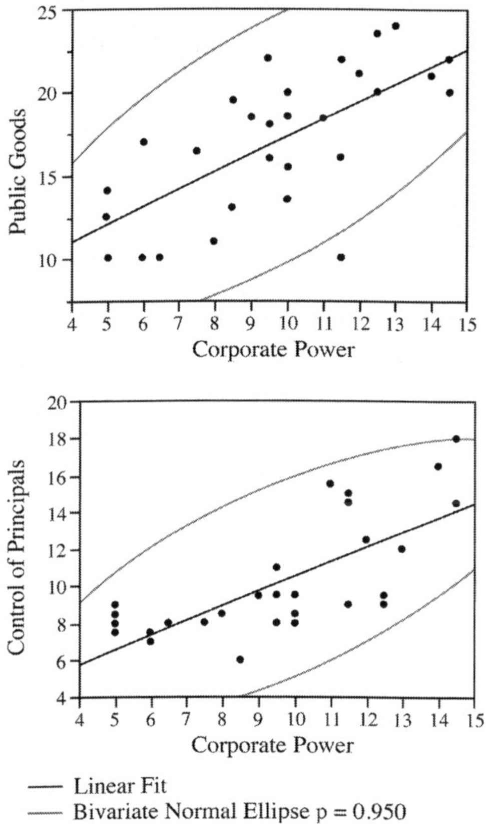

FIGURE 15.1. *Correlation plots for corporate power with (top) public goods and (bottom) control of principals.*

(Blanton et al. 1996). Principals who operate outside the official structure usurp and undermine corporate political agents and can cause collapse if these activities are sufficiently persistent. For example in dynastic China, a persistent problem for corporate state builders (*literati*) was the use of eunuchs as personal retainers by the emperor to bypass the official bureaucratic structure (e.g., Twitchett 1979:16). Thus, the personal ambitions of powerful individuals create a constant tension between corporate power strategies and interpersonal power strategies (e.g., network strategies).

CORPORATE POWER STRATEGIES IN TWO DIACHRONIC CASES

The cross-cultural patterns in corporate power strategies indicate two pathways for sharing power: (1) horizontal functional specialization and the development of precise-appeal hierarchies, and (2) collegiality. Each pathway is coupled with the development of corporate cognitive codes and bureaucratization in the selection of political agents. In this section, I propose that, in some cases, state builders implement corporate strategies to reduce the personal power of elites (e.g., aristocrats) and exclusionary strategies as pathways to power. In the following discussion, I examine this hypothesis with two diachronic cases, Sui-Tang China and the Valley of Oaxaca. Like Feinman and Nicholas (chapter 13, this volume) and Earle (chapter 14, this volume), my case selection focuses on strategies and financing and does not seek to make typological distinctions between pristine and secondary states. States are constantly reconstructed, reorganized, and reformed as principals confront ever-changing problems associated with setting revenue policies, responding to voice, controlling agency, and achieving political goals. Thus, at the very moment a state is funded it is essentially "secondary," and comparisons between any "types" of states in terms of political strategies and financing provides insights into human agency, cooperation, and political relationships.

Sui-Tang Dynasty

Following the disintegration of the Later Han (AD 220–581), political power fragmented among numerous petty kingdoms dispersed across the length and breadth of China. In the north, "powerful local magnates with personal control over extensive lands and numerous clients, families of cultivators and military retainers," dominated local politics (Twitchett 1979:2–3). Kings cobbled together decentralized states by linking these magnates into loose networks. In Southern China, a similar pattern emerged where aristocratic lineages and their generals dominated the courts of petty states (Twitchett 1979:5). Rule during this period was highly unstable and coup after coup placed one lineage after another on the many thrones of China (Twitchett 1979:5, 9).

After AD 581, the Sui dynasty began to consolidate political power through military conquest (Twitchett 1979). However, early on the Tang lineage took control of this process through a coup and completed the unification and centralization of power. In order to bring about integration and political stability, the Tang sought to reduce the power of other aristocratic linages and elite clans (Twitchett 1979:9–12). They did this by implementing a political structure consistent with Confucian political philosophy (Wechsler 1979a:190–191). Central government was divided into three departments (Chancellery, Secretariat, and State Affairs), which in turn were subdivided into numerous bureaus and boards (Twitchett 1979). Provincial and local governments were brought into this administrative structure, creating a "precise appeal hierarchy" (Twitchett 1979:13). An examination system, modeled on that of the Han, was instituted along with state schools and a national university to prepare Confucian scholars for political service (Twitchett 1979:15; Wechsler 1979b:179). This system began to supply Confucian scholar officials by the middle of the second Tang emperor's reign (Twitchett 1979:9, 14–15). These officials were initially placed in prefecture and local government positions, displacing local magnates, and later came to dominate high offices.

All officials, in both the central and local administrations, were subject to monitoring by an independent department, the Censorate (Hwa 2008:9). The second Tang emperor, Tiazong (626–649), also reorganized provincial government, reducing the power of governor-generals by placing much of their supervisory and monitoring responsibilities in "circuits" (Wechsler 1979a:188, 205). Thus, central officials were regularly dispatched from the capital to directly monitor prefectures grouped into circuits. Finally, the emperor personally appointed and reviewed all prefects (Wechsler 1979a:205–206).

Initially, Tiazong worked closely with his ministers to develop and evaluate political policy. A relatively large number of high-level ministers participated in daily council meetings (Twitchett 1979:12–13). Officials were encouraged to submit policies, participate in the debate over them, and to criticize (remonstrate) the emperor (Hwa 2008). The emperor also expected his ministers to accept criticism in turn (Hwa 2008:13). He rewarded officials who remonstrated for the benefit of the state and the people, but severely punished those who agreed with the emperor solely to curry favor or used their positions to enrich/empower themselves. This system of power-sharing was no mere show to legitimate the emperor's rule. All edicts had to be cosigned by both a high minister and the emperor before they could be published (Hwa 2008:10).

Tiazong and his ministers worked to create just and equal legal and administrative procedures. They reformed and rationalized criminal codes, removing harsh punishments (execution or amputation) for many lesser crimes (Twitchett 1979:18; Wechsler 1979b:178). They developed and implemented administrative codes (statues, regulations, and ordinances) to govern the conduct and activities of officials. Both the criminal and administrative codes were regularly updated about every 15 years (Twitchett 1979:18). These reforms created an ideology of equality within Chinese society. For example, landowners and renters entered into contracts on equal terms and responsibilities were limited to the economic sphere, as in rental contracts today in the United States (Twitchett 1979:26).

The system they developed at the beginning of the dynasty persisted with little change until about AD 755. During the reign of Hsuan-tzung (713–755), the central government was modified and the Chancellery and Secretariat were merged into a single department (Twitchett 1979:15). Hsuan-tzung reduced the number of chief ministers and increased their administrative powers, but he excluded them from policy formation. The emperor began using eunuchs and special commissions to bypass the regular bureaucracy, thereby increasing his personal power (Twitchett 1979:15). Finally, after a prolonged rebellion that ended in AD 763, military and financial power were transferred to the governor-generals, who by the end of the eighth century became largely autonomous (Twitchett 1979:17–21). They controlled revenue collection, appointed officials based on personal connections, and directly oversaw military personal. At the local and prefect levels, powerful landlords began to replace Confucian scholar-officials.

Valley of Oaxaca

As described in Joyce and Barber's chapter (chapter 2, this volume), Monte Albán experienced a shift from a corporate political economy in the Middle–Late Formative (500–150 BC) to an exclusionary network-based political economy in the Terminal Formative (150 BC–AD 250) (see also Joyce 2010:ch.5; Winter 1984). At the height of the Terminal Formative, principals at Monte Albán were probably receiving imperial revenues from conquered regions outside the Valley of Oaxaca and funneling them into a prestige economy centered on the hilltop capital (Blanton et al. 1996; Redmond 1981; Spencer 1982). Perhaps one of the most visible material correlates of this pattern are elaborate cream wares decorated with elite iconography, such as step frets, which were produced in and around the North Platform and distributed to powerful provincial elites, probably during elaborate feasts that linked these elites with Monte Albán (Elson 2007:61–66; Elson and Sherman 2007; Joyce 2010:144, 157; Kowalewski et al. 1989:180, 181–182).

However, at the outset of the Classic period (AD 250–750), Monte Albán's exclusionary political economy came under threat. Probably in the face of Teotihuacan's expansion during the Early Classic, Monte Albán's imperial domain outside the Valley of Oaxaca apparently collapsed, and with it, imperial revenues (Redmond 1981; Spencer 1982). At the same time, principals at Monte Albán faced an increasingly powerful regional elite that dominated emerging secondary centers across the valley (Elson and Sherman 2007; Kowalewski et al. 1989:180, 181–182). Faced with shrinking revenues and rising competition from provincial elites, principals at Monte Albán implemented corporate political strategies to reestablish control over the valley. Here, I postulate that the central political strategy of the Early Classic was the development of bureaucratic structures, which centralized power but divided it among many officials (cf. the assembly government of the Late Formative [see Joyce and Barber. chapter 2, this volume]). Essentially, comparatively faceless and impersonal infrastructural power replaced the personal power and prestige-goods system (exclusionary strategies) of the Terminal Formative. These changes were coupled with a shift from external revenues to internal revenues and, probably, a more collective state. In the following discussion, I marshal archaeological and epigraphic data to support my conjecture.

Terminal Formative settlement patterns indicate a relatively decentralized regional organization. Large settlement nodes (segments) dominated the eastern and northern sections of the valley and smaller nodes proliferated at the edges of the valley, in the southern section and around Monte Albán (Kowalewski et al. 1989:162, figure 7.2). Settlement nucleation increased and relatively more people lived in defensive hilltop sites than in previous periods (Kowalewski et al. 1989:153–156). High-status individuals distributed small amounts of luxury cream ware ceramics through political channels (Kowalewski et al. 1989:171, 173, 180–181). Mound groups throughout the Valley of Oaxaca were relatively closed but showed little uniformity in orientation or layout (Blanton 1989:444). The impression generated by these data is of a segmentary state integrated through elite interaction, including the exchange of prestige goods, the construction of elaborate residences, two-room temples, and tombs, as well as frequent warfare (cf. Kowalewski et al. 1989:199).

Excavations at three of the most important secondary sites occupied during this period indicate that political agents enclosed formerly open public spaces and placed burials either under central temples (e.g., Monte Albán, Dainzu; see also offerings in Structure 35 at San José Mogote) or possibly under private "plazas" (e.g., Cerro Tilcajete) (Bernal and Oliveros 1988; Elson 2007:40; Marcus and Flannery 1996:185; Martínez López et al. 1995:237). Apparently, an emergent hereditary elite "walled off" public spaces and appropriated them as personal and ancestral property. Thus, these elites were claiming large, extremely elaborate architectural complexes that included palaces, temple mounds, and plazas as personal patrimony.

The transition from the Terminal Formative to the Early Classic (AD 250–500) in the Valley of Oaxaca was marked by a massive political reform that extinguished the careers of the provincial elite. The decline of these elites is evidenced by the abandonment of the elite architectural complexes controlled by these individuals and a major shift in settlement locations (e.g., Dainzu, Magdalena Apasco, Tejalapan, San José Mogote, Cerro Tilcajete, Tlatinango, Tlalixtac), probably as a result of state policy that created new bureaucratic centers in their place (Kowalewski et al. 1989:156, 206, 229, table 8.4). Thus, the most important Early Classic centers were newly founded (or were only occupied by hamlets during the preceding Terminal Formative) (e.g., Jalieza, Tlacochahuaya, El Palmillo, Santa Cruz Mixtepec, Tilquiapan) and none were seats of entrenched local power (Kowalewski et al. 1989:229, table 8.5). A massive area encompassing Monte Albán, all of the southern arm of the valley, and the central portion of the eastern arm of the valley now formed the core of a highly integrated regional system (Kowalewski et al. 1989:210, figure 8.1).

The new system deemphasized elite lineages, as evidence by the paucity of highly elaborate Early Classic tombs in excavations at Monte Albán, El Palmillo, San Jose Mogoté, and Dainzu-Macuilxochitl (one possible exception is Monte Albán Tomb 112) (Bernal and Oliveros 1988; Feinman and Nicholas 2003, 2007, 2011a, 2011b; Marcus and Flannery 1996:230). Construction on the Main Plaza at Monte Albán continued and the architectural arrangement became more formal and more closed (Joyce 2010:figure 7.2). Several high-status residences were built along the edges of the Main Plaza at this time (Joyce 2010:218). However tombs remained confined to household contexts and were not placed in monumental platforms (cf. the Terminal Formative and Epiclassic tombs in Oaxaca).

Ruler glorification during the Early Classic was also minimal. Urcid (2005:22) identified one small shrine that showed a named ruler (Lord 5 Jaguar) and his father with 14 unnamed "secondary personages" (possibly bureaucrats governing sections of Monte Albán). At the end of the Early Classic, Lord 13 Night dismantled the shrine and used the carved stones in a resplendent construction commemorating his enthronement and military success (Joyce 2010:213; Urcid 2005:22). However, this building was also dismantled, sometime during the Late Classic, and yet another ruler reused the stones in the foundation of the South Platform (Urcid 2005:23). I argue that repeated dismantling of these monuments suggests a lack of hereditary continuity in rule throughout the Classic. I would expect that persistence in lineages would have resulted in preservation of previous monuments and the construction of new ones alongside them or the careful burying of previous monuments (cf. the North Acropolis and Temples 1 and 2 at Tikal [Sharer and Traxler 2006:302–303]).

Away from Monte Albán, public architecture, although exhibiting a high degree of conformity in orientation in accordance with centralized control, emphasized numerous open and accessible plazas possibly associated with functionally specialized buildings that were not personal residences of political officials (Blanton 1989:444, 446). Consistent with the deemphasis of tombs and more open public architecture, elaborate cream wares were replaced by standardized and relatively plainer gray wares (e.g., G-23 and G-35) that were widely distributed across the Valley, as well as across socioeconomic classes (Caso et al. 1967:311; Kowalewski et al. 1989:213–214).

Hereditary power only again became a dominant theme in architecture and iconography with the transition to the Late Classic/Epiclassic (AD 500–900). High-status lineages began to commemorate genealogies in mural paintings within elaborate residential tombs constructed below houses (e.g., Tombs 104 and 105). Highly elaborate tombs abound from this period at Monte Albán (e.g., Tombs 103, 104, 105, 112, Atzompa Edificio 6) as well as many other sites (e.g., Reyes Etla, El Palmillo, Lambityeco, Yagul) (Bernal and Gamio 1974; Bernal and Oliveros 1988; Feinman and Nicholas 2003, 2007, 2011a, 2011b; Joyce 2010:204; Lind and Urcid 2010:ch. 7; Marcus and Flannery 1996:211–216; Miller 1995:68–73; Robles García et al. 2014; Urcid 2005:50, 63–64, table 1-1). At provincial sites, political elites begin to encroach on the open plazas of the Early Classic, reducing their area and closing them off with buildings containing subfloor tombs (e.g., El Palmillo [Feinman and Nicholas 2003, 2007, 2011a, 2011b]). Also, Terminal Formative elite complexes at Dainzu and Cerro Tilcajete were reoccupied and new tombs were constructed (Bernal and Oliveros 1988; Elson 2007:89, 92).

I argue that the patterns in architecture, epigraphy, and artifacts suggest that power was centralized during the Early Classic using corporate strategies that broke down traditional provincial elite domains. Based on the cross-cultural comparative data presented above as well as the specific case of the Tang dynasty, I hypothesize that corporate strategies centered on the establishment of horizontal divisions of power and the development of precise-appeal hierarchies that bypassed traditional elite power bases. Within this context, open and competitive selection for "bureaucratic" positions would have been an important strategy to reduce hereditary power. The decline of venerable elite lineages throughout the valley (evidenced by the abandonment of elite architectural complexes), the dramatic reorganization of the regional political system with entirely new administrative seats, and the largely faceless nature of power throughout the Early Classic correspond with this hypothesis.

CONCLUSIONS

Power-sharing is an important feature of many states across cultures and through time and is not limited to modern European nation states. In this chapter, I used power-sharing philosophies from two distinct cultural traditions (Europe and China) to guide the development of a cross-cultural measure of corporate (shared) power. I concluded that corporate strategies involve the division of power among functional specialized hierarchies or assemblies, as well as open selection and monitoring of officials and the development of corporate cognitive codes that emphasize egalitarian administrative and judicial philosophies (Unfortunately extant data are insufficient to evaluate corporate cognitive codes, indicating an area that requires more intensive cross-cultural research). I then evaluated the correlation among corporate power strategies and collective action, as well as control of the behavior of principals, using the 30-state sample coded for the collective action project (Blanton and Fargher 2008). The results indicate that corporate political strategies are strongly correlated with both collective action and control of principals. I then went on to examine the deployment of corporate power strategies and their results in Sui-Tang–dynasty China and Terminal Formative/Early Classic Valley of Oaxaca. In both cases, corporate strategies helped to centralize power and reduce the influence of hereditary provincial elites. The outcome in both cases was a high degree of power-sharing that lasted for about 200 to 250 years.

References

Acemoglu, D., and J. A. Robinson. 2012. *Why Nations Fail: The Origins of Power, Prosperity, and Poverty*. New York: Crown Business.

Acuña, R. 1988. *Relaciones Geográficas del Siglo XVI: Nueva Galicia*. México DF: Universidad Autónoma de México.

Adams, J. W., and A. B. Kasakoff. 1975. "Factors Underlying Endogenous Groups Size." In *Regional Analysis: Volume II, Social Systems*, ed. C.A. Smith, 149–173. New York: Academic Press.

Adams, R. McC. 1966. *The Evolution of Urban Societ;. Early Mesopotamia and Prehispanic Mexico*. Chicago: Aldine.

Adler, M. A. 1989. "Ritual Facilities and Social Integration in Nonranked Societies." In *The Architecture of Social Integration in Prehistoric Pueblos*, ed. W. D. Lipe and M. Hegmon, 35–52. Cortez: Crow Canyon Archaeological Center.

Adler, M. A., and R. H. Wilshusen. 1990. "Large-Scale Integrative Facilities in Tribal Societies: Cross-Cultural and Southwestern U.S. Examples." *World Archaeology* 22(2):133–146. http://dx.doi.org/10.1080/00438243.1990.9980136.

Agertz, J. 2000. "Ortnamn och Bebyggelsehistoria i Gränna-Visingsöbygden (Placenames and Settlement History in the Gränna-Visingsö District)." In *Visingsöartiklar. tolv Artiklar om Visingsö från Bronsålder till Medeltid*, ed. P. Nicklasson, 111–118. Jönköping: Jönköpings läns Museum.

Ahlquist, J. S., and M. Levi. 2011. "Leadership: What It Means, What It Does, and What We Want to Know About It." *Annual Review of Political Science* 14(1):1–24. http://dx.doi.org/10.1146/annurev-polisci-042409-152654.

DOI: 10.5876/9781607325338.c016

Aimers, J., and P. M. Rice. 2006. "Astronomy, Ritual, and the Interpretation of Maya 'E-Group' Architectural Assemblages." *Ancient Mesoamerica* 17(01):79–96. http://dx.doi.org/10.1017/S0956536106060056.

Alva Ixtlilxochitl, F. de 1965. *Obras Históricas*. 2 vols. Ed. A. Chavero. México DF: Editorial Nacional.

Ames, K. M. 1995. "Chiefly Power and Household Production on the Northwest Coast." In *Foundations of Social Inequality*, ed. T. D. Price and G. M. Feinman, 155–187. New York: Plenum. http://dx.doi.org/10.1007/978-1-4899-1289-3_6.

Anderson, A.J.O., F. F. Berdan, and J. Lockhart. 1976. *Beyond the Codices*. Berkeley: University of California Press.

Anderson, K., C. S. Beekman, and V. Y. Heredia Espinoza. 2013. "The Ex-Laguna de Magdalena and pre-Columbian Settlement in Jalisco, Mexico: The Integration of Archaeological and Geomorphological Datasets." Paper presented in "Human-Environment Interactions in the Neotropics: Historical Impact to Current Challenges" Conference, Royal Geographical Society (with the Institute of British Geographers), London.

Andrén, A. 2000. "Against War! Regional Identity Across a National Border in Late Medieval and Early Modern Scandinavia." *International Journal of Historical Archaeology* 4(4):315–334. http://dx.doi.org/10.1023/A:1009556518747.

Aoyama, K. 1999. *Ancient Maya State, Urbanism, Exchange, and Craft Specialization: Chipped Stone Evidence of the Copan Valley and the La Entrada Region, Honduras*. Pittsburgh: Memoirs in Latin American Archaeology, University of Pittsburgh.

Aoyama, K. 2011. "Socioeconomic and Political Implications of Regional Studies of Maya Lithic Artifacts: Two Case Studies of the Copan Region, Honduras, and the Aguateca Region, Guatemala." In *The Technology of Maya Civilization: Political Economy and Beyond in Lithic Studies*, ed. Z. X. Hruby, G. E. Braswell, and O. Chinchilla Mazariegos, 37–54. Oakville: Equinox Publishing, Ltd.

Apicella, C. L., F. W. Marlowe, J. H. Fowler, and N. A. Christakis. 2012. "Social Networks and Cooperation in Hunter-Gatherers." *Nature* 481(7382):497–501. http://dx.doi.org/10.1038/nature10736.

Apostel, L. 1961. "Towards the Formal Study of Models in the Non-Formal Sciences." In *The Concept and Role of the Model in Mathematics and Natural and Social Sciences*, ed. H. Freudenthal, 1–37. Dordrecht, Holland: D. Reidel Publishing Company. http://dx.doi.org/10.1007/978-94-010-3667-2_1.

Appadurai, A. 1986. "Introduction: Commodities and the Politics of Value." In *The Social Life of Things: Commodities in Cultural Perspective*, ed. A. Appadurai, 3–63. Cambridge: Cambridge University Press. http://dx.doi.org/10.1017/CBO9780511819582.003.

Appel, J. 1986. "A Central-Place Analysis of Classic and Late Postclassic Settlement Patterns in the Valley of Oaxaca." *Research in Economic Anthropology, Supplement 2: Economic Aspects of Prehispanic Highland Mexico*, 375–418.

Archer, Margaret. 1982. "Morphogenesis vs. Structuration: On Combining Structure and Action." *British Journal of Sociology* 33(4):455–483. http://dx.doi.org/10.2307/589357.

Archer, Margaret. 1995. *Realist Social Theory: The Morphogenetic Approach*. Cambridge: Cambridge University Press. http://dx.doi.org/10.1017/CBO9780511557675.

Archer, Margaret. 2000. *Being Human: The Problem of Agency*. Cambridge: Cambridge University Press. http://dx.doi.org/10.1017/CBO9780511488733.

Arhem, K., K. Homewood, and W. A. Rodgers. 1981. *A Pastoral Food System: The Ngorongoro Maasai in Tanzania*. Dar Es Salaam: University of Dar Es Salaam.

Ashmore, W., and J. A. Sabloff. 2002. "Spatial Orders in Maya Civic Plans." *Latin American Antiquity* 13(2):201–215. http://dx.doi.org/10.2307/971914.

Ashmore, W., and J. A. Sabloff. 2003. "Interpreting Ancient Maya Civic Plans: Reply to Smith." *Latin American Antiquity* 14(2) 229–236. http://dx.doi.org/10.2307/3557597.

Athukorala, P., and S. Jayasuriya. 1994. *Macroeconomic Policies, Crises, and Growth in Sri Lanka, 1969–90*. Washington, DC: The World Bank. http://dx.doi.org/10.1596/0-8213-2297-4.

Axboe, M. 1999. "Towards the Kingdom of Denmark." *Anglosaxon Studies in Archaeology and History* 10:109–118.

Baldassarri, D., and G. Grossman. 2011. "Centralized Sanctioning and Legitimate Authority Promote Cooperation in Humans." *Proceedings of the National Academy of Sciences of the United States of America* 108(27):11023–11027. http://dx.doi.org/10.1073/pnas.1105456108.

Balkansky, A. K. 2002. *The Sola Valley and the Monte Albán State: A Study of Zapotec Imperial Expansion*. Ann Arbor: Museum of Anthropology, University of Michigan.

Bandy, M. S. 2008. "Global Patterns of Early Village Development." In *The Neolithic Demographic Transition*, ed. J.-P. Bocquet-Appel and O. Bar-Yosef, 333–357. Berlin: Springer. http://dx.doi.org/10.1007/978-1-4020-8539-0_13.

Bandy, M. S., and J. W. Janusek. 2005. "Settlement Patterns, Administrative Boundaries, and Internal Migration in the Early Colonial." In *Period Advances in Titicaca Basin Archaeology*, ed. C. Stanish, A. B. Cohen, and M. S. Aldenderfer, 267–288. Los Angeles: Cotsen Institute of Archaeology, UCLA.

Barber, S. B. 2013. "Defining Community and Status at Outlying Sites During the Terminal Formative Period." In *Polity and Ecology in Formative Period Coastal Oaxaca*, ed. A. A. Joyce, 165–192. Boulder: University Press of Colorado.

Barber, S. B., and A. A. Joyce. 2007. "Polity Produced and Community Consumed: Negotiating Political Centralization in the Lower Río Verde Valley, Oaxaca." In *Mesoamerican Ritual Economy*, ed. E. C. Wells and K. L. Davis-Salazar, 221–244. Boulder: University Press of Colorado.

Barber, S. B., and M. Olvera Sánchez. 2012. "A Divine Wind: The Arts of Death and Music in Terminal Formative Oaxaca." *Ancient Mesoamerica* 23(01):9–24. http://dx.doi.org/10.1017/S0956536112000016.

Bard, K. A. 2000. "The Emergence of the Egyptian State (c. 3200–2686)." In *The Oxford History of Ancient Egypt*, ed. I. Shaw, 61–88. Oxford: Oxford University Press.

Bargatzky, T. B. 1995. "Evolution, Sequential Hierarchy, and Areal Integration: The Case of Traditional Samoan Society." In *State and Society: The Emergence and Development of Social Hierarchy and Political Centralization*, ed. J. Gledhill, B. Bender, and M. T. Larsen, 41–54. New York: Routledge.

Barrett, J., R. Beukens, I. Simpson, P. Ashmore, S. Poaps, and J. Huntley. 2000. "What Was the Viking Age and When Did It Happen? A View from Orkney." *Norwegian Archaeological Review* 33(1):1–38. http://dx.doi.org/10.1080/00293650050202600.

Barry, H., III, L. Josephson, E. Lauer, and C. Marshall. 1976. "Traits Inculcated in Childhood: Cross-Cultural Codes 5." *Ethnology* 15(1):83–106. http://dx.doi.org/10.2307/3773082.

Baus de Czitrom, C. 1982. *Tecuexes y Cocas. Dos Grupos de la Región Jalisco en el Siglo XVI.* México DF: INAH.

Beals, R. 1932. *The Comparative Ethnology of Northern Mexico before 1750*. Berkeley: University of California Press.

Beals, R. 1933. *The Acaxee: A Mountain Tribe of Durango and Sinaloa.* Berkeley: University of California Press.

Beattie, J.H.M. 1959. "Rituals of Nyoro Kingship." *Africa: Journal of the International Africa Institute* 29(02):134–145. http://dx.doi.org/10.2307/1157516.

Beattie, J.H.M. 1960. *Bunyoro: An African Kingdom*. New York: Holt, Rinehart, and Winston.

Beattie, J.H.M. 1971. *The Nyoro state*. Oxford: Clarendon Press.

Beaudry, M. 1984. *Ceramic Production and Distribution in the Southeastern Maya Periphery: Late Classic Painted Serving Vessels.* BAR International Series. Oxford: British Archaeological Reports.

Beck, R. A., ed. 2007. *The Durable House: House Society Models in Archaeology*. Carbondale: Center for Archaeological Investigations, Southern Illinois University.

Beekman, C. S. 1996. "The Long-Term Evolution of a Political Boundary: Archaeological Research in Jalisco, México." PhD diss., Department of Anthropology, Vanderbilt University. Ann Arbor, MI: University Microfilms.
Beekman, C. S. 2000. "The Correspondence of Regional Patterns and Local Strategies in Formative to Classic Period West Mexico." *Journal of Anthropological Archaeology* 19(4):385–412. http://dx.doi.org/10.1006/jaar.1999.0354.
Beekman, C. S. 2003a. "Agricultural Pole Rituals and Rulership in Late Formative Central Jalisco." *Ancient Mesoamerica* 14(02):299–318. http://dx.doi.org/10.1017/S0956536103141016.
Beekman, C. S. 2003b. "Fruitful Symmetry: Corn and Cosmology in the Public Architecture of Late Formative and Early Classic Jalisco." *Mesoamerican Voices* 1:5–22.
Beekman, C. S. 2005. "Agency, Collectivities, and Emergence: Social Theory and Agent Based Simulations." In *Nonlinear Models for Archaeology and Anthropology: Continuing the Revolution*, ed. C. Beekman and W. Baden, 51–78. Aldershot: Ashgate Press.
Beekman, C. S. 2007. "Transition and Trajectory: The Late Formative to Classic Shift in the Highland Lakes District of Central Jalisco." Paper present at the 72nd Annual Meeting of the Society for American Archaeology, Austin.
Beekman, C. S. 2008. "Corporate Power Strategies in the Late Formative to Early Classic Tequila Valleys of Central Jalisco." *Latin American Antiquity* 19:414–434.
Beekman, C. S. 2009a. "Los Sistemas Políticos del Formativo en los Valles de Tequila, Jalisco y su Relación con la Subsistencia." In *Las Sociedades Complejas del Occidente de México en el Mundo Mesoamericano: Homenaje a Dr. Phil C. Weigand*, ed. E. Williams, L. López, and R. Esparza, 75–95. Zamora: Colegio de Michoacán.
Beekman, C. S. 2009b. "Estimating Population in Pre-Columbian Central Jalisco." Unpublished manuscript on file, Department of Anthropology, University of Colorado, Denver.
Beekman, C. S., and W. W. Baden. 2011. "El Cultivo del Maíz y su Impacto Regional: Agotamiento de los Suelos en el Corredor de La Venta, Jalisco." In *Patrones de Asentamiento y Actividades de Subsistencia en el Occidente de México. Reconocimiento a la Dra. Helen P. Pollard*, ed. Eduardo Williams and Phil C. Weigand, 351–382. Zamora: Colegio de Michoacán.
Beekman, C., G. Tyndall, B. Benz, S. Jennings, K. Beekman, and B. Calgaro. 2004. *Public Architecture: Navajas, Jalisco, Mexico*. Foundation for Ancient Mesoamerican Studies, Inc. http://www.famsi.org/reports/02029/.
Beekman, C. S., and P. C. Weigand. 2000. *La Cerámica Arqueológica de la Tradición Teuchitlán, Jalisco: Tipología, Análisis Petrográfico y Cronología*. Trans. E. Williams. Guadalajara, Zamora: Secretaría del Estado de Jalisco/El Colegio de Michoacán.

Beekman, C. S., and P. Weigand. 2008. "Conclusiones, Cronología y un Intento a Síntesis." In *Tradición Teuchitlán*, ed. P. Weigand, C. Beekman, and R. Esparza, 303–337. Zamora: Colegio de Michoacán.

Bentley, J. W. 1990. "Wouldn't You like to Have All of Your Land in One Place? Land Fragmentation in Northwest Portugal." *Human Ecology* 18(1):51–79. http://dx.doi.org/10.1007/BF00889072.

Bentley, R. A., and S. J. Shennan. 2003. "Cultural Transmission and Stochastic Network Growth." *American Antiquity* 68(3):459–485. http://dx.doi.org/10.2307/3557104.

Berdan, F. F. 1986. "The Tributary Provinces." In *Aztec Imperial Strategies,* by F. F. Berdan et al., 115–135. Washington, DC: Dumbarton Oaks.

Berdan, F. F. 1992. "The Imperial Tribute Roll of the *Codex Mendoza*." In *The Codex Mendoza*, ed. F. F. Berdan and P. R. Anawalt, 55–79. Berkeley: University of California Press.

Berdan, F. F. 2005. *The Aztecs of Central Mexico: An Imperial Society*. Belmont: Wadsworth.

Berdan, F. F. 2007. "The Technology of Ancient Mesoamerican Mosaics: An Experimental Investigation of Alternative Super Glues." Foundation for the Advancement of Mesoamerican Studies, Inc. http://www.famsi.org/reports/06015/06015Berdan01.pdf.

Berdan, F. F. 2012. "Turquoise in the Aztec Imperial World." In *Turquoise in Mexico and North America*, ed. J.C.H. King, M. Carocci, C. Cartwright, C. McEwan, and R. Stacey, 91–102. London: Archetype Publications.

Berdan, F. F. 2014. *Aztec Archaeology and Ethnohistory*. Cambridge: Cambridge University Press. http://dx.doi.org/10.1017/CBO9781139017046.

Berdan, F. F., and P. R. Anawalt, eds. 1992. *The Codex Mendoza*. 4 vols. Berkeley: University of California Press.

Berdan, F. F., R. E. Blanton, E. H. Boone, M. G. Hodge, M. E. Smith, and E. Umberger. 1996. *Aztec Imperial Strategies*. Washington, DC: Dumbarton Oaks.

Bernal, I., and L. Gamio. 1974. *Yagul: El palacio de los seis patios*. México DF: IIA, UNAM.

Bernal, I., and A. Oliveros. 1988. *Exploraciones arqueológicas en Dainzú, Oaxaca*. México DF: INAH.

Bernard, H. R., and P. D. Killworth. 1973. "On the Social Structure of an Ocean-Going Research Vessel and Other Important Things." *Social Science Research* 2(2):145–184. http://dx.doi.org/10.1016/0049-089X(73)90016-1.

Berry, B.J.L. 1991. *Long-Wave Rhythms in Economic Development and Political Behavior*. Baltimore: Johns Hopkins University Press.

Bikunya, P. 1927. *Ky'Abakama ba Bunyoro*. London: Sheldon Press.

Binford, L. R. 1962. "Archaeology as Anthropology." *American Antiquity* 28(2):217–225. http://dx.doi.org/10.2307/278380.

Birley, A. R. 2000. "Hadrian to the Antonines." In *The Cambridge Ancient History, Volume 11: The High Empire, A.D. 70–192*, ed. A.K. Bowman, P. Garnsey, and D. Rathbone, 132–194. Cambridge: Cambridge University Press. http://dx.doi.org/10.1017/CHOL9780521263351.004.

Blackman, M. J., and M. Vidale. 1992. "The Production and Distribution of Stoneware Bangles at Mohenjo-daro and Harappa as Monitored by Chemical Characterization Studies." In *South Asian Archaeology 1989*, ed. C. Jarrige, 37–44. Madison: Prehistory Press.

Blanco, E. 2009. *El Juego de Pelota en la Tradición Teuchitlán: Hacia una Propuesta Sobre su Función Social.* Tesis de Maestría, Centro de Estudios Arqueológicos, El Colegio de Michoacán, La Piedad.

Blanton, R. E. 1972. "Prehispanic Adaptation in the Ixtapalapa Region, Mexico." *Science* 175(4028):1317–1326. http://dx.doi.org/10.1126/science.175.4028.1317.

Blanton, R. E. 1975. "Texcoco Region Archaeology." *American Antiquity* 40(2):227–235. http://dx.doi.org/10.2307/279620.

Blanton, R. E. 1976. "Anthropological Studies of Cities." *Annual Review of Anthropology* 5(1):249–264. http://dx.doi.org/10.1146/annurev.an.05.100176.001341.

Blanton, R. E. 1978. *Monte Alban: Settlement Patterns at the Ancient Zapotec Capital.* New York: Academic Press.

Blanton, R. E. 1983a. "Advances in the Study of Cultural Evolution in Prehispanic Highland Mesoamerica." *Advances in World Archaeology* 2:245–288.

Blanton, R. E. 1983b. "Factors Underlying the Origin and Evolution of Market Systems." In *Economic Anthropology: Topics and Theories*, ed. S. Ortiz, 51–66. Lanham, MD: University Press of America.

Blanton, R. E. 1985. "A Comparison of Early Market Systems." In *Markets and Marketing*, ed. S. Plattner, 399–416. Lanham, MD: University Press of America.

Blanton, R. E. 1989. "Continuity and Change in Public Architecture: Periods I through V of the Valley of Oaxaca, Mexico." In *Monte Albán's Hinterland, Part II: Prehispanic Settlment Patterns in Tlacolula, Etla, and Ocotlan, the Valley of Oaxaca, Mexico*, vol. 1, edited by S. A. Kowalewski et al., 409–447. Ann Arbor: Museum of Anthropology, University of Michigan.

Blanton, R. E. 1994. *Houses and Households: A Comparative Study*. New York: Plenum Press. http://dx.doi.org/10.1007/978-1-4899-0990-9.

Blanton, R. E. 1995. "The Cultural Foundations of Inequality in Households." In *Foundations of Social Inequality*, ed. T. D. Price and G. M. Feinman, 105–127. New York: Plenum Press. http://dx.doi.org/10.1007/978-1-4899-1289-3_4.

Blanton, R. E. 1996. "The Basin of Mexico Market System and the Growth of Empire." In *Aztec Imperial Strategies*, by F. F. Berdan, R. E. Blanton, E. H. Boone, M. G. Hodge, M. E. Smith, and E. Umberger. 47–84. Washington, DC: Dumbarton Oaks.

Blanton, R. E. 1998a. "Beyond Centralization: Steps Toward a Theory of Egalitarian Behavior in Archaic States." In *Archaic States*, ed. G. M. Feinman and J. Marcus, 135–172. Santa Fe: School of American Research.

Blanton, R. E. 1998b. "Comment." *Current Anthropology* 39:463–464.

Blanton, R. E. 2000. *Hellenistic, Roman and Byzantine Settlement Patterns of the Coast Lands of Western Rough Cilicia.* BAR International Series. Oxford: British Archaeological Reports.

Blanton, R. E. 2010. "Collective Action and Adaptive Socioecological Cycles in Premodern States." *Cross-Cultural Research* 44(1):41–59. http://dx.doi.org/10.1177/1069397109351684.

Blanton, R. E. 2013. "Cooperation and the Moral Economy of the Marketplace." In *Merchants, Markets, and Exchange in the Pre-Columbian World*, ed. K. G. Hirth and J. Pillsbury, 23–48. Washington, DC: Dumbarton Oaks.

Blanton, R. E., J. Appel, L. M. Finsten, S. K. Kowalewski, G. M. Feinman, and E. Fisch. 1979. "Regional Evolution in the Valley of Oaxaca, Mexico." *Journal of Field Archaeology* 6:369–390.

Blanton, R. E., and L. F. Fargher. 2008. *Collective Action in the Formation of Pre-Modern States*. New York: Springer. http://dx.doi.org/10.1007/978-0-387-73877-2.

Blanton, R. E., and L. F. Fargher. 2009. "Collective Action in the Evolution of Pre-Modern States." *Social Evolution & History* 8:133–166.

Blanton, R. E., and L. F. Fargher. 2010. "Evaluating Causal Factors in Market Development in Premodern States: A Comparative Study." In *Archaeological Approaches to Market Exchange in Ancient Societies*, ed. C. P. Garraty and B. L. Stark, 207–226. Boulder: University Press of Colorado.

Blanton, R. E., and L. F. Fargher. 2011. "The Collective Logic of Pre-Modern Cities." *World Archaeology* 43(3):505–522. http://dx.doi.org/10.1080/00438243.2011.607722.

Blanton, R. E., and L. F. Fargher. 2012. "Neighborhoods and the Civic Constitutions of Premodern Cities as Seen from the Perspective of Collective Action." In *The Neighborhood as a Social and Spatial Unit in Mesoamerican Cities*, ed. M. C. Arnauld, L. R. Manzanilla, and M. E. Smith, 27–52. Tucson: The University of Arizona Press.

Blanton, R. E., and L. F. Fargher. 2016. *How Humans Cooperate: Confronting the Challenges of Collective Action*. Boulder. University Press of Colorado.

Blanton, R. E., L. F. Fargher, and V. Y. Heredia Espinoza. 2005. "The Mesoamerican World of Goods and its Transformations." In *Settlement, Subsistence, and Social*

Complexity: Essays Honoring the Legacy of Jeffrey R. Parsons, ed. R. E. Blanton, 260–294. Los Angeles: Cotsen Institute of Archaeology, UCLA.
Blanton, R. E., and G. M. Feinman. 1984. "Mesoamerican World System." *American Anthropologist* 86(3):673–682. http://dx.doi.org/10.1525/aa.1984.86.3.02a00100.
Blanton, R. E., G. M. Feinman, S. A. Kowalewski, and L. M. Nicholas. 1999. *Ancient Oaxaca*. Cambridge: Cambridge University Press. http://dx.doi.org/10.1017/CBO9780511607844.
Blanton, R. E., G. M. Feinman, S. A. Kowalewski, and P. N. Peregrine. 1996. "A Dual-Processual Theory for the Evolution of Mesoamerican Civilization." *Current Anthropology* 37(1):1–14. http://dx.doi.org/10.1086/204471.
Blanton, R. E., S. A. Kowalewski, G. M. Feinman, and L. M. Finsten. 1993. *Ancient Mesoamerica: A Comparison of Change in Three Regions*. 2nd ed. New York: Cambridge University Press.
Blanton, R. E., S. A. Kowalewski, G. M. Feinman, and J. Appel. 1982. *Monte Albán's Hinterland, Part I: The Prehispanic Settlement Patterns of the Central and Southern Parts of the Valley of Oaxaca, Mexico*. Ann Arbor: Museum of Anthropology, University of Michigan.
Blanton, R. E., S. A. Kowalewski, G. M. Feinman, and L. M. Finsten. 1981. *Ancient Mesoamerica: A Comparison of Change in Three Regions*. New York: Cambridge University Press.
Blanton, R. E., and J. Taylor. 1995. "Patterns of Exchange and the Social Production of Pigs in Highland New Guinea: Their Relevance to Questions About the Origins and Evolution of Agriculture." *Journal of Archaeological Research* 3(2):113–145. http://dx.doi.org/10.1007/BF02231435.
Blarel, B., P. Hazell, F. Place, and J. Quiggin. 1992. "The Economics of Farm Fragmentation: Evidence from Ghana and Rwanda." *World Bank Economic Review* 6(2):233–254. http://dx.doi.org/10.1093/wber/6.2.233.
Block, F., and P. Evans. 2005. "The State and the Economy." In *The Handbook of Economic Sociology*, ed. N. J. Smelser and R. Swedberg, 505–526. Princeton: Princeton University Press.
Bodley, J. H. 2003. *The Power of Scale: A Global History Approach*. Armonk: M. E. Sharpe.
Boehm, C. 1993. "Egalitarian Behavior and Reverse Dominance Hierarchy." *Current Anthropology* 34(3):227–240. http://dx.doi.org/10.1086/204166.
Bolitho, H. 1991. "The Han." In *The Cambridge History of Japan, Volume 4, Early Modern Japan*, ed. J.W. Hall, 183–234. Cambridge: Cambridge University Press.
Bondarenki, D. M., and A. A. Nemirovskiy. 2007. "Alternativeness in Cultural History: Heterarchy and Homoarchy as Evolutionary Trajectories." In *Proceeding of*

Third International Conference "Hierarchy and Power in the History of Civilizations"' Moscow: Russian Academy of Sciences Press.

Bonnet, C. 1990. *Kerma Royaume de Nubie: L'Antiquité Africaine au Temps des Pharaohs*. Geneva: Mission Archéologique de L'Université de Genève au Soudan.

Bonnet, C. 1992. "Excavations at the Nubian Royal Town of Kerma: 1975–91." *Antiquity* 66(252):611–625. http://dx.doi.org/10.1017/S0003598X00039338.

Boone, E. H. 2000. *Stories in Red and Black: Pictorial Histories of the Aztecs and Mixtecs*. Austin: University of Texas Press.

Borstein, J. A. 2001. "Tripping over Colossal Heads: Settlement Patterns and Population Development in the Upland Olmec Heartland." Unpublished PhD dissertation, Department of Anthropology, Pennsylvania State University, University Park.

Borstein, J. A. 2005. "Epiclassic Political Organization in Southern Veracruz, Mexico: Segmentary Versus Centralized Integration." *Ancient Mesoamerica* 16(01):11–21. http://dx.doi.org/10.1017/S095653610505008X.

Boulakia, J.D.C. 1971. "Ibn Khald⊠n: A Fourteenth-Century Economist." *Journal of Political Economy* 79(5):1105–1118. http://dx.doi.org/10.1086/259818.

Bourdieu, P. 1986. "The Forms of Capital." In *Handbook of Theory of Research for the Sociology of Education*, ed. J. Richardson, 241–258. Westport: Greenword Press.

Bourdieu, P. 1990. *The Logic of Practice*. Palo Alto: Stanford University Press.

Bourdieu, P. 1993. *The Field of Cultural Production: Essays on Art and Literature*. New York: Columbian University Press.

Bourdieu, P. 1998. *The State Nobility: Elite Schools in the Field of Power*. Palo Alto: Stanford University Press.

Bourdieu, P. 2004. [1997]. "From the King's House to the Reason of State: A Model of the Genesis of the Bureaucratic Field." *Constellations (Oxford, England)* 11(1):16–36. http://dx.doi.org/10.1111/j.1351-0487.2004.00359.x.

Bramm, A. 2001. "Cross-Cultural Evaluation of the Corporate-Network Strategy: A Study of Human Sociopolitical Organization." BA honor's thesis, Lawrence University, Appleton.

Braswell, G. E. 2003. *The Maya and Teotihuacan: Reinterpreting Early Classic Interaction*. Austin: University of Texas Press.

Braswell, G. E. 2010. "The Rise and Fall of Market Exchange: A Dynamic Approach to Ancient Maya Economy." In *Archaeological Approaches to Market Exchange in Ancient Societies*, ed. C. P. Garraty and B. L. Stark, 127–140. Boulder: University Press of Colorado.

Braswell, G. E., and M. D. Glascock. 2011. "Procurement and Production of Obsidian at Calakmul." In *The Technology of Maya Civilization: Political Economy and Beyond in Lithic Studies*, ed. Z. X. Hruby, O. Chinchilla, and G. Braswell, 119–129. Oakville: Equinox Press.

Braudel, F. 1970. "History and the Social Sciences: The Long Run." *Social Sciences Information. Information Sur les Sciences Sociales* 9(1):144–174. http://dx.doi.org/10.1177/053901847000900107.

Brier, B., and A. H. Hobbs. 1999. *Daily Life of the Ancient Egyptians*. Westport: Greenwood Press.

Brink, S. 1984. "Ortnamn Som Källa I Historisk Forskning (Placenames as Sources for Historic Research)." *Fornvännen* 79:165–172.

Brink, S. 1990. *Sockenbildning Och Socknnamn. Studier I Äldre Territoriell Indelning I Norden (Parish Formation and Parish Names. Studies in Early Territorial Divisions in Scandinavia)*. Uppsala: Acta Academiae Regiae Gustavi Adolphi LVII.

Brink, S. 2002. "Law and Legal Customs in Viking Age Scandinavia." In *Scandinavians from the Vendel Period to the Tenth Century*, ed. J. Jesch, 87–127. Woodbridge: Boydell Press.

Brink, S. 2003. "Legal Assemblies and Judicial Structure in Early Scandinavia." In *Political Assemblies in the Earlier Middle Ages*, ed. P. Barnwell and M. Mostert, 61–72. Turhout: Brepols. http://dx.doi.org/10.1484/M.SEM-EB.3.1930.

Brink, S. 2004. "Legal Assembly Sites in Early Scandinavia." In *Assembly Places and Practices in Medieval Europe*, ed. A. Pantos and S. Semple, 205–216. Dublin: Four Courts Press.

Brown, P., and A. Podolefsky. 1976. "Population Density, Agricultural Intensity, Land Tenure, and Groups Size in the New Guinea Highlands." *Ethnology* 15(3):211–238. http://dx.doi.org/10.2307/3773132.

Brumfiel, E. M. 1987. "Consumption and Politics at Aztec Huexotla." *American Anthropologist* 89(3):676–686. http://dx.doi.org/10.1525/aa.1987.89.3.02a00090.

Brumfiel, E. M. 1991. "Agricultural Development and Class Stratification in the Southern Valley of Mexico." In *Land and Politics in the Valley of Mexico: A Two Thousand Year Perspective*, ed. H. Harvey, 43–62. Albuquerque: University of New Mexico Press.

Brumfiel, E. M. 1992. "Breaking and Entering the Ecosystem: Gender, Class and Faction Steal the Show." *American Anthropologist* 94(3):551–567. http://dx.doi.org/10.1525/aa.1992.94.3.02a00020.

Brumfiel, E. M., and J. Fox, eds. 1994. *Factional Competition and Political Development in the New World*. Cambridge: Cambridge University Press. http://dx.doi.org/10.1017/CBO9780511598401.

Buchanan, C. A. 1974. *The Kitara Complex: The Historical Tradition of Western Uganda to the Sixteenth Century.* Ann Arbor: UMI Dissertation Services.

Burton, G. P. 1996. "Senate, Imperial Age." In *The Oxford Classical Dictionary*, ed. S. Hornblower and A. Spawforth, 1386–1387. Oxford: Oxford University Press.

Brunton, R. 1975. "Why do the Trobriands Have Chiefs?" *Man* 10(4):544–558. http://dx.doi.org/10.2307/2800132.

Butler, B., and P. Welch, eds. 2006. *Leadership and Polity in Mississippian Society*. Carbondale: Southern Illinois University Press.

Butler, M. 2011. "Excavaciones en Charco Redondo, 2009." In *Proyecto Río Verde, 2009*, ed. S. B. Barber and A. A. Joyce, 185–221. Final report submitted to the Consejo de Arqueología, Instituto Nacional de Antropología e Historia, Mexico City.

Butterwick, K. 1998. "Food for the Dead: The West Mexican Art of Feasting." In *Ancient West Mexico: Art and Archaeology of the Unknown Past*, ed. R. Townsend, 89–106. Chicago: Art Institute of Chicago.

Butterwick, K. 2005. Heritage of Power: Ancient Sculpture from West Mexico. The Andrall E. Pearson Family Collection. New Haven: Yale University Press.

Byland, B. D., and J.M.D. Pohl. 1994. *In the Realm of Eight Deer: The Archaeology of the Mixtec Codices*. Norman: University of Oklahoma Press.

Byland, B. E. 1980. "Political and Economic Evolution in the Tamazulapan Valley, Mixteca Alta, Oaxaca, México: A Regional Approach." PhD dissertation, Department of Anthropology, Pennsylvania State University, University Park.

Cam, H. M. 1950. "Shire Officials: Coroners, Constables, and Bailiffs." In *The English Government at Work, 1327–1336, Volume 3: Local Administration and Justice*, ed. J. F. Willard, W. A. Morris, and W. H. Dunham Jr., 143–183. Cambridge: The Mediaeval Academy of America.

Cap, B. 2011. "Investigating an Ancient Marketplace at Buenavista del Cayo, Belize." *Research Reports in Belizean Archaeology* 8:251–253.

Carballo, D. M., ed. 2013. *Cooperation and Collective Action: Archaeological Perspectives*. Boulder: University Press of Colorado.

Carmack, R. M. 1973. *Quichean Civilization: The Ethnohistoric, Ethnographic, and Archaeological Sources*. Berkeley: University of California Press.

Carmack, R. M. 1981. *The Quiché Mayas of Utatlán: The Evolution of a Highland Guatemala Kingdom*. Norman: University of Oklahoma Press.

Carneiro, R. L. 1962. "Scale Analysis as an Instrument for the Study of Cultural Evolution." *Southwestern Journal of Anthropology* 18(2):149–169. http://dx.doi.org/10.1086/soutjanth.18.2.3629014.

Carneiro, R. L. 1967. "On the Relationship Between Size of Population and Complexity of Social Organization." *Southwestern Journal of Anthropology* 23(3):234–243. http://dx.doi.org/10.1086/soutjanth.23.3.3629251.

Carneiro, R. L. 1968. "Ascertaining, Testing, and Interpreting Sequences of Cultural Development." *Southwestern Journal of Anthropology* 24(4):354–374. http://dx.doi.org/10.1086/soutjanth.24.4.3629356.

Carneiro, R. L. 1970. "A Theory of Origin of the State." *Science* 169(3947):733–738. http://dx.doi.org/10.1126/science.169.3947.733.
Carneiro, R. L. 1987. "Cross-Currents in the Theory of State Formation." *American Ethnologist* 14(4):756–770. http://dx.doi.org/10.1525/ae.1987.14.4.02a00110.
Carneiro, R. L. 1988. "The Circumscription Theory: Challenge and Response." *American Behavioral Scientist* 31(4) 497–511. http://dx.doi.org/10.1177/000276488031004010.
Carrasco Pizana, P. 1959. *Land and Polity in Tibet.* Seattle: University of Washington Press.
Carrasco Vargas, R., V. A. Vázquez López, and S. Martin. 2009. "Daily Life of the Ancient Maya Recorded on Murals at Calakmul, Mexico." *Proceedings of the National Academy of Sciences of the United States of America* 106(46):19245–19249. http://dx.doi.org/10.1073/pnas.0904374106.
Caso, A., I. Bernal, and J. R. Acosta. 1967. *La Cerámica de Monte Albán.* México DF: INAH.
Castilleja, A. 2011. "La Configuración del Sistema de Intercambio entre los Purépecha como Factor de Intercambio y Persistencia." In *Patrones de Asentamiento y Actividades de Subsistencia en el Occidente de México: Reconocimiento a la Doctora Helen P. Pollard*, ed. E. Williams and P. C. Weigand, 265–280. Zamora: El Colegio de Michoacán.
Castillo, L. J. 2007. "Moche Politics in the Jequetepeque Valley: A Case for Political Opportunism." In *New Perspectives on Moche Political Organization*, ed. J. Quilter and L. J. Castillo, 83–109. Washington, DC: Dumbarton Oaks.
Cederholm, M. 2007. "De värjde sin rätt. Senmedeltida Bondemotstånd i Skåne och Småland (They Defended Their Rights: Late Medieval Peasant Resistance in Scania and Småland)." PhD dissertation, Department of History, Lunds Universitet, Lund.
Chan, A.K.L. 2002. "Introduction." In *Mencius: Contexts and Intepretations*, ed. A.K.L. Chan, 1–16. Honolulu: University of Hawai'i Press.
Chandler, T. 1987. *Four Thousand Years of Urban Growth: An Historical Census.* Lewiston, NY: St. David's University Press.
Chapman, R. 2003. *Archaeologies of Complexity.* London: Routledge. http://dx.doi.org/10.4324/9780203451779.
Charsley, S. 1970. "Mobility and Village Composition in Bunyoro." *Uganda Journal* 34:15–27.
Chase, A. F., and D. Z. Chase. 1987. *Investigations at the Classic Maya City of Caracol, Belize: 1985–1987.* San Francisco: Pre-Columbian Art Research Institute.
Chase, A. F., and D. Z. Chase. 1996. "More than Kin and King: Centralized Political Organization among the Late Classic Maya." *Current Anthropology* 37(5):803–110. http://dx.doi.org/10.1086/204564.

Chase, A. F., and D. Z. Chase. 2012. "Belize Red Ceramics and their Implications for Trade and Exchange in the Eastern Maya Lowlands." *Research Reports in Belizean Archaeology* 9:3–14.
Chase, A. F., and D. Z. Chase, eds. 1992. *Mesoamerican Elites: An Archaeological Assessment.* Norman: University of Oklahoma Press.
Chase, D. Z., and A. F. Chase. 2006. "Framing the Maya Collapse: Continuity, Discontinuity, Method, and Practice in the Classic to Postclassic Southern Maya Lowlands." In *After Collapse: The Regeneration of Complex Societies*, ed. G. M. Schwartz and J. J. Nichols, 168–187. Tucson: University of Arizona Press.
Cheek, C. D. 2003. "Maya Community Buildings: Two Late Classic *Popal Nahs* at Copan, Honduras." *Ancient Mesoamerica* 14(01):131–138. http://dx.doi.org/10.1017/S0956536103141028.
Chesson, M. S. 2003. "Households, Houses, Neighborhoods and Corporate Villages: Modeling the Early Bronze Age as a House Society." *Journal of Mediterranean Archaeology* 16(1):79–102. http://dx.doi.org/10.1558/jmea.v16i1.79.
Childe, V. G. 1934. *New Light on the Most Ancient East: The Oriental Prelude to European Prehistory.* London: K. Paul, Trench, Trubner & Co., Ltd.
Childe, V. G. 1950. "The Urban Revolution." *Town Planning Review* 21(1):3–17. http://dx.doi.org/10.3828/tpr.21.1.k853061t614q42qh.
Childe, V. G. 1951. *Man Makes Himself.* London: Watts.
Childe, V. G. 1964. *What Happened in History?* New York: Penguin Books.
Chomsky, N. 2003. *Hegemony or Survival.* New York: Henry Holt and Company.
Christaller, W. 1966. *Central Places in Southern Germany.* Trans. C. Baskin. New Jersey: Prentice-Hall, Englewood Cliffs.
Christophersen, A. 1982. "Drengs, Thegns, Landmen, and Kings: Some Aspects on the Forms of Social Relations in Viking Society during the Transition to Historic Times." *Meddelanden från Lunds Universitets Historiska Museum* 1981–1982:115–134.
Clarke, D. L., ed. 1972. *Models in Archaeology.* London: Methuen.
Coburn, N. 2011. *Bazaar Politics: Power and Pottery in an Afghanistan Market Town.* Stanford: Stanford University Press.
Coe, M. D. 1962. *Mexico.* New York: Praeger.
Collis, J. 1997. "Dynamic, Descriptive and Dead-End Models: Views of an Ageing Revolutionary." In *Reconstructing Iron Age Societies*, ed. A. Gwilt and C. Haselgrove, 297–302. Oxford: Oxbow.
Cook, S. 1966. "The Obsolete 'Anti-Market' Mentality: A Critique of the Substantive Approach to Economic Anthropology." *American Anthropologist* 68(2):323–345. http://dx.doi.org/10.1525/aa.1966.68.2.02a00010.

Coppock, D., J. Ellis, and D. Swift. 1986. "Livestock Feeding Ecology and Resource Utilisation in a Nomadic Pastoral Ecosystem." *Journal of Applied Ecology* 23(2):573–583. http://dx.doi.org/10.2307/2404037.

Coria, D. de 1937. "Vesitación que se hizo en la Conquista, Donde fue por Capitán Francisco Cortés." *Boletín del Archivo General de la Nación* 8:556–572.

Cortés, H. 1928. *Five Letters of Cortes to the Emperor*, trans. J. Bayard Morris. New York: W.W. Norton.

Cossins, N. J., and M. Upton. 1988. "The Impact of Climatic Variation of the Borana Pastoral System." *Agricultural Systems* 27(2) 117–135. http://dx.doi.org/10.1016/0308-521X(88)90025-X.

Cowgill, G. C. 1983. "Rulership and the Ciudadela: Political Inferences from Teotihuacan Architecture." In *Civilizations in the Ancient Americas: Essays in Honor of Gordon R. Willey*, ed. R. M. Leventhal and A. L. Kolata, 313–343. Albuquerque: University of New Mexico Press.

Cowgill, G. C. 1997. "State and Society at Teotihuacan." *Annual Review of Anthropology* 26(1):129–161. http://dx.doi.org/10.1146/annurev.anthro.26.1.129.

Crouch, D. P., D. J. Garr, and A. I. Mundigo. 1982. *Spanish City Planning in North America*. Cambridge: MIT Press.

Crumley, C. L. 1979. "Three Locational Models: An Epistemological Assessment for Anthropology and Archaeology." *Advances in Archaeological Method and Theory* 2:141–173.

Crumley, C. L. 1994. "The Ecology of Conquest: Contrasting Agropastoral and Agricultural Societies' Adaptation to Climatic Change." In *Historical Ecology: Cultural Knowledge and Changing Landscapes*, ed. C. L. Crumley, 183–201. Santa Fe: School of American Research Press.

Crumley, C. L. 1995. "Heterarchy and the Analysis of Complex Societies." In *Heterarchy and the Analysis of Complex Societies*, ed. R. M. Ehrenreich, C. L. Crumley, and J. E. Levy, 1–5. Arlington: American Anthropological Society. http://dx.doi.org/10.1525/ap3a.1995.6.1.1.

Crumley, C. L. 2003. "Alternative Forms of Societal Order. Heterarchy, Political Economy, and the Ancient Maya." In *The Three Rivers Region of the East-Central Yucatan Peninsula*, ed. L. Vernon, F. Scarborough, and N. Dunning, 136–145. Tucson: University of Arizona Press.

Crumley, C. L. 2005. "Remember How to Organize: Heterarchy Across Disciplines." In *Nonlinear Models for Archaeology and Anthropology*, ed. C. S. Beekman and W. S. Baden, 35–50. Hampshire: Ashgate Press Aldershot.

Culbert, T. P., L. J. Kosakowsky, R. E. Fry, and W. A. Haviland. 1990. "The Population of Tikal, Guatemala." In *Precolumbian Population History in the Maya*

Lowlands, ed. T. P. Culbert and D. S. Rice, 103–121. Albuquerque: University of New Mexico Press.

Curry, O., and R. I. M. Dunbar. 2011. "Altruism in Networks: The Effect of Connections." *Biology Letters* 7(5):651–653. http://dx.doi.org/10.1098/rsbl.2010.1202.

D'Altroy, T. N. 2002. *The Incas*. Malden: Blackwell.

D'Altroy, T. N., T. K. Earle. 1985. "Staple Finance, Wealth Finance, and Storage in the Inka Political Economy." *Current Anthropology* 26(2):187–206. http://dx.doi.org/10.1086/203249.

D'Altroy, T. N., and C. Hastorf. 2001. *Empire and Domestic Economy*. New York: Kluwer Academic/Plenum Press.

Dahlin, B. H., C. T. Jensen, R. E. Terry, D. R. Wright, and T. Beach. 2007. "In Search of an Ancient Maya Market." *Latin American Antiquity* 18(4):363–384. http://dx.doi.org/10.2307/25478193.

Daneels, A. 1997. "Settlement History in the Lower Cotaxtla Basin." In *Olmec to Aztec: Settlement Patterns in the Ancient Gulf Lowlands*, ed. B. L. Stark and P. J. Arnold, III, 206–252. Tucson: University of Arizona Press.

Daneels, A. 2002. *"El Patrón de Asentamiento del Periodo Clásico en la Cuenca Baja del Río Cotaxtla, Centro de Veracruz: Un Estudio de Caso de Desarrollo de Sociedades Complejas en Tierras Bajas Tropicales."* Tesis Doctoral, IIA. UNAM.

Daneels, A. 2008a. "Monumental Earthen Architecture at La Joya, Veracruz, Mexico." Foundation for the Advancement of Mesoamerican Studies, Inc. http://www.famsi.org/reports/07021.

Daneels, A. 2008b. "Ballcourts and Politics in the Lower Cotaxtla Valley: A Model to Understand Classic Central Veracruz?" In *Cultural Currents in Classic Veracruz*, ed. P. J. Arnold, III, and C. A. Pool, 197–223. Washington, DC: Dumbarton Oaks.

Daneels, A. 2011. "Arquitectura Cívica-Ceremonial de Tierra en la Costa del Golfo: El Sitio de La Joya y el Urbanismo del Periodo Clásico." In *Monte Albán en la Crucijada Regional y Disciplinaria: Memoria de la Quinta Mesa Redonda de Monte Albán*, ed. N. M. Robles and A. I. Rivera, 445–477. México DF: INAH.

Daneels, A., and G. Gutiérrez Mendoza, eds. 2012. *El Poder Compartido*. Zamora: Colegio de Michoacan.

David, A. R. 1998. *Handbook to Life in Ancient Egypt*. New York: Facts on File.

Davies, G. 2002. *A History of Money: From Ancient Times to the Present Day*. Cardiff: University of Wales Press.

Davies, N. 1987. *The Aztec Empire: The Toltec Resurgence*. Norman: University of Oklahoma Press.

Day, J. S. 1998. "The West Mexican Ballgame." In *Ancient West Mexico: Art and Archaeology of the Unknown Past*, ed. R. Townsend, 151–168. Chicago: Art Institute of Chicago.

Day, J. S., K. Butterwick, and R. B. Pickering. 1996. "Archaeological Interpretations of West Mexican Ceramic Art from the Late Preclassic Period." *Ancient Mesoamerica* 7(01):149–161. http://dx.doi.org/10.1017/S0956536100001358.

De Montmollín, O. 1995. *Settlement and Politics in Three Classic Maya Polities*. Madison: Prehistory Press.

DeMarrais, E., L. J. Castillo, and T. Earle. 1996. "Ideology, Materialization, and Power Strategies." *Current Anthropology* 37(1):15–31. http://dx.doi.org/10.1086/204472.

Díaz del Castillo, B. 1963. *The Conquest of New Spain*. Trans. J. M. Cohen. London: Penguin.

Dilley, R. 2005. "The Visibility and Invisibility of Production: In Different Social Contexts among Senegalese Craftsman." In *Commodification: Things, Agency, and Identity: The Social Life of Things Revisited*, ed. W. van Binsbergen and P. Geschiere, 227–242. Berlin/Muster: LIT Verlag.

Dirks, N. B. 1987. *The Hollow Crown: Ethnohistory of an Indian Kingdom*. New York: Cambridge University Press.

Domínguez Covarrubias, E. 2001. "La Arquitectura Monumental del Período Clásico en el Sur de Veracruz: Un Enfoque Regional." Tesis de Licenciatura, Departamento de Antropología, Universidad de las Américas-Puebla, Puebla.

Doyle, S. 2006. *Crisis and Decline in Bunyoro: Population and Environment in Western Uganda 1860–1955*. Oxford: James Currey.

Drennan, R. D. 1987. "Regional Demography in Chiefdoms." In *Chiefdoms in the Americas*, ed. R. D. Drennan and C. A. Uribe, 307–324. Lanham, MD: University Press of America.

Drennan, R. D. 1988. "Household Location and Compact versus Dispersed Settlement in Prehispanic Mesoamerica." In *Household and Community in the Mesoamerican Past*, ed. R. Wilk and W. Ashmore, 273–293. Albuquerque: University of New Mexico Press.

Drennan, R. D. 1989. "The Mountains North of the Valley." In *Monte Albán's Hinterland, Part II: Prehispanic Settlement Patterns in Tlacolula, Etla, and Ocotlán, the Valley of Oaxaca, Mexico*, ed. S. A. Kowalewski et al., 367–384. Ann Arbor: Museum of Anthropology, University of Michigan.

Dubreuil, B. 2008. "Strong Reciprocity and the Emergence of Large-Scale Societies." *Philosophy of the Social Sciences* 38(2):192–210. http://dx.doi.org/10.1177/0048393108315509.

Dubreuil, B. 2010. *Human Evolution and the Origins of Hierarchies: The State of Nature*. Cambridge: Cambridge University Press. http://dx.doi.org/10.1017/CBO9780511780035.

Dunbar, R.I.M. 1993. "Co-Evolution of Neocortex Size as a Constraint on Group Size in Primates." *Journal of Human Evolution* 20:469–493.

Dunbar, R.I.M. 1998. "The Social Brain Hypothesis." *Evolutionary Anthropology* 6(5):178–190. http://dx.doi.org/10.1002/(SICI)1520-6505(1998)6:5<178::AID-EVA N5>3.0.CO;2-8.

Dunbar, R.I.M. 2011. "Constraints on the Evolution of Social Institutions and Their Implications of Information Flow." *Journal of Institutional Economics* 7(03):345–371. http://dx.doi.org/10.1017/S1744137410000366.

Durán, D. 1971. *Book of the Gods and Rites and the Ancient Calendar*. Ed. and trans. F. Horcasitas and D. Heyden. Norman: University of Oklahoma Press.

Durán, D. 1994. *The History of the Indies of New Spain*. Trans. D. Heyden. Norman: University of Oklahoma Press.

Durand-Forest, J. 1994. "The Aztec Craftsman and the Economy." In *Chipping Away on Earth*, ed. E. Quiñones, 173–176. Lancaster: Labyrinthos.

duVall, S. 2007. "Shared Symbolism in Powerful Places: Cosmological Principles Displayed through the Ceremonial Public Architecture of the Teuchitlán Tradition in the Tequila Valleys of Jalisco, West Mexico." Unpublished MA thesis, Department of Anthropology, University of Colorado, Denver.

Earle, T. K. 1976. "A Nearest-Neighbor Analysis of Two Formative Settlement Systems." In *Early Mesoamerican Village*, ed. K. Flannery, 196–223. New York: Academic Press.

Earle, T. K. 1978. *Economic and Social Organization of a Complex Chiefdom, the Halelea District, Kaua'i, Hawaii*. Ann Arbor: Museum of Anthropology, University of Michigan.

Earle, T. K. 1980. "Prehistoric Irrigation in the Hawaiian Islands: An Evaluation of Evolutionary Significance." *Archaeology and Physical Anthropology in Oceania* 15:1–28.

Earle, T. K. 1987. "Specialization and the Production of Wealth: Hawaiian Chiefdoms and the Inka Empire." In *Specialization, Exchange, and Complex Societies*, ed. E. Brumfiel and T. Earle, 73–81. Cambridge: Cambridge University Press.

Earle, T. K. 1994. "Wealth Finance in the Inka Empire: Evidence from the Calchaqui Valley, Argentina." *American Antiquity* 59(3):443–460. http://dx.doi.org/10.2307/282457.

Earle, T. K. 1997. *How Chiefs Come to Power: The Political Economy in Prehistory*. Stanford: Stanford University Press.

Earle, T. K. 2002. *Bronze Age Economics*. Boulder: Westview Press.

Earle, T. K. 2004. "Culture Matters: The Neolithic Transition and Emergence of Hierarchy in Thy, Denmark." *American Anthropologist* 106(1):111–125. http://dx.doi.org/10.1525/aa.2004.106.1.111.

Earle, T. K. 2011. "Redistribution in Aegean Palatial Societies: Redistribution and the Political Economy: The Evolution of an Idea." *American Journal of Archaeology* 115(2):237–244. http://dx.doi.org/10.3764/aja.115.2.237.
Earle, T. K. 2012. "Taro Irrigation and Primary State Formation in Hawai'i." In *Irrigated Taro (Colocasia esculenta) in the Indo-Pacific*, ed. M. Spriggs, D. Addison, and P. Matthews, 95–114. Osaka: National Museum of Ethnology.
Earle, T. K., and T. D'Altroy. 1989. "The Political Economy of the Inka Empire: The Archaeology of Power and Finance." In *Archaeological Thought in America*, ed. C. C. Lamberg, 183–204. Cambridge: Cambridge University Press. http://dx.doi.org/10.1017/CBO9780511558221.013.
Earle, T. K., and D. Doyle. 2008. "The Engineered Landscapes of Irrigation." In *Economics and the Transformation of Landscape*, ed. L. Cliggett and C. A. Pool, 19–46. Lanham, MD: AltaMira Press.
Earle, T. K., and K. Kristiansen, eds. 2010. *Organizing Bronze Age Societies*. Cambridge: Cambridge University Press. http://dx.doi.org/10.1017/CBO9780511779282.
Earle, T. K., R. Preucel. 1987. "Processual Archaeology and the Radical Critique." *Current Anthropology* 28(4):501–537. http://dx.doi.org/10.1086/203551.
Earle, T. K., and M. Spriggs. 2015. "Political Economy in Prehistory: A Marxist Approach to Pacific Sequences." *Current Anthropology* 56(4):515–544. http://dx.doi.org/10.1086/682284.
Eck, W. 2000a. "The Growth of Administrative Posts." In *The Cambridge Ancient History, Volume 11: The High Empire, A.D. 70–192*, ed. A.K. Bowman, P. Garnsey, and D. Rathbone, 238–265. Cambridge: Cambridge University Press. http://dx.doi.org/10.1017/CHOL9780521263351.007.
Eck, W. 2000b. "Provincial Administration and Finance." In *The Cambridge Ancient History, Volume 11: The High Empire, A.D. 70–192*, ed. A. K. Bowman, P. Garnsey, and D. Rathbone, 266–292. Cambridge: Cambridge University Press. http://dx.doi.org/10.1017/CHOL9780521263351.008.
Eck, W. 2000c. "Emperor, Senate, and Magistrates." In *The Cambridge Ancient History, Volume 11: The High Empire, A.D. 70–192*, ed. A. K. Bowman, P. Garnsey, and D. Rathbone, 214–237. Cambridge: Cambridge University Press. http://dx.doi.org/10.1017/CHOL9780521263351.006.
Ellen, R. 2010. "Theories in Anthropology and 'Anthropological Theory.'" *Journal of the Royal Anthropological Institute* 16(2):387–404. http://dx.doi.org/10.1111/j.1467-9655.2010.01631.x.
Elson, C. M. 2007. *Excavations at Cerro Tilcajete: A Monte Albán II Administrative Center in the Valley of Oaxaca*. Ann Arbor: Museum of Anthropology, University of Michigan.

Elson, C. M., and R. J. Sherman. 2007. "Crema Ware and Elite Power at Monte Alban: Ceramic Production and Iconography in the Oaxaca Valley, Mexico." *Journal of Field Archaeology* 32(3):265–282. http://dx.doi.org/10.1179/009346907791071548.

Ember, C. R., and M. Ember. 1992a. "Resource Unpredictability, Mistrust, and War: A Cross-Cultural Study." *Journal of Conflict Resolution* 36(2):242–262. http://dx.doi .org/10.1177/0022002792036002002.

Ember, C. R., and M. Ember. 1992b. "Warfare, Aggression, and Resource Problems: Cross-Cultural Codes." *Cross-Cultural Research* 26(1-4):169–226. http://dx.doi.org /10.1177/106939719202600108.

Emerson, T. E. 1997. *Cahokia and the Archaeology of Power*. Tuscaloosa: University of Alabama Press.

Enzensberger, H. M. 1982. *Svensk höst*. Stockholm: Dagens Nyheter.

Fargher, L. F. 2007. "A Microscopic View of Ceramic Production: An Analysis of Thin-Sections from Monte Albán." *Latin American Antiquity* 18(3):313–332. http://dx.doi.org/10.2307/25478183.

Fargher, L. F. 2009. "A Comparison of the Spatial Distribution of Agriculture and Craft Specialization in Five State-Level Societies." *Journal of Anthropological Research* 65(3):353–387. http://dx.doi.org/10.3998/jar.0521004.0065.301.

Fargher, L. F. 2012. "Sangre, Humo y Ojos Estelares: Iconografía, Ideología y la Construcción de la República de Tlaxcallan." Paper presented at the Conference: Tlaxcala Pasado y Presente: Investigaciones Antropológicas Recientes. Tlaxcala: Gobierno del Estado de Tlaxcala and Universidad Autónoma de México.

Fargher, L. F., and R. E. Blanton. 2007. "Revenue, Voice, and Public Goods in Three Premodern States." *Comparative Studies in Society and History* 49(04):848–882. http://dx.doi.org/10.1017/S0010417507000783.

Fargher, L. F., R. E. Blanton, and V. Y. Heredia Espinoza. 2010. "Egalitarian Ideology and Political Power in Prehispanic Central Mexico: The Case of Tlaxcallan." *Latin American Antiquity* 21(3): 227–251. http://dx.doi.org/10.7183/1045-6635.21.3.227.

Fargher, L. F., V. Y. Heredia Espinoza, and R. E. Blanton. 2011. "Alternative Pathways to Power in Late Postclassic Highland Mesoamerica." *Journal of Anthropological Archaeology* 30(3):306–326. http://dx.doi.org/10.1016/j.jaa.2011.06.001.

Fargher, L. F., R. E. Blanton, V. Y. Heredia Espinoza, J. Millhauser, N. Xiuhtecutli, and L. Overholtzer. 2011. "Tlaxcallan: The Archaeology of an Ancient Republic in the New World." *Antiquity* 85(327):172–186. http://dx.doi.org/10.1017/S0003598 X0006751X.

Fargher, L. F., R. E. Blanton, V. Y. Heredia Espinoza, J. Millhauser, N. Xiuhtecutli, and L. Overholtzer. 2011. "Tlaxcallan: The Archaeology of an Ancient Republic in the New World." *Antiquity* 85(327) 172–186. http://dx.doi.org/10.1017/S0003598 X0006751X.

Fargher, L. F., V. Y. Heredia Espinoza, and A. Uriarte Torres. 2014. "Collective Action, Intermediate Spatial Units and Urban Organization in Pre-Hispanic Mesoamerica: An Interregional Cross-Scale Comparison." Paper presented at the 79th annual meeting of the Society for American Archaeology, Austin.

Farriss, N. M. 1985. *Maya Society Under Colonial Rule: The Collective Enterprise of Survival.* Princeton: Princeton University Press.

Fash, B., W. L. Fash, S. Lane, R. Larios, L. Schele, J. Stomper, and D. Stuart. 1992. "Investigations of a Classic Maya Council House at Copán, Honduras." *Journal of Field Archaeology* 19 (4):419–442. http://dx.doi.org/10.2307/530426.

Fash, W. L. 1991. *Scribes, Warriors, and Kings: The City of Copán and the Ancient Maya.* New York: Thames and Hudson.

Feinman, G. M. 1995. "The Emergence of Inequality: A Focus on Strategies and Processes." In *Foundations of Social Inequality*, ed. T. D. Price and G. M. Feinman, 255–279. New York: Plenum Press. http://dx.doi.org/10.1007/978-1-4899-1289-3_10.

Feinman, G. M. 1998. "Scale and Social Organization: Perspectives on the Archaic State." In *Archaic States*, ed. G. M. Feinman and J. Marcus, 95–133. Santa Fe: School of American Research Press.

Feinman, G. M. 2001. "Mesoamerican Political Complexity: The Corporate-Network Dimension." In *From Leaders to Rulers*, ed. J. Haas, 151–175. New York: Kluwer Academic/Plenum Publishers. http://dx.doi.org/10.1007/978-1-4615-1297-4_8.

Feinman, G. M. 2011. "Size, Complexity, and Organizational Variation: A Comparative Approach." *Cross-Cultural Research* 45(1):37–58. http://dx.doi.org/10.1177/1069397110383658.

Feinman, G. M. 2012. "Comparative Frames for the Diachronic Analysis of Complex Societies, Next Steps." In *The Comparative Archaeology of Complex Societies*, ed. M. E. Smith, 21–43. Cambridge: Cambridge University Press.

Feinman, G. M., K. G. Lightfoot, and S. Upham. 2000. "Political Hierarchies and Organizational Strategies in the Puebloan Southwest." *American Antiquity* 65(3):449–470. http://dx.doi.org/10.2307/2694530.

Feinman, G. M., and J. Marcus, eds. 1998. *Archaic States*. Santa Fe: School of American Research Press.

Feinman, G. M., and J. Neitzel. 1984. "Too Many Types: An Overview of Sedentary Prestate Societies in the Americas." *Advances in Archaeological Method and Theory* 7:39–102.

Feinman, G. M., and L. M. Nicholas. 1990. "At the Margins of the Monte Albán State: Settlement Patterns in the Ejutla Valley, Oaxaca, Mexico." *Latin American Antiquity* 1(3):216–246. http://dx.doi.org/10.2307/972162.

Feinman, G. M., and L. M. Nicholas. 2000. "High-Intensity Household-Scale Production in Ancient Mesoamerica: A Perspective from Ejutla, Oaxaca." In *Cultural*

Evolution: Contemporary Viewpoints, ed. G. M. Feinman and L. Manzanilla, 119–142. New York: Kluwer Academic/Plenum. http://dx.doi.org/10.1007/978-1-4615-4173-8_5.
Feinman, G. M., and L. M. Nicholas. 2003. "Top-Notch Tomb." *Archaeology* 56:14.
Feinman, G. M., and L. M. Nicholas. 2004. *Hilltop Terrace Sites of Oaxaca, Mexico: Intensive Surface Survey at Guirún, El Palmillo, and the Mitla Fortress*. Chicago: Field Museum of Natural History. http://dx.doi.org/10.5962/bhl.title.3393.
Feinman, G. M., and L. M. Nicholas. 2007. "The Socioeconomic Organization of the Classic Period Zapotec State: A Bottom-Up Perspective from El Palmillo." In *The Political Economy of Ancient Mesoamerica: Transformations During the Formative and Classic Periods*, ed. V. Scarborough and J. Clark, 135–152. Albuquerque: University of New Mexico Press.
Feinman, G. M., and L. M. Nicholas. 2011. "Poder y Desigualdad: Variación y Cambio en la Mesoamérica Prehispánica." In *Mesoamérica: Debates y Perspectivas*, ed. E. Williams, M. García, P. C. Weigand, and M. Gándara, 133–153. Zamora, Mexico: El Colegio de Michoacán.
Feinman, G. M., and L. M. Nicholas. 2011a. "Domestic Craft Production and the Classic Period Economy of Oaxaca." In *Producción Artesanal y Especializada en Mesoamérica: Areas de Actividad y Procesos Productivos*, ed. L. Manzanilla and K. Hirth, 29–58. México DF: INAH and IIA, UNAM.
Feinman, G. M., and L. M. Nicholas. 2011b. "The Ballcourt at El Palmillo: Implications for Late Classic Oaxaca, Mexico." *Mexicon* 33:98–104.
Feinman, G. M., L. M. Nicholas, and F. Hui. 2010. "The Imprint of China's First Emperor on the Distant Realm of Eastern Shandong." *Proceedings of the National Academy of Sciences of the United States of America* 107(11):4851–4856. http://dx.doi.org/10.1073/pnas.0914961107.
Feinman, G. M., L. M. Nicholas, and H. R. Haines. 2007. "Classic Period Agricultural Intensification and Domestic Life at El Palmillo, Valley of Oaxaca, Mexico." In *Seeking a Richer Harvest: The Archaeology of Subsistence Intensification, Innovation, and Change*, ed. T. J. Thurston and C. T. Fisher, 23–61. New York: Springer. http://dx.doi.org/10.1007/978-0-387-32762-4_2.
Feinman, G. M., S. Upham, and K. G. Lightfoot. 1981. "The Production Step Measure: An Ordinal Index of Labor Input in Ceramic Manufacture." *American Antiquity* 46(4):871–884. http://dx.doi.org/10.2307/280113.
Feinman, G. M., S. A. Kowalewski, and R. E. Blanton. 1984. "Modelling Ceramic Production and Organizational Change in the Pre-Hispanic Valley of Oaxaca, Mexico." In *The Many Dimensions of Pottery*, ed. S. E. van der Leeuw, 297–337. Amsterdam: Universiteit Van Amsterdam Alb Voor Prae-En Protohistorie.

Feinman, G. M., S. A. Kowalewski, L. Finsten, R. E. Blanton, and L. M. Nicholas. 1985. "Long-Term Demographic Change: A Perspective from the Valley of Oaxaca, Mexico." *Journal of Field Archaeology* 12:333–362.

Feldman, L. H. 1978. "Inside a Mexican Market." In *Mesoamerican Communication Routes and Cultural Contacts*, ed. T. A. Lee and C. Navarrete, 219–222. Provo: Brigham Young University.

Ferguson, R. B., and N. L. Whitehead. 1992. "The Violent Edge of Empire." In *War in the Tribal Zone: Expanding States and Indigenous Warfare*, ed. R. B. Ferguson and N. L. Whitehead, 1–30. Santa Fe: School of American Research Press.

Ferguson, Y. H., and R. W. Mansbach. 1996. *Polities: Authority, Identities, and Change.* Columbia: University of South Carolina Press.

Filloy Nadal, L., F. Solís Olguín, and L. Navarijo. 2006. "Un Eccezionale Mosaico di Piume Azteco: il 'Copri-calice' del Museo Nacional de Antropología." In *Gli Aztechi tra Passato e Presente*, 105–115. Rome: Carocci Editore.

Finsten, L. M. 1996. "Periphery and Frontier in Southern Mexico: The Mixtec Sierra in Highland Oaxaca." In *Pre-Columbian World-systems*, ed. P. N. Peregrine and G. M. Feinman, 77–96. Madison: Prehistory Press.

Fisher, A. B. 1911. *Twilight Tales of the Black Baganda.* London: Marshall Brothers.

Flannery, K. V. 1972. "The Cultural Evolution of Civilizations." *Annual Review of Ecology and Systematics* 3(1):399–426. http://dx.doi.org/10.1146/annurev.es.03.110172.002151.

Flannery, K. V. 1973. "Archeology with a Capital S." In *Research and Theory in Current Archaeology*, ed. C. L. Redman, 47–53. New York: John Wiley.

Flannery, K. V. 1998. "The Ground Plans of Archaic States." In *Archaic States*, ed. G. M. Feinman and J. Marcus, 15–57. Santa Fe: School of American Research.

Fleisher, J., and S. Wynne-Jones. 2010. "Authorisation and the Process of Power: The View from African Archaeology." *Journal of World Prehistory* 23(4):177–193. http://dx.doi.org/10.1007/s10963-010-9038-y.

Fleming, D. E. 2004. *Democracy's Ancient Ancestors: Mari and Early Collective Governance.* Cambridge: Cambridge University Press. http://dx.doi.org/10.1017/CBO9780511499623.

Fletcher, R. 2012. "Low-Density, Agrarian-Based Urbanism: Scale, Power, and Ecology." In *The Comparative Archaeology of Complex Societies*, ed. M. E. Smith, 285–320. Cambridge: Cambridge University Press.

Florén, A., G. Rydén, L. Dashkevich, D. V. Gavrilo, and S. Ustiantsev. 2003. "The Social Organisation of Work at Mines, Furnaces and Forges." In *Iron-Making Societies: Early Industrial Development in Sweden and Russia, 1600–1900*, ed. M. Ågren, 61–140. New York: Berghahn Books.

Foias, A. E., and R. L. Bishop. 1997. "Changing Ceramic Production and Exchange in the Petexbatun Region, Guatemala: Reconsidering the Classic Maya Collapse." *Ancient Mesoamerica* 8(02):275–291. http://dx.doi.org/10.1017/S0956536100001735.

Foias, A. E., and R. L. Bishop. 2007. "Pots, Sherds, and Glyphs: Pottery Production and Exchange in the Petexbatun Polity, Peten, Guatemala." In *Pottery Economics in Mesoamerica*, ed. C. A. Pool and G. J. Bey, 212–236. Tucson: The University of Arizona Press.

Forge, A. 1972. "Normative Factors in Settlement Size of Neolithic Cultivators (New Guinea)." In *Man, Settlement, and Urbanism*, ed. P. J. Ucko, R. Tringham, and G. W. Dimbleby, 363–376. London: Duckworth.

Foster, B. R. 1995. *From Distant Days: Myths, Tales, and Poetry of Ancient Mesopotamia*. Bethesda: Capital Decisions, Ltd.

Fowler, J. H., and N. A. Christakis. 2010. "Cooperative Behavior Cascades in Human Social Networks." *Proceedings of the National Academy of Sciences of the United States of America* 107(12):5334–5338. http://dx.doi.org/10.1073/pnas.0913149107.

Fowler, W. R., C. S. Beekman, and R. B. Pickering. 2006. "Introduction: Special Section on Mortuary Practices and Shaft Tombs in Western Mexico: An *Homenaje* to Phil Weigand." *Ancient Mesoamerica* 17(02):231–233. http://dx.doi.org/10.1017/S0956536106060202.

Fox, J. G., W. Ashmore, J. H. Blitz, S. D. Gillespie, S. D. Huston, T. J. J. Leyenaar, J. Marcus, J. D. Moore, P. A. Urban, E. M. Schortman, and David Webster. 1996. "Playing with Power: Ballcourts and Political Ritual in Southern Mesoamerica [and Comments and Reply]." *Current Anthropology* 37(3):483–509. http://dx.doi.org/10.1086/204507.

Fox, J. W. 1978. *Quiche Conquest: Centralism and Regionalism in Highland Guatemalan State Development*. Albuquerque: University of New Mexico Press.

Fox, J. W. 1987. *Maya Postclassic State Formation: Segmentary Lineage Migration in Advancing Frontiers*. New York: Cambridge University Press.

Frachetti, M. 2008a. *Pastoralist Landscapes and Social Interaction in Bronze Age Eurasia*. Berkeley: University of California Press.

Frachetti, M. 2008b. "Variability of Mobile Pastoralism." In *Variability and Dynamic Landscapes of Mobile Pastoralism in Ethnography and Prehistory*, ed. H. Barnard and W. Wendrick, 366–392. Los Angeles: Cotsen Institute of Archaeology, UCLA.

Franke-Vogt, U. 1989. "Inscribed Objects from Mohenjodaro: Some Remarks on Stylistic Variability and Distribution Patterns." *South Asian Archaeology* 1:103–112.

Frankfort, H. 1956. *The Birth of Civilization in the Near East*. Garden City: Doubleday.

Fratkin, E. 1989. "Household Variation and Gender Inequality in Ariaal Pastoral Production: Results of a Stratified Time-Allocation Survey." *American Anthropologist* 91(2):430–440. http://dx.doi.org/10.1525/aa.1989.91.2.02a00100.

Fratkin, E., and K. Smith. 1994. "Labor, Livestock and Land: The Organization of Pastoral Production." In *African Pastoralist Systems: New Frontiers in Method and Theory*, ed. E. Fratkin, K. A. Galvin, and E. A. Roth, 91–112. Boulder: Lynne Reiner.
Frenez, D., and M. Tosi. 2005. "The Lothal Sealings: Records from an Indus Civilization Port Town at the Eastern End of the Maritime Trade Network Across the Arabia Sea." In *Studi in Onore di Enrica FIandra, Contributi di Archeologia Egea e Vicinorientale*, ed. M. Pernaa, 65–103. Paris: Diffusion de Boccard.
Fried, M. H. 1967. *The Evolution of Political Society: An Essay in Political Anthropology*. New York: Random House.
Fried, M. H. 1975. *The Notion of Tribe*. Menlo Park: Cummings.
Friedman, J. 1981. "Notes on Structure and History in Oceania." *Folk (Kobenhavn)* 23:275–295.
Friedman, J., and M. Rowlands. 1977. "Notes Towards an Epigenetic Model of the Evolution of 'Civilization." In *The Evolution of Social Systems*, ed. J. Friedman and M. Rowlands, 201–276. London: Duckworth.
Fry, R. E. 1980. "Models of Exchange for Major Shape Classes of Lowland Maya Pottery." In *Models and Methods in Regional Exchange*, ed. R. E. Fry, 3–18. Washington, DC: Society for American Archaeology.
Fujita, M., P. R. Krugman, and A. J. Venables, eds. 1999. *The Spatial Economy, Cities, Region, and International Trade*. Cambridge: MIT Press.
Furst, P. T. 1975. "House of Darkness and House of Light: Sacred Functions of West Mexican Funerary Art." In *Death and the Afterlife in Pre-Columbian America*, ed. E. P. Benson, 33–68. Washington, DC: Dumbarton Oaks.
Galsterer, H. 2000. "Local and Provincial Institutions and Government." In *The Cambridge Ancient History, Volume 11: The High Empire, A.D. 70–192*, ed. A. K. Bowman, P. Garnsey, and D. Rathbone, 344–360. Cambridge: Cambridge University Press. http://dx.doi.org/10.1017/CHOL9780521263351.011.
Galván Villegas, L. J. 1991. *Las Tumbas de Tiro del Valle de Atemajac, Jalisco*. México DF: INAH.
Galván Villegas, L. J. 1983. *Proyecto Valle de Atemajac: Rescate del Fraccionamiento Bugambilias, Zapopan, Jalisco*. México DF: INAH.
Garcia, K. 2008. "Maya Market Economics: Petrographic Analysis of Common Late and Terminal Classic Ceramics Groups in the Upper Belize Valley." Unpublished Master's thesis, Department of Anthropology, University of Alabama, Tuscaloosa.
García Ayala, G. 2011. *Urbanismo en la Mixteca Alta, Oaxaca: El Caso de Naduza-Ñiaxugue en la Región de Coixtlahuaca*. México DF: Tesis de Licenciatura, Escuela Nacional de Antropología e Historia.
Garfinkle, S. J. 2002. "Turan-ili and the Community of Merchants in the Ur III Period." *Journal of Cuneiform Studies* 54:29–48. http://dx.doi.org/10.2307/1360041.

Garraty, C. P. 2000. "Ceramic Indices of Aztec Eliteness." *Ancient Mesoamerica* 11(02):323–340. http://dx.doi.org/10.1017/S0956536100112052.
Garraty, C. P. 2010. "Investigating Market Exchange in Ancient Societies: A Theoretical Review." In *Archaeological Approaches to Market Exchange in Ancient Societies*, ed. C. P. Garraty and B. L. Stark, 3–32. Boulder: University Press of Colorado.
Garvin, R. D. 1994. "Modern and Prehispanic Agriculture in the Sierra Mixteca, Oaxaca, Mexico." PhD dissertation, Department of Archaeology, University of Calgary, Calgary.
Geertz, C. 1980. *Negara: The Theatre State in Nineteenth-Century Bali*. Princeton: Princeton University Press.
Gerring, J. 2012. *Social Science Methodology: A Unified Framework*. Cambridge: Cambridge University Press.
Giddens, A. 1984. *The Constitution of Society: Outline of the Theory of Structuration*. Cambridge: Polity Press.
Giles, M. 2007. "Making Metal and Forging Relations: Ironworking in the British Iron Age." *Oxford Journal of Archaeology* 26(4):395–413. http://dx.doi.org/10.1111/j.1468-0092.2007.00290.x.
Gillespie, S. D. 1991. "Ballgames and Boundaries." In *The Mesoamerican Ballgame*, ed. V. L. Scarborough and D. R. Wilcox, 317–345. Tucson: University of Arizona Press.
Gilman, A. 1991. "Trajectories Towards Social Complexity in the Later Prehistory of the Mediterranean." In *Chiefdoms: Power, Economy and Ideology*, ed. T. Earle, 146–168. Cambridge: Cambridge University Press.
Gilman, A. 1995. "Prehistoric European Chiefdoms: Rethinking 'Germanic' Societies." In *Foundations of Social Inequality*, ed. T. D. Price and G. M. Feinman, 235–251. New York: Plenum. http://dx.doi.org/10.1007/978-1-4899-1289-3_9.
Glassow, M. 1967. "The Ceramics of Huistla, a West Mexican Site in the Municipality of Etzatlán, Jalisco." *American Antiquity* 32(1):64–83. http://dx.doi.org/10.2307/278779.
Gluckman, M. 1941. *Economy of the Central Barotse Plain*. Livingstone: Rhodes-Livingstone Institute.
Gluckman, M. 1943. *Essays on Lozi land and Royal Property*. Livingstone: Rhodes-Livingstone Institute.
Gluckman, M. 1961. "The Lozi of Barotseland in North-Western Rhodesia." In *Seven Tribes of British Central Africa*, ed. E. Colson and M. Gluckman, 1–93. Manchester: Manchester University Press.
Golden, C., and A. K. Scherer. 2013. "Territory, Trust, Growth, and Collapse in Classic Period Maya Kingdoms." *Current Anthropology* 54(4):397–435. http://dx.doi.org/10.1086/671054.

Goldstone, J. A. 1998. "Initial Conditions, General Laws, Path Dependence, and Explanation in Historical Sociology." *American Journal of Sociology* 104(3):829–845. http://dx.doi.org/10.1086/210088.

González-Ruibal, A. 2006. "House Societies vs. Kinship-based Societies: An Archaeological Case from Iron Age Europe." *Journal of Anthropological Archaeology* 25(1):144–173. http://dx.doi.org/10.1016/j.jaa.2005.09.002.

Gordon, R. B., and T. S. Reynolds. 1986. "Medieval Iron in Society-Norberg, Sweden, May 6–10, 1985." *Technology and Culture* 27(1):110–117. http://dx.doi.org/10.2307/3104948.

Graeber, D. 2011. *Debt: The First 5,000 Years*. Brooklyn: Melville House.

Gramsci, A. 1971. "The Intellectuals." In *Selections from the Prison Notebooks*, ed. G. N. Smith, 3–23. New York: International Publishers.

Grandin, B. E., S. Bekure, and P. Nestel. 1991. "Livestock Transactions, Food Consumption and Household Budgets." In *Maasai Herding: An Analysis of the Livestock Production System of Maasai Pastoralists in Eastern Kajiado District, Kenya*, ed. S. Bekure, P. N. de Leeuw, B. E. Grandin, and P. J. H. Neate, 103–114. Addis Ababa: International Livestock Centre for Africa.

Granovetter, M. 1985. "Economic Action and Social Structure: The Problem of Embeddedness." *American Journal of Sociology* 91(3):481–510. http://dx.doi.org/10.1086/228311.

Green, A. 2009. "Carving Indus Seals: An Archaeological Methodology for Analyzing Formal Variation to Reconstruct Social Relations in the Indus Civilization." PhD dissertation, Department of Anthropology, New York University, New York.

Green, A. 2016. "Finding Harappan Carvers: An Operational Chain Approach and Identifying People in the Past." *Journal of Archaeological Research* 72: 128–141.

Green, R. 2002. "Rediscovering the Social Aspects of Ancestral Oceanic Societies through Archaeology, Linguistics, and Ethnology." In *Fifty Years in the Field: Essays in Honour and Celebration of Richard Shutler Jr's Archaeological Career*, ed. S. Bedford, C. Sand, and D. Burley, 21–35. Auckland: New Zealand Archaeological Association.

Griffin, M. 2000. "The Flavians." In *The Cambridge Ancient History, Volume 11: The High Empire, A.D. 70–192*, ed. A.K. Bowman, P. Garnsey, and D. Rathbone, 1–83. Cambridge: Cambridge University Press. http://dx.doi.org/10.1017/CHOL9780521263351.002.

Grove, D. C., and S. D. Gillespie. 1984. "Chalcatzingo's Portrait Figurines and the Cult of the Ruler." *Archaeology* 37:27–33.

Gulick, C. B. 1973. *The Life of the Ancient Greeks, with Special Reference to Athens*. New York: Cooper Square Publishers.

Gullick, J. M. 1958. *Indigenous Political Systems of Western Malaya*. London: University of London and Athlone Press.

Gutting, G. 2012. "How Reliable are the Social Sciences?" *The New York Times*, May 17.

Guyer, J. I. 1993. "Wealth in People and Self-Realization in Equatorial Africa." *Man* 28(2):243–265. http://dx.doi.org/10.2307/2803412.

Guyer, J. I., and S. M. Belinga. 1995. "Wealth in People as Wealth in Knowledge: Accumulation and Composition in Equatorial Africa." *Journal of African History* 36(01):91–120. http://dx.doi.org/10.1017/S0021853700026992.

Haas, J. 1982. *The Evolution of the Prehistoric State*. New York: Columbia University Press.

Habib, I. 1963. *The Agrarian System of Mughal India, 1556–1707*. Bombay: Dept. of History, Aligarh Muslim University.

Häggström, L. 2004. "Att datera agrara lämningar med OSL (Dating agricultural remains with OSL)." In *Om OSL-metoden och dess tillämpning I Öggestorp, norra Småland*, ed. Lönn. Stockholm: *Aktuella metodfrågor*, RAÄ UV skrifter.

Hagstrum, M. B. 1988. "Ceramic Production in the Central Andes, Peru: An Archaeological and Ethnographic Comparison." In *A Pot for All Reasons: Ceramic Ecology Revisited*, ed. C. C. Kolb and L. M. Lackey, 127–145. Philadelphia: Temple University Press.

Hall, J. W. 1991. "The Bakuhan System." In *The Cambridge History of Japan, Volume 4, Early Modern Japan*, ed. J. W. Hall, 128–182. Cambridge: Cambridge University Press.

Hallenberg, H., J. Holm, and D. Johansson. 2008. "Organization, Legitimation, Participation: State Formation as a Dynamic Process—the Swedish Example c. 1523–1680." *Scandinavian Journal of History* 33(3):247–268. http://dx.doi.org/10.1080/03468750802150242.

Hanks, B. 2010. "Archaeology of the Eurasian Steppes and Mongolia." *Annual Review of Anthropology* 39(1):469–486. http://dx.doi.org/10.1146/annurev.anthro.012809.105110.

Hansard. 1943. "House of Commons Rebuilding." *Parliamentary Debates: House of Commons*: vol. 393, columns 403–473.

Hansen, M. H. 1999. *The Athenian Democracy in the Age of Demosthenes: Structure, Principles, and Ideology*. Norman: University of Oklahoma Press.

Hasan, I. 1936. *The Central Structure of the Mughal Empire and its Practical Working Up to the Year 1657*. London: Oxford University Press.

Hassane, B. 2001. "Pastoral and Agro-pastoral Organizations' Collective Action Approach to Devolution of Natural Resource Management in West and Central

African Countries." In *Collective Action, Property Rights and Devolution of Natural Resource Management: Exchange of Knowledge and Implications for Policy*, ed. R. Meinzen-Dick, A. Knox, M. Di Gregorio, and M. Feldafing, 249–254. Germany: DSE/ZEL.

Haug, G. H., D. Günther, L. C. Peterson, D. M. Sigman, K. A. Hughen, and B. Aeschlimann. 2003. "Climate and the Collapse of Maya Civilization." *Science* 299(5613):1731–1735. http://dx.doi.org/10.1126/science.1080444.

Haven Putnam, B. 1950. "Shire Officials: Keepers of the Peace and Justices of the Peace." In *The English Government at Work, 1327–1336, Volume 3: Local Administration and Justice*, ed. J. F. Willard, W. A. Morris, and W. H. Dunham Jr., 185–217. Cambridge: The Mediaeval Academy of America.

Hayden, B. D. 1983. "Social Characteristics of Early Austronesian Colonisers." *Indo-Pacific Prehistory Bulletin* 4:123–134.

Healan, D. M. 2011. "El Aprovechamento de la Obsidiana en el Occidente de México en la Epoca Prehispánica." In *Patrones de Asentamiento y Actividades de Subsistencia en el Occidente de México: Reconocimiento a la Doctora Helen P. Pollard*, ed. E. Williams and P. C. Weigand, 189–208. Zamora: Colegio de Michoacán.

Hechter, M., and S. Kanazawa. 1997. "Sociological Rational Choice Theory." *Annual Review of Sociology* 23(1):191–214. http://dx.doi.org/10.1146/annurev.soc.23.1.191.

Hedeager, L. 1992. *Iron Age Societies. From Tribe to State in Northern Europe, 500 BC to AD 700*. Oxford: Blackwell.

Hedström, P., and R. Swedberg. 1996. "Social Mechanisms." *Acta Sociologica* 39(3):281–308. http://dx.doi.org/10.1177/000169939603900302.

Helander, A., and P. Zetterlund. 1998. "*Linnebergsgravfältet från Romersk Järnålder* (Linnebergs Grave Field from the Roman Iron Age)." *UV Rapport 1998* no. 52. Linköping: Arkeologiska Uppdragsverksamheten Öst, Riksantikvarieämbetet.

Helmke, C., J. Awe, and N. Grube. 2010. "The Carved Monuments and Inscriptions of Xunantunich: Implications for Terminal Classic Sociopolitical Relationships in the Belize Valley." In *Classic Maya Provincial Politics: Xunantunich and Its Hinterlands*, ed. L. J. LeCount and J. Yaeger, 97–121. Tucson: University of Arizona Press.

Hendon, J. A. 1996. "Archaeological Approaches to the Organization of Domestic Labor: Household Practice and Domestic Relations." *Annual Review of Anthropology* 25(1):45–61. http://dx.doi.org/10.1146/annurev.anthro.25.1.45.

Heredia Espinoza, V. Y. 2007. *Cities on Hills: Classic Society in Mesoamerica's Mixteca Alta.* BAR International Series. Oxford: British Archaeological Reports.

Heredia Espinoza, V. Y. 2008. *The Agave Landscape and its Archaeological Context in the Tequila Volcano area*. Foundation for Ancient Mesoamerican Studies, Inc. Accessed March 5, 2010. http://www.famsi.org/reports/07012/.

Heredia Espinoza, V. Y. 2011. *Beyond the Teuchitlán Polity: Late Classic and Postclassic Sociopolitical Developments in the Central Valleys of Jalisco*. Paper presented at the 77th annual meeting of the Society for American Archaeology, Memphis.

Heredia Espinoza, V. Y. 2014. *Reflections on the use of Full-Coverage Archaeological Survey in the Tequilla Valleys of Jalisco, Mexico*. Paper presented at the 79th annual meeting of the Society for American Archaeology, Austin.

Hers, M.-A. 1977. "Mission Archeologique Belge dans la Sierra del Nayar (Mexique)." *Reune des Archeoloques et Historiens d'Art de Louvain* 10:274–282.

Hill, J. D. 1989. "Re-Thinking the Iron Age." *Scottish Archaeological Review* 6:16–24.

Hill, J. D. 1993. "Can We Recognise a Different European Past? A Contrastive Archaeology of Later Prehistoric Settlements in Southern England." *Journal of European Archaeology* 1(1):57–75. http://dx.doi.org/10.1179/096576693800731226.

Hill, R. A., and R.I.M. Dunbar. 2003. "Social Network Size in Humans." *Human Nature (Hawthorne, N.Y.)* 14(1):53–72. http://dx.doi.org/10.1007/s12110-003-1016-y.

Hillier, B., A. Leaman, P. Stansall, and M. Bedford. 1976. "Space Syntax." *Environment & Planning B* 3(2):147–185. http://dx.doi.org/10.1068/b030147.

Hintzman, M. W. 2000. "Scarce-Resource Procurement and Use: The Technological Analysis of an Obsidian Blade Workshop in the Lowlands of Belize." Unpublished Master's thesis, Department of Anthropology, University of California, Riverside.

Hirth, K. G. 1998. "The Distributional Approach: A New Way to Identify Marketplace Exchange in the Archaeological Record." *Current Anthropology* 39(4):451–476. http://dx.doi.org/10.1086/204759.

Hirth, K. G. 2003. "The Altepetl and Urban Structure in Prehispanic Mesoamerica." In *El Urbanismo en Mesoamérica/Urbanism in Mesoamerica*, ed. W. T. Sanders, A. G. Mastache, and R. H. Cobean, 57–84. México DF: INAH.

Hirth, K. G. 2013. "Economic Consumption and Domestic Economy in Cholula's Rural Hinterland, Mexico." *Latin American Antiquity* 24(2):123–148. http://dx.doi.org/10.7183/1045-6635.24.2.123.

Hirth, K. G., ed. 2009. *Housework: Craft Production and Domestic Economy in Ancient Mesoamerica*. Hoboken: Archaeological Papers of the American Anthropological Association.

Hobbes, T. 1651. *Leviathan*. London: Green Dragon in St. Pauls Church-Yard.

Hodge, M. G. 1984. *Aztec City-States*. Ann Arbor: Museum of Anthropology, University of Michigan.

Hodos, T. 2006. *Local Responses to Colonization in the Iron Age Mediterranean*. New York: Routledge.

Hoebel, E. A. 1988. *The Cheyennes: Indians of the Great Plains*. Belmont: Wadsworth/Thompson.

Hoedl, L. 2013. "The Many Shades of Black: Obsidian Distribution and Social Organization in the Tequila Valley Region of Jalisco, Mexico." Master's thesis, Department of Anthropology, University of Colorado, Denver.

Hole, F. 1991. "Middle Khabur Settlement and Agriculture in the Ninevite 5 Period." *Bulletin of the Canadian Society for Mesopotamian Studies* 21:17–29.

Hole, F. 2000. "The Prehistory of the Khabur." In *La Djéziré et l'Euphrate Syriens de la Protohistoire à la fin du Second Millénaire av. J.C: Tendances dans l'Interprétation Historique des Données Nouvelles (Subartu VII)*, ed. O. Rouault and M. Wäfler, 17–27. Turnhout: Brepols.

Holmes, G. 1962. *The Later Middle Ages, 1272–1485*. vol. 3. A History of England. Edinburgh: T. Nelson.

Hommon, R. 2013. *The Ancient Hawaiian State*. Oxford: Oxford University Press. http://dx.doi.org/10.1093/acprof:oso/9780199916122.001.0001.

Honeychurch, W., and C. Amartuvshin. 2007. "Hinterlands, Urban Centers, and Mobile Settings: The 'New' Old World Archaeology from the Eurasian Steppe." *Asian Perspective* 46 1):36–64. http://dx.doi.org/10.1353/asi.2007.0005.

Hooper, P. L., H. S. Kaplan, and J. L. Boone. 2010. "A Theory of Leadership in Human Cooperative Groups." *Journal of Theoretical Biology* 265(4):633–646. http://dx.doi.org/10.1016/j.jtbi.2010.05.034.

Hopkins, K. 1980. "Taxes and Trade in the Roman Empire (200 B.C.–A.D. 400)." *Journal of Roman Studies* 70:101–125. http://dx.doi.org/10.2307/299558.

Horton, M., and J. Middleton. 2000. *The Swahili: The Social Landscape of a Mercantile Society*. Malden: Blackwell.

Houston, S. D., and T. Inomata. 2009. *The Classic Maya*. Cambridge: Cambridge University Press.

Huang, R. 1998. "The Ming Fiscal Administration." In *The Cambridge History of China, Volume 8: The Ming Dynasty, 1368–1644, Part 2*, ed. D. Twitchett and F.W. Mote, 106–171. Cambridge: Cambridge University Press. http://dx.doi.org/10.1017/CHOL9780521243339.004.

Hucker, C. O. 1998. "Ming Fovernment." In *The Cambridge History of China, Volume 8: The Ming Dynasty, 1368–1644, Part 2*, ed. D. Twitchett and F. W. Mote, 9–105. Cambridge: Cambridge University Press. http://dx.doi.org/10.1017/CHOL9780521243339.003.

Hull, K. 2010. "An Epigraphic Analysis of Classic-Period Maya Foodstuffs." In *Pre-Columbian Foodways: Interdisciplinary Approaches to Food, Culture, and Markets in Ancient Mesoamerica*, ed. J. E. Staller and M. Carrasco, 235–256. New York: Springer. http://dx.doi.org/10.1007/978-1-4419-0471-3_9.

Hunt, R. C., and A. Gilman, eds. 1998. *Property in Economic Context*. Lanham, MD: University Press of America.

Hurgronje, C. S. 1906. *The Achehnese*. London: E. J. Brill.
Hwa, L. 2008. "State Building in the Government of Tang Taizong." In *Forum on Public Policy Online*, vol. 2. Accessed October 2, 2014. http://forumonpublicpolicy.com/summer08papers/archivesummer08/hwa.rev.pdf.
Iles, L. E. 2010. "Reconstructing the Iron Production Technologies of Western Uganda: Reconciling Archaeometallurgical and Ethnoarchaeological Approaches." Unpublished PhD dissertation, University College, London.
Inalcik, H. 1994. "The Ottoman State: Economy and Society, 1300–1600." In *An Economic and Social History of the Ottoman empire, 1300–1914*, ed. H. Inalcik and D. Quataert, 9–410. Cambridge: Cambridge University Press.
Inomata, T., and L. S. Coben, eds. 2006. *Archaeology of Performance: Theaters of Power, Community, and Politics*. Lanham, MD: Altamira Press.
Irwin, G. 1992. *The Prehistoric Exploration and Colonisation of the Pacific*. Cambridge: Cambridge University Press. http://dx.doi.org/10.1017/CBO9780511518225.
Ivanhoe, P., ed. 2009. *Mencius*. New York: Columbia University Press.
Iversen, F. 2013. "Concilium and Pagus-Revisiting the Early Germanic Thing System of Northern Europe." *Journal of the North Atlantic* 5:5–17.
Jacobs, J. 1969. *The Economy of Cities*. New York: Vintage.
Jacobsen, T. 1943. "Primitive Democracy in Ancient Mesopotamia." *Journal of Near Eastern Studies* 2 (3): 159–172. http://dx.doi.org/10.1086/370672.
James, S. 1999. *The Atlantic Celts: Ancient People or Modern Invention?* Madison: University of Wisconsin Press.
James, S. 2000. "A Reply to Amy Hale." *Folklore* 111:307–313.
James, T.G.H. 1984. *Pharaoh's People: Scenes from Life in Imperial Egypt*. London: Bodley Head.
Jansson, K., and A. Kristensson. 2004. *Knutpunkt Bredestad Undersökningar av Järnåldershärdar och ett Vikingatida hus, Fornlämningarna 76 och 78* (Investigation of Iron Age Hearths and a Viking Age House) 2004:36. Jönköpings: Läns Museum, Jönköping.
Janusek, J. W. 2008. *Ancient Tiwanaku*. Cambridge: Cambridge University Press.
Jefferson, T. 1999. *Political Writings*. Ed. J. Appleby and T. Ball. Cambridge: Cambridge University Press. http://dx.doi.org/10.1017/CBO9781139164351.
Jennings, J. 2011. *Globalizations and the Ancient World*. Cambridge: Cambridge University Press.
Johns, C. 2014. "Ceramic Activity Analysis of Navajas Circle 5 and the Need for Practice Theory in Unusual Monumental Architecture." Master's thesis, Department of Anthropology, University of Colorado, Denver.
Johnson, A., and T. Earle. 2000. *The Evolution of Human Societies*. Stanford: Stanford University Press.

Johnson, G. A. 1982. "Organizational Structure and Scalar Stress." In *Theory and Explanation in Archaeology: The Southampton Conference*, ed. C. Renfrew, M. T. Rowlands, and B. A. Seagraves-Whallon, 389–421. New York: Academic Press.

Johnson, G. A. 1983. "Decision-Making Organization and Pastoral Nomad Camp Size." *Human Ecology* 11(2):175–199. http://dx.doi.org/10.1007/BF00891742.

Jones, C. 1996. *Excavations in the East Plaza of Tikal*. Philadelphia: University of Pennsylvania Museum of Archaeology and Anthropology.

Joyce, A. A. 1991. "Formative Period Social Change in the Lower Río Verde Valley, Oaxaca, Mexico." *Latin American Antiquity* 2(2):126–150. http://dx.doi.org/10.2307/972274.

Joyce, A. A. 2000. "The Founding of Monte Albán: Sacred Propositions and Social Practices." In *Agency in Archaeology*, ed. M. A. Dobres and J. Robb, 71–91. London: Routledge Press.

Joyce, A. A. 2004. "Sacred Space and Social Relations in the Valley of Oaxaca." In *Mesoamerican Archaeology*, ed. J. Hendon and R. Joyce, 192–216. Oxford: Blackwell.

Joyce, A. A. 2006. "The Inhabitation of Río Viejo's Acropolis." In *Space and Spatial Analysis in Archaeology*, ed. E. C. Robertson, J. D. Seibert, D. C. Fernández, and M. U. Zeder, 83–96. Calgary: University of New Mexico and University of Calgary Presses.

Joyce, A. A. 2010. *Mixtecs, Zapotecs, and Chatinos: Ancient Peoples of Southern Mexico*. Malden: Wiley-Blackwell.

Joyce, A. A. 2014. "Warfare in Late/Terminal Formative Period Oaxaca." In *Embattled Bodies, Embattled Places: War in Pre-Columbian Mesoamerica and the Andes*, ed. A. K. Scherer and J. W. Verano, 117–141. Washington, DC: Dumbarton Oaks Research Library and Collection.

Joyce, A. A., L. A. Bustamante, and M. N. Levine. 2001. "Commoner Power: A Case Study from the Classic Period Collapse on the Oaxaca Coast." *Journal of Archaeological Method and Theory* 8(4):343–385. http://dx.doi.org/10.1023/A:1013786700137.

Joyce, A. A., and S. B. Barber. 2011. "Excavating the Acropolis at Río Viejo, Oaxaca, Mexico." *Mexicon* 33(1):15–20.

Joyce, A. A., M. N. Levine, and S. B. Barber. 2013. "Place-Making and Power in the Terminal Formative Period: Excavations on Río Viejo's Acropolis." In *Polity and Ecology in Formative Period Coastal Oaxaca*, ed. A. A. Joyce, 135–163. Boulder: University Press of Colorado.

Joyce, A. A., M. Winter, and R. G. Mueller. 1998. *Arqueología de la Costa de Oaxaca: Asentamientos del Periodo Formativo en el Valle del Río Verde Inferior*. Estudios de Antropología e Historia No. 40. Oaxaca: Centro INAH.

Joyce, R. A., and S. D. Gillespie, eds. 2000. *Beyond Kinship: Social and Material Reproduction in House Societies*. Philadelphia: University of Pennsylvania Press.

Julien, C. J. 1982. "Inca Decimal Administration in the Lake Titicaca Region." In *The Inca and Aztec States, 1400–1800: Anthropology and History*, ed. G. A. Collier, R. I. Rosaldo, and J. D. Wirth, 119–151. New York: Academic Press.

Junker, L. 1990. "The Organization of Intra-Regional and Long-Distance Trade in Prehispanic Philippine Complex Societies." *Asian Perspective* 29:167–209.

Junker, L. 1999. *Raiding, Trading, and Feasting: The Political Economy of Philippine Chiefdoms*. Honolulu: University of Hawai'i Press.

K. W. [Sir Tito Winyi]. 1935. "The Kings of Bunyoro-Kitara." *Uganda Journal* 3:155–160.

K. W. [Sir Tito Winyi]. 1936. "The Kings of Bunyoro-Kitara. Part II." *Uganda Journal* 4:75–83.

K. W. [Sir Tito Winyi]. 1937. "The Procedure in Accession to the Throne of a Nominated King in the Kingdom of Bunyoro-Kitara." *Uganda Journal* 4:289–299.

Kaliff, A. 2003. *Slättbygdsprojektet. Västra Östragötlands Kulturlandskap i ett Långtidsperspektiv (Western Östergötlands Cultural Landscape in Long-Term Perspective)*. Linköping: Riksantikvarieämbetet.

Karunatilake, H. N. S. 1988. *The Accelerated Mahaweli Programme and Its Impact*. Colombo: Centre for Demographic and Socioeconomic Studies.

Katajala, K. 2004. "Against Tithes and Taxes, for King and Province. Peasant Unrest and Medieval Scandinavian Political Culture." In *Northern Revolts: Medieval and Early Modern Unrest in the Nordic Countries*, ed. K. Katajala, 32–52. Helsinki: Finnish Literature Society.

Keller, A. H. 2006. "Roads to the Center: The Design, Use, and Meaning of the Roads of Xunantunich, Belize." Unpublished PhD dissertation, University of Pennsylvania. University, Ann Arbor.

Kelley, J. C. 1974. "Speculations on the Culture History of Northwestern Mesoamerica." In *The Archaeology of West Mexico*, ed. B. Bell, 19–39. Ajijic: Sociedad de Estudios Avanzados del Occidente de México.

Kelly, I. 1943. "Notes on a West Coast Survival of the Ancient Mexican Ball Game." *Notes on Middle American Archaeology, Carnegie Institution of Washington Division of Historical Research* 6:163–175.

Kelly, M. 1992. "The 1992 Campaign." *New York Times*, October 31. Accessed October 14, 2013. http://www.nytimes.com/1992/10/31/us/1992-campaign-democrats-clinton-bush-compete-be-champion-change-democrat-fights.html.

Kemp, B. J. 1989. *Ancient Egypt: Anatomy of a Civilization*. London: Routledge.

Kepecs, S., and P. Kohl. 2003. "Conceptualizing Macroregional Interaction: World-Systems Theory and the Archaeological Record." In *The Postclassic Mesoamerican World*, ed. M. E. Smith and F. F. Berdan, 14–20. Salt Lake City: University of Utah Press.

Khaldûn, I. 1967. *The Muqaddimah: An Introduction to History.* Trans. F. Rosenthal. London: Routledge & Kegan Paul.

Kihumuro-Apuuli, D. 1994. *A Thousand Years of Bunyoro-Kitara Kingdom: The People and the Rulers.* Kampala: Fountain Publishers.

Killion, T. W., and J. Urcid. 2001. "The Olmec Legacy: Cultural Continuity and Change in Mexico's Southern Gulf Coast Lowlands." *Journal of Field Archaeology* 28:3–25.

Kirch, P. 1991. "Prehistoric Exchange in Western Melanesia." *Annual Review of Anthropology* 20(1):141–165. http://dx.doi.org/10.1146/annurev.an.20.100191.001041.

Kirch, P. 1994. *The Wet and the Dry: Irrigation and Agricultural Intensification in the Pacific.* Chicago: University of Chicago Press.

Kirch, P. 1997. *The Lapita People: Ancestors of the Oceanic World.* Oxford: Blackwell.

Kirch, P. 2000. *On the Roads of the Winds: An Archaeological History of the Pacific Islands Before European Contact.* Berkeley: University of California Press.

Kirch, P. V. 2010. *How Chiefs Became Kings: Divine Kingship and the Rise of Archaic States in Ancient Hawai'i.* Berkeley: University of California Press. http://dx.doi.org/10.1525/california/9780520267251.001.0001.

Kiser, E., and Y. Cai. 2003. "War and Bureaucratization in Qin China: Exploring an Anomalous Case." *American Sociological Review* 68(4):511–539. http://dx.doi.org/10.2307/1519737.

Kiser, E., and M. Hechter. 1991. "The Role of General Theory in Comparative-Historical Sociology." *American Journal of Sociology* 97(1):1–30. http://dx.doi.org/10.1086/229738.

Kiser, E., and A. Linton. 2002. "The Hinges of History: State-Making and Revolt in Early Modern France." *American Sociological Review* 67(6):889–910. http://dx.doi.org/10.2307/3088975.

Kiser, E., and S. Pfaff. 2010. "Comparative-Historical Methodology in Political Sociology." In *Handbook of Politics: State and Society in Global Perspective*, ed. K. T. Leight and J. C. Jenkins, 571–587. New York: Springer. http://dx.doi.org/10.1007/978-0-387-68930-2_30.

Kitchen, K. A. 1982. *Pharaoh Triumphant: The Life and Times of Ramesses II King of Egypt.* Warminster: Aris & Phillips.

Klusáková, L., and S. G. Ellis. 2006. "Introduction." In *Frontiers and Identities: Exploring the Research Area*, ed. L. Klusáková and S. G. Ellis, 13–15. Pisa: Edizioni Plus, Pisa University.

Knapp, A. B. 1993. "Social Complexity: Incipience, Emergence, and Development on Prehistoric Cyprus." *Bulletin of the American Schools of Oriental Research. American Schools of Oriental Research* 292(292):85–106. http://dx.doi.org/10.2307/1357250.

Koenig, W. J. 1990. *The Burmese Polity, 1752–1819: A Study of Kon Baung Politics, Administration, and Social Organization*. Ann Arbor: Center for South and Southeast Asian Studies, University of Michigan.

Kohl, P. L. 2009. "The Maikop Singularity: The Unequal Accumulation of Wealth on the Bronze Age Eurasian Steppe." In *Social Complexity in Prehistoric Eurasia: Monuments, Metals, and Mobility*, ed. B. K. Hanks and K. M. Linduff, 91–104. Cambridge: Cambridge University Press. http://dx.doi.org/10.1017/CBO97805 11605376.007.

Köhler, E. C. 2010. "Theories of State Formation." In *Egyptian Archaeology*, ed. W. Wendrich, 36–54. Chichester: Wiley-Blackwell.

Kohler, T. A., D. Cockburn, P. L. Hooper, R. K. Bocinsky, and Z. Kobti. 2012. "The Coevolution of Group Size and Leadership: An Agent-Based Public Goods Model for Prehispanic Pueblo Societies." *Advances in Complex Systems* 15 (01n02): 1150007. http://dx.doi.org/10.1142/S0219525911003256.

Kohler, T. A., M. W. Van Pelt, and L. Y. L. Yap. 2000. "Reciprocity and Its Limits: Considerations for a Study of the Prehispanic Pueblo World." In *Alternative Leadership Strategies in the Prehispanic Southwest*, ed. B. J. Mills, 180–206. Tucson: University of Arizona Press.

Kolb, M. 1994. "Monumental Grandeur and the Rise of Religious Authority in Precontact Hawaii." *Current Anthropology* 34:9–37.

Kosakowsky, L. J. 2012. "Ceramics and Chronology at Chan." In *Chan: An Ancient Maya Farming Community*, ed. C. Robin, 42–70. Gainesville: University Press of Florida.

Kosse, K. 1990. "Group Size and Societal Complexity: Thresholds in the Long-Term Memory." *Journal of Anthropological Archaeology* 9(3):275–303. http://dx.doi.org/10 .1016/0278-4165(90)90009-3.

Kosse, K. 1994. "The Evolution of Large, Complex Groups: A Hypothesis." *Journal of Anthropological Archaeology* 13(1):35–50. http://dx.doi.org/10.1006/jaar.1994.1003.

Kowalewski, S. A. 1990. "The Evolution of Complexity in the Valley of Oaxaca." *Annual Review of Anthropology* 19(1):39–58. http://dx.doi.org/10.1146/annurev.an .19.100190.000351.

Kowalewski, S. A. 1995. "Spatial and Temporal Variation in the Porfirian Economy." Paper presented at the 15th Annual Meeting of the Society for Economic Anthropology, Santa Fe.

Kowalewski, S. A. 2006. "Coalescent Societies." In *Light on the Path: Essays in the Anthropology and History of the Southeastern Indians*, ed. T. J. Pluckhahn and R. Ethridge, 94–122. Tuscaloosa: University of Alabama Press.

Kowalewski, S. A. 2012. "A Theory of the Ancient Mesoamerican Economy." *Research in Economic Anthropology* 32:187–224. http://dx.doi.org/10.1108/S0190-1281(2012)0000032012.

Kowalewski, S. A., A. K. Balkansky, L. R. Stiver Walsh, T. J. Pluckhahn, J. F. Chamblee, V. Pérez Rodríguez, V. Y. Heredia Espinoza, and C. A. Smith. 2009. *Origins of the Ñuu: Archaeology in the Mixteca Alta, Mexico.* Boulder: University Press of Colorado.

Kowalewski, S. A., L. Barba Pingarrón, G. García Ayala, B. A. Steere, J. Blancas, M. Y. Cortés Vilchis, L. López Zárate, A. Ortiz, T. J. Pluckhahn, and B. Vilchis Flores. 2010. "La Presencia Azteca en Oaxaca: La Provincia de Coixtlahuaca." *Anales de Antropología* 44:77–103.

Kowalewski, S. A., L. A. Barba Pingarrón, J. Blancas, M. Y. Cortés, Vilchis, G. García Ayala, L. López Zárate, A. Ortíz, T. J. Pluckhahn, B. A. Steere and B. Vilchis Flores. 2008. *Proyecto Urbanismo Temprano y Tardío en Coixtlahuaca, Oaxaca.* Mexico DF : INAH.

Kowalewski, S. A., R. E. Blanton, G. M. Feinman, and L. Finsten. 1983. "Boundaries, Scale, and Internal Organization." *Journal of Anthropological Archaeology* 2(1):32–56. http://dx.doi.org/10.1016/0278-4165(83)90007-7.

Kowalewski, S. A., G. M. Feinman, L. Finsten, and R. E. Blanton. 1991. "Pre-Hispanic Ballcourts from the Valley of Oaxaca, Mexico." In *The Mesoamerican Ballgame*, ed. V. L. Scarborough and D. R. Wilcox, 25–44. Tucson: University of Arizona Press.

Kowalewski, S. A., G. M. Feinman, L. Finsten, R. E. Blanton, and L. M. Nicholas. 1989. *Monte Albán's Hinterland, Part II: Prehispanic Settlement Patterns in Tlacolula, Etla, and Ocotlan, the Valley of Oaxaca, Mexico.* Ann Arbor: Museum of Anthropology, University of Michigan.

Kramer, C. 1997. *Pottery in Rajasthan: Ethnoarchaeology in Two Indian Cities.* New York: Academic Press.

Kristiansen, K. 1991. "Chiefdoms, States, and Systems of Social Evolution." In *Chiefdoms: Power, Economy, and Ideology*, ed. T. Earle, 16–43. Cambridge: Cambridge University Press.

Kristiansen, K. 1994. "The Emergence of the European World System in the Bronze Age: Divergence, Convergence, and Social Evolution during the First and Second Millennia BC in Europe." In *Europe in the First Millennium B.C*, ed. K. Kristiansen and J. Jensen, 7–30. Sheffield: J. R. Collis Publications.

Kristiansen, K. 1998. *Europe Before History.* Cambridge: Cambridge University Press.

Kristiansen, K. 1999. *Europe Before History.* Cambridge: Cambridge University Press.

Kristiansen, K. 2005. "Chiefdoms, States, and Systems of Social Evolution." In *Social Transformations in Archaeology: Global and Local Perspectives*, ed. K. Kristiansen and M. Rowlands, 236–259. New York: Routledge.

Kristiansen, K., and T. B. Larsson. 2005. *The Rise of Bronze Age Society-Travels, Transmissions, and Transformations*. Cambridge: Cambridge University Press.

Krugman, P. 1980. "Scale Economies, Product Differentiation, and the Pattern of Trade." *American Economic Review* 70:950–959.

Ladefoged, T., P. Kirch, S. Gon, III, O. Chadwick, A. Hartshorn, and P. Vitousek. 2009. "Opportunities and Constraints for Intensive Agriculture in the Hawaiian Archipelago Prior to European Contact." *Journal of Archaeological Science* 36(10):2374–2383. http://dx.doi.org/10.1016/j.jas.2009.06.030.

Ladrón de Guevara, S. 1999. *Imagen y Pensamiento en El Tajín*. Xalapa: Universidad Veracruzana.

Landa, D. 1941. *Landa's Relación de las Cosas de Yucatán, a Translation*. Translated by A.M. Tozzer. Cambridge: Papers of the Peabody Museum of American Archaeology and Ethnology, Harvard University.

Lane, F. C. 1973. *Venice, a Maritime Republic*. Baltimore: Johns Hopkins University Press.

Lanning, E. C. 1959. "The Death of Chieftain Kateboha." *Uganda Journal* 23:186–188.

Lanning, E. C. 1966. "Excavations at Mubende Hill." *Uganda Journal* 30:153–163.

Lansing, J. S. 1991. *Priests and Programmers: Technologies of Power in the Engineered Landscape of Bali*. Princeton: Princeton University Press.

Lansing, J. S. 2006. *Perfect Order: Recognizing Complexity in Bali*. Princeton: Princeton University Press.

Laursen, S. T. 2010. "The Westward Transmission of Indus Valley Sealing Technology: Origin and Development of the 'Gulf Type' Seal and Other Administrative Technologies in Early Dilmun, c. 2100–2000 BC." *Arabian Archaeology and Epigraphy* 21(2):96–134. http://dx.doi.org/10.1111/j.1600-0471.2010.00329.x.

LaViolette, A., and J. Fleisher. 2005. "The Archaeology of Sub-Saharan Urbanism: Cities and Their Countrysides." In *African Archaeology: A Critical Introduction*, ed. A. Stahl, 327–352. Malden: Blackwell Publishing.

Law, R. 1977. *The Oyo Empire, c.1600–c.1836: A West African Imperialism in the Era of the Atlantic Slave Trade*. Oxford: Clarendon Press.

Leach, J. 1994. *Survival Psychology*. New York: New York University Press. http://dx.doi.org/10.1057/9780230372719.

LeCount, L. J. 1996. "Pottery and Power: Feasting, Gifting, and Displaying Wealth among the Late and Terminal Classic Lowland Maya." PhD diss., Department of Anthropology, University of California, Los Angeles.

LeCount, L. J. 1999. "Polychrome Pottery and Political Strategies in Late and Terminal Classic Lowland Maya." *Latin American Antiquity* 10(3):239–258. http://dx.doi.org/10.2307/972029.

LeCount, L. J. 2001. "Like Water for Chocolate: Feasting and Political Ritual among the Late Classic Maya of Xunantunich, Belize." *American Anthropologist* 103(4):935–953. http://dx.doi.org/10.1525/aa.2001.103.4.935.

LeCount, L. J. 2010. "Mount Maloney People? Domestic Pots, Every Day Practice, and the Social Formation of the Xunantunich Polity." In *Classic Maya Provincial Politics: Xunantunich and its Hinterlands*, ed. L. J. LeCount and J. Yaeger, 209–232. Tucson: University of Arizona Press.

LeCount, L. J., and J. Yaeger. 2010. "Conclusions: Placing Xunantunich and Its Hinterland Settlements in Perspective." In *Classic Maya Provincial Politics: Xunantunich and Its Hinterlands*, ed. L. J. LeCount and J. Yaeger, 337–369. Tucson: University of Arizona Press.

LeCount, L. J., J. Yaeger, R. M. Leventhal, and W. Ashmore. 2002. "Dating the Rise and Fall of Xunantunich, Belize: A Late and Terminal Classic Lowland Maya Regional Center." *Ancient Mesoamerica* 13(1):41–63. http://dx.doi.org/10.1017/S0956536102131117.

Lejju, B. J., D. Taylor, and P. Robertshaw. 2005. "Late Holocene Environmental Variability at Munsa Archaeological Site, Uganda: A Multi-Core, Multi-Proxy Approach." *Holocene* 15(7):1044–1061. http://dx.doi.org/10.1191/0959683605hl877ra.

Lekson, S. 1985. "Largest Settlement Size and the Interpretation of Socio-Political Complexity at Chaco Canyon, New Mexico." *Haliksa'i: UNM Contributions to Anthropology* 4:68–75.

León-Portilla, M. 1985. *Tonalamatl de los Pochtecas (Códice Mesoamericano Fejérváry-Mayer).* México DF: Celanese Mexicana, S.A.

Levi, M. 1981. "The Predatory Theory of Rule." *Politics & Society* 10(4):431–465. http://dx.doi.org/10.1177/003232928101000403.

Levi, M. 1988. *Of Rule and Revenue*. Berkeley: University of California Press.

Levick, B. M. 1996. "Senators, Patterns of Recruitment." In *The Oxford Classical Dictionary*, ed. S. Hornblower and A. Spawforth, 1387–1388. Oxford: Oxford University Press.

Levine, M. N. 2002. "Ceramic Change and Continuity in the Lower Río Verde Region of Oaxaca Mexico: The Late Formative to Early Terminal Formative Transition." Unpublished MA thesis, Department of Anthropology, University of Colorado, Boulder.

Levine, M. N. 2011. "Negotiating Political Economy at Late Postclassic Tututepec (Yucu Dzaa), Oaxaca, Mexico." *American Anthropologist* 113(1):22–39. http://dx.doi.org/10.1111/j.1548-1433.2010.01304.x.

Levy, J. 2009. "Montesquieu's Constitutional Legacies." In *Montesquieu and his Legacy*, ed. R. E. Kingston, 115–137. Albany: State University of New York Press.

Leyenaar, T.J.J. 2001. "The Modern Ballgames of Sinaloa: A Survival of the Aztec Ullamaliztli." In *The Sport of Life and Death: The Mesoamerican Ballgame*, ed. E. M. Whittington, 122–130. New York: Thames and Hudson.

Lichbach, M. I. 1995. *The Rebel's Dilemma. Ann Harbor*. University of Michigan.

Lichbach, M. I. 1996. *The Cooperator's Dilemma*. Ann Arbor: University of Michigan Press.

Lind, M., and J. Urcid. 2010. *The Lords of Lambityeco and their Nearest Neighbors: Political Evolution in the Valley of Oaxaca during the Xoo Phase*. Boulder: University Press of Colorado.

Lindow, J. 1976. *Comitatus, Individual, and Honor: Studies in Northern Germanic Institutional Vocabulary*. Berkeley: University of California Publications in Linguistics.

Linduff, K. M., and K. S. Rubinson, eds. 2008. *Are all Warriors Male?* Lanham, MD: Altamira Press.

Little, D. 2000. "Explaining Large-Scale Historical Change." *Philosophy of the Social Sciences* 30(1):89–112. http://dx.doi.org/10.1177/004839310003000105.

Little, D. 2008. "Eurasian Historical Comparisons: Conceptual Issues in Comparative Historical Inquiry." *Social Science History* 32:235–261.

Ljungkvist, J. 2008. "Dating Two Royal Mounds of Old Uppsala Evaluating the Elite of the 6th–7th Century in Middle Sweden." *Archäologisches Korrespondenzblatt* 38:263–282.

Lloyd, P. C. 1971. *The Political Development of Yoruba Kingdoms in the Eighteenth and Nineteenth Centuries*. London: Royal Anthropological Institute.

Locke, J. 2003. *Two Treatises of Government and a Letter Concerning Toleration*. Ed. I. Shapiro. New Haven: Yale University.

Lockhart, J., F. F. Berdan, and A. J. O. Anderson. 1986. *The Tlaxcalan Actas: A Compendium of the Records of the Cabildo of Tlaxcala (1545–1627)*. Salt Lake City: University of Utah Press.

Long, S. V. 1966. "Archaeology of the Municipio de Etzatlán, Jalisco." PhD dissertation, Department of Anthropology, University of California at Los Angeles.

Loughlin, M. L. 2012. "El Mesón Regional Survey: Settlement Patterns and Political Economy in the Eastern Papaloapan Basin, Veracruz, Mexico." Unpublished PhD diss., Department of Anthropology, University of Kentucky.

Lucero, L. J., J. D. Gunn, and V. L. Scarborough. 2011. "Climate Change and Classic Maya Water Management." *Water (Basel)* 3(4):479–494. http://dx.doi.org/10.3390/w3020479.

Lunagómez, R. 2011. *Los Patrones Arquitectónicos Prehispánicos del Sur de Veracruz: Epoca Clásica*. Xalapa: Universidad Veracruzana.

Lybyer, A. H. 1966. *The Government of the Ottoman Empire in the Time of Suleiman the Magnificent*. New York: Russell & Russell.
MacDonald, K. C., and S. Camara. 2012. "Segou, Slavery, and *Sifinso*." In *Power and Landscape in Atlantic West Africa*, ed. J. C. Monroe and A. Ogundiran, 169–190. Cambridge: Cambridge University Press. http://dx.doi.org/10.1017/CBO9780511921032.010.
MacDonald, W. L. 1986. *An Urban Appraisal*. vol. 2. The Architecture of the Roman Empire. New Haven: Yale University Press.
Mackay, E. J. H. 1943. *Chanhu-daro Excavations, 1935–1936*. New Haven: American Oriental Society.
MacSweeney, N. 2004. "Social Complexity and Population: A Study in the Early Bronze Age Aegean." *Papers from the Institute of Archaeology* 15:82–89. http://dx.doi.org/10.5334/pia.225.
Magnusson, G. 1984. "Lapphyttan en Masugn i Karbennings Socken (Lapphyttan a Blast Furnace in Karbennings Parish)." In *Karbenning en Bergslagssocken* (Karbenning, a Mining Parish), ed. L. Bygge, K. Bergman, and V. Melin, 60–85. Norberg: Norbergs Kommun.
Magnusson, G. 2002. "Bergslagens Bebyggelse Under 1000 år (A Millennium of Human Settlement in Bergslagen)." *Bebyggelsehistorisk Tidskrift* 43:7–30.
Magnusson, G. 2009. "Om Järnets Roll (On Iron's Role)." In Järnets roll Skånelands och södra Smålands Järnframställning under Förhistorisk och Historisk Tid (*Iron's Role: Scania and Southern Småland's Iron Industry in Prehistoric and Historic Times*), ed. B. Helgesson, 7–18. Sweden: Regionmuseet, Kristianstad.
Malinowski, B. 1922. *Argonauts of the Western Pacific*. London: Routledge.
Malpass, M. A. 1996. *Daily Life in the Inca Empire*. Westport: Greenwood Press.
Mann, M. 1986. *The Sources of Social Power: A History of Power from the Beginning to AD 1760*. vol. 1. Cambridge: Cambridge University Press. http://dx.doi.org/10.1017/CBO9780511570896.
Manzanilla, L. 1999. "The Emergence of Complex Urban Societies in Central Mexico: The Case of Teotihuacan." In *Archaeology in Latin America*, ed. G. G. Politis and B. Alberti, 96–126. London: Routledge.
Marcus, J. 1992. *Mesoamerican Writing Systems*. Princeton: Princeton University Press.
Marcus, J. 1993. "Ancient Maya Political Organization." In *Lowland Maya Civilization in the Eighth Century A.D*, ed. J. A. Sabloff and J. S. Henderson, 65–89. Washington, DC: Dumbarton Oaks.
Marcus, J. 1998. "The Peaks and Valleys of Ancient States: An Extension of the Dynamic Model." In *Archaic States*, ed. G. M. Feinman and J. Marcus, 59–94. Santa Fe: School of American Research Press.

Marcus, J. 2003. "Recent Advances in Maya Archaeology." *Journal of Archaeological Research* 11(2):71–148. http://dx.doi.org/10.1023/A:1022919613720.
Marcus, J. 2008. "The Archaeological Evidence of Social Evolution." *Annual Review of Anthropology* 37(1):251–266. http://dx.doi.org/10.1146/annurev.anthro.37.081407.085246.
Marcus, J., and K. V. Flannery. 1996. *Zapotec Civilization: How Urban Society Evolved in Mexico's Oaxaca Valley*. London: Thames and Hudson.
Markens, R., and C. Martínez. 2009. "El Sistema de Producción Cerámica en Monte Albán durante el Preclásico Tardío y el Clásico Tardío." In *Bases de la Complejidad Social en Oaxaca: Memoria de la Cuarta Mesa Redonda de Monte Albán*, ed. N. Robles, 123–152. México DF: INAH.
Markman, C. W. 1981. *Prehistoric Settlement Dynamics in Central Oaxaca, Mexico: A View from the Miahuatlán Valley*. Nashville: Vanderbilt University.
Martin, S., and N. Grube. 2008. *Chronicle of the Maya Kings and Queens*. London: Thames and Hudson.
Martínez López, C., M. Winter, and P. A. Juárez. 1995. "Entierros Humanos del Proyecto Especial Monte Albán 1992–1994." In *Entierros Humanos de Monte Albán: Dos Estudios*, ed. M. Winter, 79–247. México DF: Instituto Nacional de Anthropología e Historia.
Masson, M. A., and D. A. Freidel. 2012. "An Argument for Classic Era Maya Market Exchange." *Journal of Anthropological Archaeology* 31(4):455–484. http://dx.doi.org/10.1016/j.jaa.2012.03.007.
Mayr, E. 1985. *The Growth of Biological Thought: Diversity, Evolution, and Inheritance*. Cambridge, MA: Harvard University Press.
McAnany, P. A. 2004. "Appropriate Economies: Labor Obligations and Luxury Goods in Ancient Maya Societies." In *Archaeological Perspectives on Ancient Economies*, ed. G. Feinman and L. Nicholas, 145–165. Salt Lake City: University of Utah Press.
McAnany, P. A., and N. Yoffee, eds. 2010. *Questioning Collapse: Human Resilience, Ecological Vulnerability, and the Aftermath of Empire*. Cambridge: Cambridge University Press.
McCafferty, S. D., and G. G. McCafferty. 2000. "Textile Production in Postclassic Cholula, Mexico." *Ancient Mesoamerica* 11(01):39–54. http://dx.doi.org/10.1017/S0956536100111071.
McCann, L. 2000. "Seasons of Labor: Family, Work, and Land in a Nineteenth-Century Nova Scotia Shipbuilding Community." *History of the Family* 4(4):485–527. http://dx.doi.org/10.1016/S1081-602X(99)00026-3.
McCaskie, T. C. 1995. *State and Society in Pre-Colonial Asante*. New York: Cambridge University Press.

McGuire, R. H. 1983. "Breaking Down Cultural Complexity: Inequality and Heterogeneity." In *Advances in Archaeological Method and Theory*, vol. 6. ed. M. B. Schiffer, 91–142. New York: Academic Press. http://dx.doi.org/10.1016/B978-0-12-003106-1.50008-0.

McIntosh, R. 2005. *Ancient Middle Niger: Urbanism and the Self-Organizing Landscape*. Cambridge: Cambridge University Press.

McIntosh, S. K. 1999. "Pathways to Complexity: An African Perspective." In *Beyond Chiefdoms: Pathways to Complexity*, ed. S. McIntosh, 1–30. Cambridge: Cambridge University Press. http://dx.doi.org/10.1017/CBO9780511558238.002.

Meadow, R. H., and J. M. Kenoyer. 2000. "The 'Tiny' Steatite Seals (Incised Steatite Tablets) of Harappa: Some Observations on their Context and Dating." In *South Asian Archaeology 1997*, ed. M. Taddei and G. DeMarco, 2–20. Rome: Istituto Italiano per L'Africa e'L'Oriente.

Mierow, C. C. 1908. "The Origin and Deeds of the Goths." English Version. Princeton: Princeton University Press.

Miller, A. G. 1995. *The Painted Tombs of Oaxaca, Mexico: Living with the Dead*. Cambridge: Cambridge University Press.

Miller, D. 1985a. "Ideology and the Harappan Civilization." *Journal of Anthropological Archaeology* 4(1):34–71. http://dx.doi.org/10.1016/0278-4165(85)90013-3.

Miller, D. 1985b. *Artifacts as Categories: A Study of Ceramic Variability in Central India*. Cambridge: Cambridge University Press.

Miller, M. E. 1991. "Rethinking the Classic Sculptures of Cerro de las Mesas, Veracruz, Mexico." In *Settlement Archaeology of Cerro de las Mesas, Veracruz, Mexico*, ed. B. L. Stark, 26–38. Los Angeles: University of California, Institute of Archaeology.

Millon, R. 1973. *The Teotihuacán Map, Part 1: Text*. Austin: University of Texas Press.

Millon, R. F. 1955. "When Money Grew on Trees: A Study of Cacao in Ancient Mesoamerica." PhD dissertation, Department of Anthropology, Columbia University, New York.

Mills, B., ed. 2000. *Alternative Leadership Strategies in the Prehispanic Southwest*. Tucson: University of Arizona Press.

Minc, L. D. 2006. "Monitoring Regional Market Systems in Prehistory: Models, Methods, and Metrics." *Journal of Anthropological Archaeology* 25(1):82–116. http://dx.doi.org/10.1016/j.jaa.2005.09.003.

Mintz, S. W. 1961. "Pratik: Haitian Personal Economic Relationships." In *Proceedings of the 1961 Annual Spring Meeting of the American Ethnological Society*, ed. Viola E. Garfield, 54–63. Seattle: University of Washington Press.

Mixter, D., T. Jamison, and L. LeCount. 2013. "Actuncan's Noble Court: New Insights into Classic-Period Hierarchies, Polities, and their Failure in the Upper Belize River Valley." *Research Reports in Belizean Archaeology* 10:91–103.

Mjøset, L. 2009. "Contextualist Approach to Social Science Methodology." In *The Sage Handbook of Case-Based Methods*, ed. D. S. Byrne and C. C. Ragin, 39–68. London: Sage Publications. http://dx.doi.org/10.4135/9781446249413.n3.

Modelski, G. 1997. "Cities of the Ancient World: An Inventory (–3500 to –1200)." Manuscript on file, Integrative Research Center, Field Museum of Natural History, Chicago.

Modelski, G. 1999. "Ancient World Cities 4000–1000 BC: Centre/Hinterland in the World System." *Global Society* 13 (4): 383–392.

Moertono, S. 1981. *State and Statecraft in Old Java: A Study of the Later Mataram Period, 16th to 19th Century*. Ithaca: Modern Indonesia Project, Southeast Asia Program, Cornell University.

Moholy-Nagy, H. 2003. "Source Attribution and the Utilization of Obsidian in the Maya Area." *Latin American Antiquity* 14(3):301–310. http://dx.doi.org/10.2307/3557561.

Moholy-Nagy, H. 2008. *The Artifacts of Tikal: Ornamental and Ceremonial Artifacts and Unworked Material.* Tikal Report No. 27, Part A. Philadelphia: University of Pennsylvania Museum of Archaeology and Anthropology.

Monaghan, J. 1995. *The Covenants with Earth and Rain: Exchange, Sacrifice, and Revelation in Mixtec Sociality*. Norman: University of Oklahoma Press.

Monaghan, J. 2009. "Sacrificio y Poder en Mesoamérica." In *Bases de la Complejidad Social en Oaxaca: Memoria de la Cuarta Mesa Redonda de Monte Albán*, ed. N. Robles, 181–197. México DF: INAH.

Montelius, S. 1953. "The Burning of Forest Land for the Cultivation of Crops, 'Svedjebruk' in Central Sweden." *Geografiska Annaler* 35(1):41–54. http://dx.doi.org/10.2307/520265.

Montet, P. 1964. *Eternal Egypt*. New York: New American Library.

Montet, P. 1981. *Everyday Life in Egypt in the Days of Ramesses the Great*. Philadelphia: University of Pennsylvania Press.

Moore, T., and X. L. Armada, eds. 2011. *Atlantic Europe in the First Millennium BC: Crossing the Divide*. Oxford: Oxford University Press.

Moreno Guzmán, M., and M. Korn. 2012. "Construcción y Técnicas." In *El Penacho del México Antiguo*, ed. S. Haag, A. de M. y Campos, L. Rivero and C. Feest, 61–82. Altenstadt: ZKF Publishers.

Morgan, L.H.L. 1901. *League of the Ho-De'-No-Sau-Nee or Iroquois*. 2 vols. New York: Dodd, Mead, and Company.

Morris, I. 1999. "Negotiated Peripherality in Iron Age Greece: Accepting and Resisting the East." In *World-Systems Theory in Practice: Leadership, Production, and Exchange*, ed. P. N. Kardulias, 63–84. Lanham, MD: Rowman and Littlefield Publishers, Inc.

Morris, I. 2010. "Social Development." October. htttp://ianmorris.org.
Morris, W. A., ed. 1940. *The English Government at Work, 1327–1336, Volume 1: Central and Prerogative Administration*. Cambridge: The Mediaeval Academy of America.
Morris, W. A., and J. R. Strayer, eds. 1947. *The English Government at Work, 1327–1336, Volume 2: Fiscal Administration.* Cambridge: The Mediaeval Academy of America.
Murnane, W. J. 1998. "The Organization of Government under Amenhotep III." In *Amenhotep III: Perspectives on his Reign*, ed. D. O'Connor and E. H. Cline, 173–221. Ann Arbor: University of Michigan Press.
Murra, J. V. 1980. "The Economic Organization of the Inka State." *Research in Economic Anthropology* Supplement 1.
Myrberg, N. 2008. "Room for All? Spaces and Places for Thing Assemblies: The Case of the All-Thing on Götland, Sweden." *Viking and Medieval Scandinavia* 4:133–157. http://dx.doi.org/10.1484/J.VMS.1.100308.
Nadel, S. F. 1942. *A Black Byzantium: The Kingdom of Nupe in Nigeria*. London: Oxford University Press.
Nance, C. R., S. L. Whittington, and B. E. Jones-Borg. 2003. *Archaeology and Ethnohistory of Iximché*. Gainesville: University Press of Florida.
Nance, R. C. 2013. "Ceramic Descriptions." In *Correspondence Analysis and West Mexico Archaeology*, ed. C. R. Nance, J. de Leeuw, P. C. Weigand, K. Prado, and D. S. Verity, 101–144. Albuquerque: University of New Mexico Press.
Nelson, B. A. 2004. "Elite Residences in West Mexico." In *Palaces of the Ancient New World*, ed. S. T. Evans and J. Pillsbury, 59–82. Washington, DC: Dumbarton Oaks.
Netting, R. 1989. "Smallholders, Householders, Freeholders: Why the Family Farm Works Well Worldwide." In *The Household Economy: Reconsidering the Domestic Mode of Production*, ed. R. R. Wilk, 221–244. Boulder: Westview Press.
Nichols, D. L., and T. H. Charlton, eds. 1997. *The Archaeology of City-States: Cross-Cultural Approaches*. Washington, DC: Smithsonian Institution Press.
Norwich, J. J. 1982. *A History of Venice*. New York: Knopf.
Nyakatura, J. (Original work published 1947) 1973. *Anatomy of an African Kingdom: A History of Bunyoro-Kitara*. Edited by. G. Uzoigwe, trans. T. Muganwa. New York: Doubleday.
O'Gorman, R., J. Henrich, and M. Van Vugt. 2009. "Constraining Free-Riding in Public Goods Games: Designated Solitary Punishers Can Sustain Human Cooperation." *Proceedings. Biological Sciences* 276(1655):323–329. http://dx.doi.org/10.1098/rspb.2008.1082.
Ofek, H. 2001. *Second Nature: Economic Origins of Human Evolution*. Cambridge: Cambridge University Press. http://dx.doi.org/10.1017/CBO9780511754937.

Ohnersorgen, M. A., and M. D. Varien. 1996. "Formal Architecture and Settlement Organization in Ancient West Mexico." *Ancient Mesoamerica* 7(01):103–120. http://dx.doi.org/10.1017/S0956536100001322.
Ojeda Gastélum, S., B. F. Benz, and L. López Mestas C. 2008. "La Tradición Teuchitlán, el Maguey y el Tequila. Un Debate Acerca de su Confluencia Espacial, Histórica y Cultural. Integración o Destrucción?" In Tradición *Teuchitlán*, ed. Phil C. Weigand, Christopher Beekman, and Rodrigo Esparza, 275–302. Zamora: Colegio de Michoacán.
Oliveros Morales, J. A. 2004. *Hacedores de Tumbas en El Opeño, Michoacán*. Zamora: Ayuntamiento de Jacona, Colegio de Michoacán.
Olson, M. 1965. *The Logic of Collective Action: Public Goods and the Theory of Groups*. Cambridge, MA: Harvard University Press.
Oosthuizen, S. 2013. "Beyond Hierarchy: The Archaeology of Collective Governance." *World Archaeology* 45(5):714–729. http://dx.doi.org/10.1080/00438243.2013.847634.
Österberg, E. 1993. "Vardagens Sträva Samförstånd: Bondepolitik i den Svenska Modellen från Vasatid till Frihetstid." In *Svensk Historia Underifrån: Tänka, Tycka, Tro*, ed. G. Broberg, U. Wikander, and K. Åmark, 126–146. Stockholm: Ordfronts Förlag.
Ostrom, E. 1990. *Governing the Commons: The Evolution of Institutions for Collective Action*. Cambridge: Cambridge University Press. http://dx.doi.org/10.1017/CBO9780511807763.
Ostrom, E. 2003. "How Types of Goods and Property Rights Jointly Affect Collective Action." *Journal of Theoretical Politics* 15(3):239–270. http://dx.doi.org/10.1177/0951692803015003002.
Oubiña, C. P. 2003. "Looking Forward in Anger: Social and Political Transformations in the Iron Age of the North-Western Iberian Peninsula." *European Journal of Archaeology* 6(3):267–299. http://dx.doi.org/10.1177/14619571030060030 4.
Paddock, J. 1970. "Commitment—in Archæology?" *Boletín de Estudios Oaxaqueños* 34. Mitla: Museo Frissell de Arte Zapoteca.
Parker Pearson, M. 1984. "Economic and Ideological Change: Cyclical Growth in the Pre-State Societies of Jutland." In *Ideology, Power, and Prehistory*, ed. D. Miller and C. Tilley, 69–92. Cambridge: Cambridge University Press.
Parkinson, W. A., and M. L. Galaty. 2007. "Secondary States in Perspective: An Integrated Approach to State Formation in the Prehistoric Aegean." *American Anthropologist* 109(1):113–129. http://dx.doi.org/10.1525/aa.2007.109.1.113.
Parkinson, W. A., and M. L. Galaty. 2009. "Introduction." In *Archaic State Interaction: The Eastern Mediterranean in the Bronze Age*, ed. W. A. Parkinson and M. L. Galaty, 3–28. Santa Fe: School for Advanced Research Press.

Parliament. 2012. "Living Heritage: Bomb Damage." Accessed 27 October 2012. www.parliament.uk/about/living-heritage/building/palace/architecture/palace structure/bomb-damage/.

Parpola, A. 1994. *Deciphering the Indus Script*. Cambridge: Cambridge University Press.

Parpola, S., A. Parpola, and R. H. Brunswig. 1977. "The Meluhha Village: Evidence of Acculturation of Harappan Traders in Late Third Millennium Mesopotamia?" *Journal of Economic and Social History of the Orient* 20:129–165.

Parry, W. J. 1987. *Chipped Stone Tools in Formative Oaxaca, Mexico: Their Procurement, Production, and Use*. Ann Arbor: Museum of Anthropology, University of Michigan.

Parsons, J. R. 1971. *Prehistoric Settlement Patterns in the Texcoco Region, Mexico*. Ann Arbor: Museum of Anthropology, University of Michigan.

Patterson, T. C. 1991. *The Inca Empire: The Formation and Disintegration of a Pre-Capitalist State*. New York: Berg.

Pauketat, T. R. 2001. "Practice and History in Archaeology: An Emerging Paradigm." *Anthropological Theory* 1(1):73–98. http://dx.doi.org/10.1177/146349960122228638.

Pauketat, T. R. 2007. *Chiefdoms and Other Archaeological Delusions*. Lanham, MD: AltaMira Press.

Pennisi, E. 2005. "How Did Cooperative Behavior Evolve?" *Science* 309(5731): 93–96. http://dx.doi.org/10.1126/science.309.5731.93.

Peregrine, P. N. 1998. "Corporate and Network Strategies in the Evolution of Mississippian Societies." Paper Presented at the 63rd Annual Meeting of the Society for American Archaeology, Seattle, March 25–29.

Peregrine, P. N. 2001. "Matrilocality, Corporate Strategy, and the Organization of Production in the Chacoan World." *American Antiquity* 66(1):36–46. http://dx.doi .org/10.2307/2694316.

Peregrine, P. N. 2003. "Large-Scale Processes in Cultural Evolution". Paper presented at the Santa Fe Institute Working Group on Language and Prehistory Conference, Santa Fe, April 29-May 2.

Peregrine, P. N. 2008. "Political Strategy and Cross-Cultural Variation in Games." *Cross-Cultural Research* 42(4):386–393. http://dx.doi.org/10.1177/1069397108321895.

Peregrine, P. N. 2009. "Political Participation, Political Strategies, and Variation in Warfare." Paper presented at the 38th Annual Conference of the Society for Cross-Cultural Research, Las Vegas, February 18–21.

Peregrine, P. N. 2012. "Power and Legitimation: Political Strategies, Typology, and Cultural Evolution." In *Comparative Archaeology of Complex Societies*, ed. M. E. Smith, 165–191. New York: Cambridge University Press.

Peregrine, P. N., C. R. Ember, and M. Ember. 2007. "Modeling State Origins Using Cross-Cultural Data." *Cross-Cultural Research* 41(1):75–86. http://dx.doi.org/10.1177/1069397106295445.

Perez, L. G. 2002. *Daily Life in Early Modern Japan*. Westport: Greenwood Press.

Pérez Rodríguez, V., K. C. Anderson, and M. K. Neff. 2011. "The Cerro Jazmín Archaeological Project: Investigating Prehispanic Urbanism and its Environmental Impact in the Mixteca Alta, Oaxaca, Mexico." *Journal of Field Archaeology* 36(2):83–99. http://dx.doi.org/10.1179/009346911X12991472411321.

Peterson, C. E., and R. D. Drennan. 2005. "Communities, Settlements, Sites, and Surveys: Regional-Scale Analysis of Prehistoric Human Interaction." *American Antiquity* 70(1):5–30. http://dx.doi.org/10.2307/40035266.

Peterson, C. E., and R. D. Drennan. 2012. "Patterned Variation in Trajectories of Community Societies." In *The Comparative Archaeology of Complex Societies*, ed. M. E. Smith, 88–137. Cambridge: Cambridge University Press.

Phipps, E., and L. Commoner. 2006. "Investigation of a Colonial Latin American Textile." Textile Society of America, 10th Biennial Symposium: Textile Narratives and Conversations, Toronto, Ontario, Canada.

Pinder, D., I. Shimada, and D. Gregory. 1979. "The Nearest-Neighbor Statistic: Archaeological Application and New Developments." *American Antiquity* 44(3):430–445. http://dx.doi.org/10.2307/279543.

Pittman, H. 2013. "New Evidence for Interaction between the Iranian Plateau and the Indus Valley: Seals and Sealings from Konar Sandal South." In *Connections and Complexity: New Approaches to the Archaeology of South Asia*, ed. S. A. Abraham, P. Gullapalli, T. P. Raczek, and U. Z. Rizvi, 63–89. Walnut Creek, CA: Left Coast Press.

Pluckhahn, T. J., and S. A. Kowalewski. 2003. "Plazas and Markets in the Prehispanic Mixteca Alta." Paper presented at the 68th annual meeting of the Society for American Archaeology, Milwaukee.

Plunket, P. S. 1983. "An Intensive Survey in the Yucuita Sector of the Nochixtlán Valley, Oaxaca, Mexico." PhD dissertation, Department of Anthropology, Tulane University, New Orleans. Ann Arbor: University Microfilms.

Pohl, J. M. D., and B. E. Byland. 1990. "Mixtec Landscape Perception and Archaeological Settlement Patterns." *Journal of Field Archaeology* 1:113–131.

Pool, C. A. 2008. "Architectural Plans, Factionalism, and the Protoclassic-Classic Transition at Tres Zapotes." In *Classic Period Cultural Currents in Southern and Central Veracruz*, ed. P. J. Arnold and C. A. Pool, 121–157. Cambridge, MA: Harvard University Press.

Porter, A. 2000. "Mortality, Monuments, and Mobility: Ancestor Traditions and the Transcendance of Space." Unpublished PhD Dissertation, Department of Near Eastern Languages and Civilizations, University of Chicago, Chicago.

Porter, A. 2002. "Communities in Conflict. Death and the Contest for Social Order in the Euphrates River Valley." *Near Eastern Archaeology* 65(3):156–173. http://dx.doi.org/10.2307/3210881.

Porter, A. 2012. *Mobile Pastoralism and the Formation of Near Eastern Civilization: Weaving Together Society*. Cambridge: Cambridge University Press. http://dx.doi.org/10.1017/CBO9780511895012.

Postill, J. 2011. *Localizing the Internet: An Anthropological Account*. Oxford: Berghahn.

Powell, M. A. 1996. "Money in Mesopotamia." *Journal of Economic and Social History of the Orient* 39(3):224–242. http://dx.doi.org/10.1163/1568520962601225.

Price, B. 1978. "Secondary State Formation: An Exploratory Model." In *Origin of the State: The Anthropology of Political Evolution*, ed. R. Cohen, 161–224. Philadelphia: Institute for the Study of Human Issues.

Prins, A. H. J. 1971. *Didemic Lamu: Social Stratification and Spatial Structure in a Muslim Maritime Town*. Groningen: Instituut voor Culturele Antropologie der Rijksuniversiteit Groningen.

Pryor, F. L. 1977. *The Origins of the Economy: A Comparative Study of Distribution in Primitive and Peasant Economies*. New York: Academic Press.

Quiñones Keber, E. 1995. *Codex Telleriano-Remensis*. Austin: University of Texas Press.

Rabibhadana, A. 1969. *The Organization of Thai Society in the Early Bangkok Period, 1782–1873*. Ithaca: Southeast Asia Program, Cornell University.

Rainey, L. D. 2010. *Confucius and Confucianism: The Essentials*. Malden, MA: Wiley-Blackwell.

Ramos de la Vega, J., and L. López Mestas Camberos. 1996. "Datos Preliminares sobre el Descubrimiento de una Tumba de Tiro en el Sitio de Huitzilapa, Jalisco." *Ancient Mesoamerica* 7(01):121–134. http://dx.doi.org/10.1017/S0956536100001334.

Rands, R. L., and R. L. Bishop. 1980. "Resource Procurement Zones and Patterns of Ceramic Exchange in the Palenque Region, Mexico." In *Models and Methods in Regional Exchanged*, ed. R. E. Fry, 19–46. Washington, DC: Society for American Archaeology.

Randsborg, K. 1980. *The Viking Age in Denmark: The Formation of a State*. London: Duckworth.

Rapoport, A. 1969. *House Form and Culture*. Milwaukee: University of Wisconsin.

Rapoport, A. 1990. *The Meaning of the Built Environment: A Nonverbal Communication Approach*. Tucson: University of Arizona Press.

Rattray, R. S. 1929. *Ashanti Law and Constitution*. Oxford: Clarendon Press.

Redmond, E. M. 1981. "A Fuego y Sangre: Early Zapotec Imperialism in the Cuicatlan Canada, Oaxaca." PhD dissertation, Department of Anthropology, Yale University, New Haven.

Redmond, E. M., and C. S. Spencer. 2006. "From Raiding to Conquest: Warfare Strategies and Early State Development in Oaxaca, Mexico." In *The Archaeology of Warfare: Prehistories of Raiding and Conquest*, ed. E. N. Arkush and M. W. Allen, 336–393. Gainesville: University Press of Florida.

Reid, A. 1975. "Trade and the Problem of Royal Power in Aceh, Three Stages: c. 1550–1700." In *Pre-Colonial State Systems in Southeast Asia: The Malay Peninsula, Sumatra, Bali-Lombok, South Celebes*, ed. A. Reid and L. Castles, 45–55. Kuala Lumpur: Malaysian Branch of the Royal Asiatic Society.

Renfrew, C. 1972. *The Emergence of Civilization: The Cyclades and the Aegean in the Third Millennium BC.* London: Methuen.

Renfrew, C. 1974. "Beyond a Subsistence Economy: The Evolution of Social Organization in Prehistoric Europe." In *Reconstructing Complex Societies: An Archaeological Colloquium*, ed. C. B. Moore, 69–85. Cambridge: American Schools of Oriental Research.

Renfrew, C., and J. Cherry. 1986. *Peer Polity Interaction and Socio-Political Change.* Cambridge: Cambridge University Press.

Reyna, S. P. 1990. *Wars without End: The Political Economy of a Precolonial African State.* Hanover: University Press of New England.

Ricardo, D. 1821. *On the Principles of Political Economy and Taxation.* 3rd ed. London: John Murray.

Riisøy, A. I. 2013. "Sacred Legal Places in Eddic Poetry: Reflected in Real Life?" *Journal of the North Atlantic* 5:28–41.

Ringle, W. M., T. Gallareta Negrón, and G. J. Bey III. 1998. "The Return of Quetzalcoatl: Evidence for the Spread of a World Religion during the Epiclassic Period." *Ancient Mesoamerica* 9: 183–232.

Rissman, P. C. 1988. "Public Displays and Private Values: A Guide to Buried Wealth in Harappan Archaeology." *World Archaeology* 20(2):209–228. http://dx.doi.org/10.1080/00438243.1988.9980068.

Rivera Guzmán, A. I. 1999. "El Patrón de Asentamiento en la Mixteca Baja de Oaxaca: Análisis del Area de Tequixtepec-Chazumba." Unpublished MA thesis, Department of Anthropology, Escuela Nacional de Antropología, México DF.

Rivero Weber, L., and C. Feest. 2012. "La Sombra de los Dioses: El Arte Plumario en el México del Siglo XVI." In *El Penacho del México Antiguo*, ed. S. Haag, A. de M. y Campos, L. Rivero, and C. Feest, 41–60. Altenstadt: ZKF Publishers.

Robertshaw, P. 1994. "Archaeological Survey, Ceramic Analysis and State Formation in Western Uganda." *African Archaeological Review* 12(1):105–131. http://dx.doi.org/10.1007/BF01953040.

Robertshaw, P. 1999a. "Women, Labor and State Formation in Western Uganda." In *Complex Polities in the Ancient Tropical World*, ed. E. A. Bacus and L. J. Lucero, 51–65. Washington, DC: American Anthropological Association Archaeological Memoirs. http://dx.doi.org/10.1525/ap3a.1999.9.1.51.

Robertshaw, P. 1999b. "Seeking and Keeping Power in Bunyoro-Kitara, Uganda." In *Beyond Chiefdoms: Pathways to Complexity in Africa*, ed. S. K. McIntosh, 124–135. Cambridge: Cambridge University Press. http://dx.doi.org/10.1017/CBO9780511558238.011.

Robertshaw, P. 2001. "The Age and Function of the Ancient Earthworks of Western Uganda." *Uganda Journal* 47(1):20–33. http://dx.doi.org/10.4314/uj.v47i1.23050.

Robertshaw, P. 2002. "The Ancient Earthworks of Western Uganda: Capital Sites of a Cwezi Empire?" *Uganda Journal* 48(1):17–32. http://dx.doi.org/10.4314/uj.v48i1.23001.

Robertshaw, P. 2003. "Explaining the Origins of the State in East Africa." In *East African Archaeology: Foragers, Potters, Smiths, and Traders*, ed. C. M. Kusimba and S. B. Kusimba, 149–166. Philadelphia: Museum of Archaeology and Anthropology, University of Pennsylvania Press.

Robertshaw, P. 2010. "Beyond the Segmentary State: Creative and Instrumental Power in Western Uganda." *Journal of World Prehistory* 23(4):255–269. http://dx.doi.org/10.1007/s10963-010-9039-x.

Robertshaw, P., K. Murphy, and S. H. Ambrose. 2012. "Life and Death in Bunyoro-Kitara, Uganda: The Analysis of Human Remains and Burials from the Early to Mid-Second Millennium CE." Paper presented at the conference of the Society of Africanist Archaeologists, Toronto, June.

Robertshaw, P., and D. Taylor. 2000. "Climate Change and the Rise of Political Complexity in Western Uganda." *Journal of African History* 41(1):1–28. http://dx.doi.org/10.1017/S0021853799007653.

Robertshaw, P., D. Taylor, S. Doyle, and R. Marchant. 2004. "Famine, Climate, and Crisis in Western Uganda." In *Past Climate Variability through Europe and Africa*, ed. R. W. Battarbee, F. Grasse, and C. E. Stickley, 535–549. Dordrecht: Springer. http://dx.doi.org/10.1007/978-1-4020-2121-3_25.

Robin, C. 2003a. "New Directions in Classic Maya Household Archaeology." *Journal of Archaeological Research* 11(4):307–356. http://dx.doi.org/10.1023/A:1026327105877.

Robin, C. 2003b. "Social Diversity and Everyday Life within Classic Maya Settlements." In *Mesoamerican Archaeology*, ed. J. Hendon and R. Joyce, 148–168. Oxford: Blackwell Publishers.

Robles García, N. M., J. Vera Estrada, and E. García Wigueras. 2014. "El Edificio Funerario de Atzompa." *Arqueología Mexicana* 21(126):48–51.
Rogers, J. D. 2012. "Inner Asian States and Empire: Theories and Syntheses." *Journal of Archaeological Research* 20(3):205–256. http://dx.doi.org/10.1007/s10814-011-9053-2.
Rojas, M. 2005. *Sweden After the Swedish Model from Tutorial State to Enabling State.* Stockholm: Timbro.
Roscoe, J. 1923. *The Bakitara or Banyoro.* Cambridge: Cambridge University Press.
Roscoe, J. 1965. *The Baganda: An Account of their Native Customs and Beliefs.* London: F. Cass.
Roscoe, P. B. 1993. "Practice and Political Centralization: A New Approach to Political Evolution." *Current Anthropology* 34(2):111–124. http://dx.doi.org/10.1086/204149.
Roscoe, P. B. 2000. "Costs, Benefits, Typologies, and Power: The Evolution of Political Hierarchy." In *Hierarchies in Action: Cui Bono?* ed. M. W. Diehl, 113–133. Carbondale: Center for Archaeological Investigations, Southern Illinois University.
Ross, M. H. 2007. *Cultural Contestation in Ethnic Conflict.* Cambridge: Cambridge University Press. http://dx.doi.org/10.1017/CBO9780511491115.
Ross, M. L. 2004. "Does Taxation Lead to Representation?" *British Journal of Political Science* 34(2):229–29. http://dx.doi.org/10.1017/S0007123404000031.
Rousseau, J. J. 1764. *A Treatise on the Social Compact; or the Principles of Politic Law.* London: T. Becket and P. A. De Hondt, in the Strand.
Rueda Smithers, S. 2009. "Ceremonial Shield (*Chimalli*)." In *Moctezuma: Aztec Ruler*, ed. C. McEwan and L. López, 108–109. London: The British Museum Press.
Rydberg, S. 1979. *Stora Kopparberg—1000 Years of an Industrial Activity.* Stockholm: Gullers International AB.
Sabloff, J. A. 2012. "Foreword." In *The Comparative Archaeology of Complex Societies*, ed. M. E. Smith, xvii–xix. Cambridge: Cambridge University Press.
Sahagún, B. de 1950–1982. *Florentine Codex: General History of the Things of New Spain.* 12 vols. Ed. and trans. A. J. O. Anderson and C. E. Dibble. Salt Lake City: University of Utah Press.
Sahagún, B. de 1993. *Primeros Memoriales.* Norman: University of Oklahoma Press.
Saitta, D. J. 2013. "Agency and Collective Action: Insights from North American Historical Archaeology." In *Cooperation and Collective Action: Archaeological Perspectives*, ed. D. Carballo, 129–148. Boulder: University Press of Colorado.
Saletore, B. A. 1934. *Social and Political Life in the Vijayanagara Empire (A.D. 1346–A.D. 1646).* Madras: B. G. Paul & Co.
Salzman, P. C. 1972. "Multi-Resource Nomadism in Iranian Baluchistan." In *Perspectives on Nomadism*, ed. W. Irons and N. Dyson-Hudson, 60–68. Leiden: E. J. Brill.

Salzman, P. C. 2002. "Pastoral Nomads: Some General Observations Based on Research in Iran." *Journal of Anthropological Research* 58(2):245–264. http://dx.doi.org/10.1086/jar.58.2.3631038.
Sanders, W. T., and S. T. Evans. 2006. "Rulership and Palaces at Teotihuacan." In *Palaces and Power in the Americas: From Peru to the Northwest Coast*, ed. J. J. Christie and P. J. Sarro, 256–284. Austin: University of Texas Press.
Sanders, W. T., J. R. Parsons, and M. Logan. 1976. "Summary and Conclusions." In *The Valley of Mexico: Studies in Pre-Hispanic Ecology and Society*, ed. E. R. Wolf, 161–178. Albuquerque: University of New Mexico Press.
Sanders, W. T., J. R. Parsons, and R. S. Santley. 1979. *The Basin of Mexico: Ecological Processes in the Evolution of a Civilization*. New York: Academic Press.
Sanders, W. T., and B. J. Price. 1968. *Mesoamerica: The Evolution of a Civilization*. New York: Random House.
Sanmark, A. 2009. "Administrative Organisation and State Formation: A Case Study of Assembly Sites in Södermanland, Sweden." *Medieval Archaeology* 53(1):205–241. http://dx.doi.org/10.1179/007660909X12457506806289.
Sanmark, A., and S. Semple. 2008. "Places of Assembly: New Discoveries in Sweden and England." *Fornvännen* 103:245–259.
Santley, R. S., and K. G. Hirth, eds. 1993. *Prehispanic Domestic Units in Western Mesoamerica: Studies of the Household, Compound, and Residence*. Boca Raton: CRC Press.
Sarkar, J. 1963. *Mughal Administration*. Calcutta: M.C. Sarkar.
Saville, M. 1920. *The Goldsmith's Art in Ancient Mexico*. New York: Museum of the American Indian, Heye Foundation.
Scarborough, V. L. 2003. *The Flow of Power: Ancient Water Systems and Landscapes*. Santa Fe: School for American Research Press.
Scarborough, V. L., and F. Valdez, Jr. 2009. "An Alternative Order: The Dual Economies of the Ancient Maya." *Latin American Antiquity* 20:207–227.
Scarborough, V. L., and D. R. Wilcox, eds. 1991. *The Mesoamerican Ballgame*. Tucson: University of Arizona Press.
Schele, L., and J. Guernsey Kappelman. 2001. "What the Heck's Coatépec? 'The Formative Roots of an Enduring Mythology." In *Landscape and Power in Ancient Mesoamerica*, ed. R. Koontz, K. Reese, and A. Headrick, 29–54. Boulder: Westview Press.
Schelling, T. C. 2006. *Micromotives and Macrobehavior*. New York: W.W. Norton.
Schneider, J. 1977. "Was There a Precapitalist World-System?" *Peasant Studies* 6:20–29.
Schoenbrun, D. 1999. "The (In)visible Roots of Bunyoro-Kitara and Buganda in the Lakes Region: AD 800–1300." In *Beyond Chiefdoms: Pathways to Complexity in Africa*, ed. S. McIntosh, 136–150. Cambridge: Cambridge University Press. http://dx.doi.org/10.1017/CBO9780511558238.012.

Schortman, E. M., and P. A. Urban. 1992. "The Place of Interaction in Archaeological Thought." In *Resources, Power, and Interregional Interaction*, ed. E. M. Schortman and P. A. Urban, 3–15. New York: Plenum Press. http://dx.doi.org/10.1007/978-1-4757-6416-1_1.

Schulte Nordholt, H. 1991. *State, Village, and Ritual in Bali: A Historical Perspective*. Amsterdam: VU University Press.

Scott, J. C. 1989. "Everyday Forms of Resistance." In *Everyday Forms of Peasant Resistance*, ed. F. D. Colburn, 3–33. Armonck: M.E. Sharpe, Inc.

Scott, J. C. 1990. *Domination and the Arts of Resistance: Hidden Transcripts*. New Haven: Yale University Press.

Scott, J. C. 1998. *Seeing Like a State: How Certain Schemes to Improve the Human Condition have Failed*. New Haven: Yale University Press.

Scott, J. C. 2009. *The Art of Not Being Governed: An Anarchist History of Upland Southeast Asia*. New Haven: Yale University Press.

Semple, S., and A. Sanmark. 2013. "Assembly in North West Europe: Collective Concerns for Early Societies?" *European Journal of Archaeology* 16(3):518–342. http://dx.doi.org/10.1179/1461957113Y.0000000035.

Service, E. R. 1975. *Origins of the State and Civilization: The Process of Cultural Evolution*. New York: W.W. Norton.

Sewell, W. H., Jr. 2005. *Logics of History: Social Theory and Social Transformation*. Chicago: University of Chicago Press. http://dx.doi.org/10.7208/chicago/9780226749198.001.0001.

Shanks, M., and C. Tilley. 1982. "Ideology, Symbolic Power and Ritual Communication: A Reinterpretation of Neolithic Mortuary Practices." In *Symbolic and Structural Archaeology*, ed. I. Hodder, 129–154. Cambridge: University of Cambridge Press. http://dx.doi.org/10.1017/CBO9780511558252.013.

Sharer, R. J., and L. P. Traxler. 2006. *The Ancient Maya*. Stanford: Stanford University Press.

Shaw, J. M. 2001. "Maya Sacbeob: Form and Function." *Ancient Mesoamerica* 12(02):261–272. http://dx.doi.org/10.1017/S0956536101121048.

Shaw, L. C. 2012. "The Elusive Maya Marketplace: An Archaeological Consideration of the Evidence." *Journal of Archaeological Research* 20(2):117–155. http://dx.doi.org/10.1007/s10814-011-9055-0.

Shipton, P. 1984. "Strips and Patches: A Demographic Dimension in Some African Land-holding and Political Systems." *Man* 19(4):613–634. http://dx.doi.org/10.2307/2802329.

Shults, S. 2012. "Uncovering Ancient Maya Exchange Networks: Using the Distributional Approach to Interpret Obsidian Exchange at Actuncan, Belize."

Unpublished Master's thesis, Department of Anthropology, University of Alabama, Tuscaloosa.
Sieff, D. F. 1997. "Herding Strategies of the Datoga Pastoralists of Tanzania: Is Household Labor a Limiting Factor." *Human Ecology* 25(4):519–544. http://dx.doi.org/10.1023/A:1021829806765.
Sigurðsson, J. V. 2006. "Tendencies in the Historiography on the Medieval Nordic States (to 1350)." In *Public Power in Europe Studies in Historical Transformations*, ed. A. K. Isaacs and G. Hálfdanarson, 1–15. Pisa: Edizioni Plus, Pisa University.
Silver, M. 1983. "Karl Polanyi and Markets in the Ancient Near East: The Challenge of the Evidence." *Journal of Economic History* 43(04):795–829. http://dx.doi.org/10.1017/S002205070003076X.
Sinha, S. 1978. "Coexistence of Multiple Scales and Networks of Civilization: India." In *Scale and Social Organization*, ed. F. Barth, 122–132. Oslo: Universitetsforlaget.
Small, D. B. 2009. "The Dual-processual Model in Ancient Greece: Applying a Post-neoevolutionary Model to a Data Rich Environment." *Journal of Anthropological Archaeology* 28(2):205–221. http://dx.doi.org/10.1016/j.jaa.2009.02.004.
Smith Marquez, S. M. 2009. *El Caso Teuchitlán-Loma Alta: Un Ejemplo de Patron de Asentamiento Mesoamericano.* Tesis de Maestría, Centro de Estudios Arqueológicos, Colegio de Michoacán, La Piedad.
Smith Marquez, S. M., and J. Herrejón Villicaña. 2004. *Las Unidades Habitacionales del Posclásico en la Zona de Teuchitlan, Jalisco.* Universidad Autónoma de Guadalajara, Guadalajara: Tesis de Licenciatura.
Smith, A. T. 2000. "Rendering the Political Aesthetic: Political Legitimacy in Urartian Representations of the Built Environment." *Journal of Anthropological Archaeology* 19(2):131–163. http://dx.doi.org/10.1006/jaar.1999.0348.
Smith, A. T. 2003. *The Political Landscape: Constellations of Authority in Early Complex Polities.* Berkeley: University of California Press.
Smith, C. 1974. "Economics of Marketing Systems: Models from Economic Geography." *Annual Review of Anthropology* 3(1):167–201. http://dx.doi.org/10.1146/annurev.an.03.100174.001123.
Smith, C. 1976a. "Exchange Systems and the Spatial Distribution of Elites: The Organization of Stratification in Agrarian Societies." In *Regional Analysis*, ed. C. Smith, 309–374. New York: Academic Press. http://dx.doi.org/10.1016/B978-0-12-652102-3.50022-0.
Smith, C. 1976b. "Regional Economic Systems: Linking Geographic Models and Socioeconomic Problems." In *Regional Analysis*, ed. C. Smith, 3–63. New York: Academic Press. http://dx.doi.org/10.1016/B978-0-12-652101-6.50008-5.
Smith, C. A., ed. 1976. *Regional Analysis.* 2 vols. New York: Academic Press.

Smith, Ch. A. 1993. "Prehispanic Mixtec Social Organization: The Architectural Evidence." MA thesis, Department of Anthropology, University of Georgia, Athens.

Smith, Ch. A. 2002. "Concordant Change and Core-periphery Dynamics: A Synthesis of Highland Mesoamerican Archaeological Survey Data." PhD dissertation, Department of Anthropology, University of Georgia, Athens.

Smith, K. P. 2004. "Patterns in Time and the Tempo of Change: A North Atlantic Perspective on the Evolution of Complex Societies." In *Exploring the Role of Analytical Scale in Archaeological Interpretation*, ed. J. R. Mathiew and R. E. Scott, 83–99. BAR International Series. Oxford: British Archaeological Reports.

Smith, L. E., and F. K. Branom. 1937. *New International Atlas of the World*. Chicago: Geographical Publishing Company.

Smith, M. E. 1998. "Comment on Hirth's 'Distributional Approach.'" *Current Anthropology* 40:528–530.

Smith, M. E. 2003a. "Can We Read Cosmology in Ancient Maya City Plans? Comment on Ashmore and Sabloff." *Latin American Antiquity* 14(2):221–228. http://dx.doi.org/10.2307/3557596.

Smith, M. E. 2003b. "Economic Change in Morelos Households." In *The Postclassic Mesoamerican World*, ed. M. E. Smith and F. F. Berdan, 249–258. Salt Lake City: University of Utah Press.

Smith, M. E. 2003c. "Key Commodities." In *The Postclassic Mesoamerican World*, ed. M. E. Smith and F. F. Berdan, 117–125. Salt Lake City: University of Utah Press.

Smith, M. E. 2005. "Did the Maya Build Architectural Cosmograms?" *Latin American Antiquity* 16(2):217–224. http://dx.doi.org/10.2307/30042813.

Smith, M. E. 2006. "How Do Archaeologists Compare Early States?" *Revista de Antropologia* 35:5–35.

Smith, M. E. 2007. "Form and Meaning in the Earliest Cities: A New Approach to Ancient Urban Planning." *Journal of Urban Planning* 6:3–47.

Smith, M. E. 2008. *Aztec City-state Capitals*. Gainesville: University Press of Florida.

Smith, M. E. 2009. "Ancient Cities." In *The Encyclopedia of Urban Studies*, ed. R. Hutchison, 24–28. London: Sage Publications.

Smith, M. E. 2012. *The Aztecs*. Chichester: Wiley-Blackwell.

Smith, M. E., J. B. Wharton, and J. M. Olson. 2003. "Aztec Feasts, Rituals, and Markets: Political Uses of Ceramic Vessels in a Commercial Economy." In *The Archaeology and Politics of Food and Feasting in Early States and Empires*, ed. B. Tamara, 235–268. New York: Kluwer Academic/Plenum. http://dx.doi.org/10.1007/978-0-306-48246-5_9.

Smith, M. H. 2013. "Thorir's Bargain: Gender, Vaðmál, and the Law." *World Archaeology* 45 (5): 730–46. http://dx.doi.org/10.1080/00438243.2013.860272.

Smith, M. L. 2005. "Networks, Territories, and the Cartography of Ancient States." *Annals of the Association of American Geographers* 95(4):832–849. http://dx.doi.org/10.1111/j.1467-8306.2005.00489.x.
Soofi, A. 1995. "Economics of Ibn Khaldun Revisited." *History of Political Economy* 27(2):387–404. http://dx.doi.org/10.1215/00182702-27-2-387.
Soto de Arechavaleta, M. de los D. 1994. "El Primer Censo Neogallego: Treslado de una visitación de 1525." In *Contribuciones a la Arqueología y Etnohistoria del Occidente de México*, ed. E. Williams, 341–353. Zamora: El Colegio de Michoacán.
Southwold, M. 1961. *Bureaucracy and Chiefship in Buganda: The Development of Appointive Office in the History of Buganda*. Kampala: East African Institute of Social Research.
Spencer, C. S. 1982. *The Cuicatlán Cañada and Monte Albán: A Study of Primary State Formation*. New York: Academic Press.
Spencer, C. S., and E. M. Redmond. 1997. *Archaeology of the Cañada de Cuicatlán, Oaxaca*. New York: Anthropological Papers, American Museum of Natural History.
Spencer, C. S., and E. M. Redmond. 2004. "A Late Monte Albán I Phase (300–100 B.C.) Palace in the Valley of Oaxaca." *Latin American Antiquity* 15(4):441–455. http://dx.doi.org/10.2307/4141587.
Spengler, O. 1926–1928. *The Decline of the West*. Translated by C. F. Atkinson. New York: Knopf.
Spores, R. 1972. *An Archaeological Settlement Survey of the Nochixtlán Valley, Oaxaca*. Nashville: Vanderbilt University.
Spores, R. 2007. *Ñuu Ñudzahui, La Mixteca de Oaxaca: La Evolución de la Cultura Mixteca desde los Primeros Pueblos Preclásicos hasta la Independencia*. Oaxaca: Instituto Estatal de Educación Pública de Oaxaca.
Spores, R., and N. Robles García. 2007. "A Prehispanic (Postclassic) Capital Center in Colonial Transition: Excavations at Yucundaa Pueblo Viejo de Teposcolula, Oaxaca, Mexico." *Latin American Antiquity* 18(3):333–353. http://dx.doi.org/10.2307/25478184.
Stanish, C. 2010. "Labor Taxes, Market Systems, and Urbanization in the Prehispanic Andes: A Comparative Approach." In *Archaeological Approaches to Market Exchange in Ancient Societies*, ed. C. P. Garraty and B. L. Stark, 185–205. Boulder: University Press of Colorado.
Stark, B. L. 1999. "Formal Architectural Complexes in South-Central Veracruz, Mexico: A Capital Zone?" *Journal of Field Archaeology* 26:197–225.
Stark, B. L. 2003. "Cerro de las Mesas: Social and Economic Perspectives on a Gulf Center." In *El Urbanismo en Mesoamérica: Urbanism in Mesoamerica*, ed. W. T. Sanders, A. G. Mastache, and R. Cobean, 391–426. Mexico DF: INAH.

Stark, B. L. 2007. "Out of Olmec." In *The Political Economy of Ancient Mesoamerica: Transformations during the Formative and Classic Periods*, ed. V. L. Scarborough and J. E. Clark, 47–63. Albuquerque: University of New Mexico Press.
Stark, B. L. 2008. "Polity and Economy in the Western Lower Papaloapan Basin." In *Classic Period Cultural Currents in Southern and Central Veracruz*, ed. P. J. Arnold, III, and C. A. Pool, 85–119. Washington, DC: Dumbarton Oaks.
Stark, B. L. 2012. "Canchas de la Mixtequilla." In *Primer Coloquio Temas Selectos del Centro-Sur de Veracruz. El Juego de Pelota*, ed. A. Daneels, N. R. Donner, and J. Hernández Arana. Manuscript submitted to editors.
Stark, B. L. 2013. "Economic Growth in Mesoamerica." Paper presented at the 78th Annual Meeting of the Society for American Archaeology, Honolulu.
Stark, B. L., ed. 2001. *Classic Period Mixtequilla, Veracruz, Mexico: Diachronic Inferences from Residential Investigations. Albany*. New York: Institute for Mesoamerican Studies, The University at Albany.
Stark, B. L., and C. P. Garraty. 2010. "Detecting Marketplace Exchange in Archaeology: A Methodological Review." In *Archaeological Approaches to Market Exchange in Ancient Societies*, ed. C. P. Garraty and B. L. Stark, 33–60. Boulder: University Press of Colorado.
Stark, B. L., L. Heller, and M. A. Ohnersorgen. 1998. "People with Cloth: Mesoamerican Economic Change from the Perspective of Cotton in South-Central Veracruz." *Latin American Antiquity* 9 (1):7–36. http://dx.doi.org/10.2307/972126.
Stark, B. L., and A. Ossa. 2007. "Ancient Settlement, Urban Gardening, and Environment in the Gulf Lowlands of Mexico." *Latin American Antiquity* 18(4):385–406. http://dx.doi.org/10.2307/25478194.
Stark, B. L., and A. Ossa. 2010. "Origins and Development of Mesoamerican Marketing: Evidence from South-Central Veracruz, Mexico." In *Archaeological Approaches to Market Exchange in Ancient Societies*, ed. C. P. Garraty and B. L. Stark, 99–126. Boulder: University of Colorado Press.
Steere, B. A., and S. A. Kowalewski. 2012. "Wealth Stratification in Ancient Mesoamerica." *Social Evolution & History* 11:20–48.
Stein, G. 1999. *Rethinking World-Systems: Diasporas, Colonies, and Interaction in Uruk Mesopotamia*. Tucson: University of Arizona Press.
Steinhart, E. I. 1977. *Conflict and Collaboration: The Kingdoms of Western Uganda, 1890–1907*. Princeton: Princeton University Press.
Sterelny, K. 2013. "Cooperation in a Complex World: The Role of Proximate Factors in Ultimate Explanations." *Biological Theory* 7(4):358–367. http://dx.doi.org/10.1007/s13752-012-0069-8.
Steward, J. H. 1955. *Theory of Culture Change*. Urbana: University of Illinois Press.

Stirling, M. W. 1943. *Stone Monuments of Southern Mexico*. Washington, DC: Bureau of American Ethnology, United States Government Printing Office.
Stomper, J. A. 1996. "The Popol Na: A Model for Ancient Maya Community Structure at Copán, Honduras." Unpublished PhD diss., Department of Anthropology, Yale University, New Haven.
Stomper, J. A. 2001. "A Model for Late Classic Community Structure at Copán, Honduras." In *Landscape and Power in Ancient Mesoamerica*, ed. R. Koontz, K. Reese-Taylor, and A. Headrick, 197–229. Boulder: Westview Press.
Stone, E. C. 2007. "The Mesopotamian Urban Experience." In *Settlement and Society*, ed. E. C. Stone, 213–234. Los Angeles: Cotsen Institute of Archaeology, UCLA.
Stone, E. C., and P. Zimansky. 2004. *The Anatomy of a Mesopotamian City: Survey and Soundings at Maskhan-shapir*. Winona Lake, Indian: Eisenbrauns.
Strathern, A. 1969. "Finance and Production: Two Strategies in New Guinea Highlands Exchange Systems." *Oceania* 40(1):42–67. http://dx.doi.org/10.1002/j.1834-4461.1969.tb01056.x.
Sugiyama, S. 1993. "Worldview Materialized in Teotihuacán, Mexico." *Latin American Antiquity* 4(2):103–129. http://dx.doi.org/10.2307/971798.
Sutton, J. 1993. "The Antecedents of the Interlacustrine Kingdoms." *Journal of African History* 34(01):33–64. http://dx.doi.org/10.1017/S0021853700032990.
Szuchman, J., ed. 2009. *Nomads, Tribes, and the State in the Ancient Near East: Cross-Disciplinary Perspectives*. Chicago: The Oriental Institute of the University of Chicago.
Taladoire, E. 1981. *Les Terrains de Jeu de Balle*. vol. 4. México DF: Mission Archeologique et Ethnologique Francaise au Mexique.
Tan, S., N. Heerink, and F. Qu. 2006. "Land Fragmentation and its Driving Forces in China." *Land Use Policy* 23(3):272–285. http://dx.doi.org/10.1016/j.landusepol.2004.12.001.
Tarkanian, M. J., and D. Hosler. 2011. "America's First Polymer Scientists: Rubber Processing, Use and Transport in Mesoamerica." *Latin American Antiquity* 22(4):469–486. http://dx.doi.org/10.7183/1045-6635.22.4.469.
Tatsuya, T. 1991. "Politics in the Eighteenth Century." In *The Cambridge History of Japan, Volume 4, Early Modern Japan*, ed. J. W. Hall, 425–477. Cambridge: Cambridge University Press.
Taylor, P. J. 2012. "Extraordinary Cities: Early 'City-ness' and the Origins of Agriculture and States." *International Journal of Urban and Regional Research* 36(3):415–447. http://dx.doi.org/10.1111/j.1468-2427.2011.01101.x.
Tello, F. A. 1968. *Crónica Miscelánea de la Sancta Provincia de Xalisco. Libro Segundo, Volumen I*. Guadalajara: Gobierno del Estado de Jalisco and Universidad de Guadalajara.

Temin, P. 2013. *Roman Market Economy*. Princeton: Princeton University Press.

Thomson, P. 2008. "Field." In *Pierre Bourdieu: Key Concepts*, ed. Michael Grenfell, 67–81. Stocksfield: Acumen.

Thurston, T. L. 2001. *Landscapes of Power, Landscapes of Conflict: State Formation in South Scandinavian Iron Age*. New York: Kluwer Academic/Plenum Publishing.

Thurston, T. L. 2009. "Unity and Diversity in the European Iron Age: Out of the Mists, Some Clarity?" *Journal of Archaeological Research* 17(4):347–423. http://dx.doi.org/10.1007/s10814-009-9032-z.

Thurston, T. L. 2010. "Bitter Arrows and Generous Gifts: What Was a King in the European Iron Age?" In *Pathways to Power New Perspectives on the Emergence of Social Inequality*, ed. T. D. Price and G. M. Feinman, 193–254. New York: Springer. http://dx.doi.org/10.1007/978-1-4419-6300-0_8.

Thurston, T. L. 2015. "Surplus from Below: Self Organization of Production in Early Sweden." In *The Politics of Production and the Strategies of Everyday Life*, ed. C. Morehart and K. De Lucia, 121–151. Boulder: University Press of Colorado. http://dx.doi.org/10.5876/9781607323808.c005.

de Torquemada, J. 1969. *Monarquia Indiana*. 3 vols. Mexico: Editorial Porrua.

Trigger, B. 1974. "The Archaeology of Government." *World Archaeology* 6(1):95–106. http://dx.doi.org/10.1080/00438243.1974.9979591.

Trigger, B. G. 2002. *The Huron: Farmers of the North*. Belmont: Wadswoth/Thomson Learning.

Trigger, B. G. 2003. *Understanding Early Civilizations*. Cambridge: Cambridge University Press. http://dx.doi.org/10.1017/CBO9780511840630.

Turchin, P. 2003. *Historical Dynamics: Why States Rise and Fall*. Princeton: Princeton University Press.

Turner, V. W. 1974. *Dramas, Fields and Metaphors: Symbolic Action in Human Society*. Ithaca: Cornell University Press.

Twitchett, D. 1979. "Introduction." In *The Cambridge History of China: Volume 3, Sui and Tang China, 589–906, Part I*, ed. D. Twitchett, 1–47. Cambridge: Cambridge University Press. http://dx.doi.org/10.1017/CHOL9780521214469.002.

Udo, R. K. 1965. "Disintegration of Nucleated Settlement in Eastern Nigeria." *Geographical Review* 55(1):53–67. http://dx.doi.org/10.2307/212855.

Underhill, A. P., G. M. Feinman, L. M. Nicholas, H. Fang, F. Luan, H. Yu, and F. Cai. 2008. "Changes in Regional Settlement Patterns and the Development of Complex Societies in Southeastern Shandong, China." *Journal of Anthropological Archaeology* 27(1):1–29. http://dx.doi.org/10.1016/j.jaa.2006.11.002.

Urban, P., and E. Schortman. 2004. "Opportunities for Advancement: Intra-community Power Contests in the Midst of Political Decentralization in Terminal Classic

Southeastern Mesoamerica." *Latin American Antiquity* 15(3):251–272. http://dx.doi.org/10.2307/4141574.

Urcid, J. 2005. *The Zapotec Scribal Tradition: Knowledge, Memory, and Society in Ancient Oaxaca*. Foundation for the Advancement of Mesoamerican Studies Inc. Accessed October 2, 2014. http://www.famsi.org/zapotecwriting/.

Urcid, J. 2011. "Los Oráculos y la Guerra: el Papel de las Narrativas Pictóricas en el Desarrollo Temprano de Monte Albán (500 a.C.–200 d.C.)." In *Monte Albán en la Encrucijada Regional y Disciplinaria: Memoria de la Quinta Mesa Redonda de Monte Albán*, ed. N. M. Robles and A. Rivera, 163–240. México DF: INAH.

Urcid, J., and A. Joyce. 2014. "Early Transformations of Monte Albán's Main Plaza and their Political Implications, 500 BC–AD 200." In *Mesoamerican Plazas*, ed. K. Tsukamoto and T. Inomata, 149–167. Tucson: University of Arizona Press.

Urcid, J., and T. W. Killion. 2008. "Social Landscapes and Political Dynamics in the Southern Gulf Coast Lowlands." In *Cultural Currents in Classic Veracruz*, ed. P. J. Arnold and C. A. Pool, 260–291. Washington, DC: Dumbarton Oaks.

Uriarte Torres, A. J. 2011. *Estrategias Políticas y Organización Espacial durante el Formativo en Ciudad Caucel, Yucatán*. Tesis de maestría, El Colegio de Michoacán, La Piedad.

Uzoigwe, G. 1972. "Precolonial Markets in Bunyoro-Kitara." *Comparative Studies in Society and History* 14(04):422–455. http://dx.doi.org/10.1017/S0010417500006794.

Uzoigwe, G. 2013. "Bunyoro-Kitara Revisited: A Reevaluation of the Decline and Diminishment of an African Kingdom." *Journal of Asian and African Studies* 48(1):16–34. http://dx.doi.org/10.1177/0021909611432094.

Van Vugt, M., S. F. Jepson, C. M. Hart, and D. de Cremer. 2003. "Autocratic Leadership in Social Dilemmas: A Threat to Group Stability." *Journal of Experimental Social Psychology* 40(1):1–13. http://dx.doi.org/10.1016/S0022-1031(03)00061-1.

van Zantwijk, R.A.M. 1985. *The Aztec Arrangement: The Social History of Pre-Spanish Mexico*. Norman: University of Oklahoma Press.

Vansina, J. 1973. *The Tio Kingdom of the Middle Congo, 1880–1892*. London: Oxford University Press.

Vansina, J. 1978a. *The Children of Woot: A History of the Kuba Peoples*. Madison: University of Wisconsin Press.

Vansina, J. 1978b. "The Kuba State." In *The Early State*, ed. H. J. M. Claessen and P. Skalník, 359–380. The Hague: Mouton. http://dx.doi.org/10.1515/9783110813326.359.

Vansina, J. 1990. *Paths in the Rainforest: Towards a History of Political Tradition in Equatorial Africa*. London: James Currey.

Vermaak, P. S. 2008. "Guabba, the Meluhhan Village in Mesopotamia." *Journal for Semitics* 17:553–570.

Vestbö-Franzén, A. 2005. *Råg och Rön: Om Mat, Människor och Landskapsförändringar i Norra Småland, ca 1550–1700* [*Research on Rye: Food, Humans, and Structural Changes in the Agrarian Landscape of Northern Småland 1550–1700*]. Jönköping: Jönköpings läns museum.
Vestergaard, T. 1979. "Class and State Development in Scandinavia: Necessary and Voluntary Relations." *Kontaktstencil* 16:25–52.
Vidale, M. 1989. "Specialized Producers and Urban Elites: On the Role of Craft Industries in Mature Harappan Urban Contexts." In *Old Problems and New Perspectives in the Archaeology of South Asia*, ed. J. M. Kenoyer, 171–182. Madison: University of Wisconsin.
Vidale, M. 2000 *The Archaeology of Indus Crafts*. Rome: IsIAO.
Vidale, M. 2005. "The Short-Horned Bull on the Indus Seals: A Symbol of the Families in the Western Trade?" In *Proceedings of the Seventeenth International Conference of the European Association of South Asian Archaeologists (7–11 July 2003, Bonn)*, ed. U. Franke-Vogt and H.-J. Weisshaar, 147–158. Aichwald: Linden Soft Verlag.
von Winning, H. 1969. *Pre-Columbian Art of Mexico and Central America*. London: Thames and Hudson.
von Winning, H. 1971. "Keramische Hausmodelle aus Nayarit, Mexiko." *Baessler-Archiv, Neue Folge* 19:343–377.
von Winning, H., and O. Hammer. 1972. *Anecdotal Sculpture of Ancient West Mexico*. Los Angeles: Ethnic Arts Council of Los Angeles.
Wacquant, L.J.D. 1989. "Towards a Reflexive Sociology: A Workshop with Pierre Bourdieu." *Sociological Theory* 7(1):26–63. http://dx.doi.org/10.2307/202061.
Wagner, J. 2014. "Circles of Glass and Grain: Economic Differences between Core and Semi-peripheral Zones, a Study of Public Center Lithics from the Tequila Valleys of West Mexico." Master's thesis, Department of Anthropology, University of Colorado Denver.
Wailes, B. 1995. "A Case Study of Heterarchy in Complex Societies: Early Medieval Ireland and Its Archeological Implications." In *Heterarchy and the Analysis of Complex Societies*, ed. R. M. Ehrenreich, C. L. Crumley, and J. E. Levy, 55–69. Arlington: Archeological Papers of the American Anthropological Association. http://dx.doi.org/10.1525/ap3a.1995.6.1.55.
Wallace, D. T. 1977. "An Intra-Site Locational Analysis of Utatlan: The Structure of an Urban Site." In *Archaeology and Ethnohistory of the Central Quiché*, ed. D. T. Wallace and R. M. Carmack, 20–54. Albany: State University of New York.
Warner, J. C. 1976. "Survey of the Market System in the Nochixtlán Valley and the Mixteca Alta." In *Markets in Oaxaca*, ed. S. Cook and M. Diskin, 107–131. Austin: University of Texas Press.

Watson, P. J., S. A. LeBlanc, and C. L. Redman. 1971. *Explanation in Archaeology: An Explicitly Scientific Approach*. New York: Columbia University Press.

Waugh, S. L. 1991. *England in the Reign of Edward III*. New York: Cambridge University Press. http://dx.doi.org/10.1017/CBO9781139167918.

Weber, M. 1947. *The Theory of Social and Economic Organization*. New York: Free Press.

Weber, M. 1978. *Economy and Society*. Berkeley: University of California Press.

Weber, S. A. 2003. "Archaeobotany at Harappa: Indications of Change." In *Indus Ethnobiology: New Perspectives from the Field*, ed. S. A. Weber and W. R. Belcher, 175–198. Lanham, MD: Lexington Books.

Webster, D. 1975. "Warfare and the Evolution of the State: A Reconsideration." *American Antiquity* 40(4):464–470. http://dx.doi.org/10.2307/279334.

Webster, D. L. 1997. "City-States of the Maya." In *The Archaeology of City-States: Cross-Cultural Approaches*, ed. D. L. Nichols and T. E. Charlton, 135–154. Washington, DC: Smithsonian Institution Press.

Wechsler, H. J. 1979a. "T'ai-tsung (reign 626–649) the Consolidator." In *The Cambridge History of China: Volume 3, Sui and Tang China, 589–906, Part I*, ed. D. Twitchett, 188–241. Cambridge: Cambridge University Press. http://dx.doi.org/10.1017/CHOL9780521214469.005.

Wechsler, H. J. 1979b. "The Founding of the T'ang Dynasty: Kao-tsu (reign 618–626)." In *The Cambridge History of China: Volume 3, Sui and Tang China, 589–906, Part I*, ed. D. Twitchett, 150–187. Cambridge: Cambridge University Press. http://dx.doi.org/10.1017/CHOL9780521214469.004.

Weigand, P. C. 1991. "The Western Mesoamerican Tlachco: A Two-Thousand-Year Perspective." In *The Mesoamerican Ballgame*, ed. V. L. Scarborough and D. R. Wilcox, 73–86. Tucson: University of Arizona Press.

Weigand, P. C. 1993. *Evolución de una Civilización Prehispánica: Arqueología de Jalisco, Nayarit y Zacatecas*. Trans. P. R. Aviñoá. Zamora: El Colegio de Michoacán.

Weigand, P. C. 1996. "The Architecture of the Teuchitlan Tradition of the Occidente of Mesoamerica." *Ancient Mesoamerica* 7(01):91–101. http://dx.doi.org/10.1017/S0956536100001310.

Weigand, P. C. 2000. "The Evolution and Decline of a Core of Civilization: The Teuchitlán Tradition and the Archaeology of Jalisco." In *Greater Mesoamerica: The Archaeology of West and Northwest Mexico*, ed. M. S. Foster and S. Gorenstein, 43–58. Salt Lake City: University of Utah Press.

Weigand, P. C. 2013. "Archaeology and Ethnohistory of Etzatlán and its Region." In *Correspondence Analysis and West Mexico Archaeology*, ed. C. R. Nance, J. de Leeuw, P. C. Weigand, K. Prado, and D. S. Verity, 17–66. Albuquerque: University of New Mexico Press.

Weigand, P. C. and A. García de Weigand. 1995. *Los Orígenes de los Caxcanes y su Relación con la Guerra de los Nayaritas: Una Hipótesis*. Zapopan: El Colegio de Jalisco.
Weigand, P. C., and A. García de Weigand. 1996. *Tenamaxtli y Guaxicar: las Raíces Profundas de la Rebelión de Nueva Galicia*. Trans. E. W. Martínez. Zamora: El Colegio de Michoacán.
Weigand, P. C., and A. García de Weigand. 1997. "Arqueología y Turismo en el Estado de Jalisco." Antropologia en Jalisco, una Visión Actual 8. Guadalajara: Secretaría de Cultura, Gobierno de Jalisco.
Weigand, P. C., and C. S. Beekman. 1998. "The Teuchitlán Tradition: Rise of a State-like Society." In *Ancient West Mexico: Art and Archaeology of the Unknown Past*, ed. R. Townsend, 35–51. Chicago: Art Institute of Chicago.
Wells, E. C., and K. L. Davis-Salazar, eds. 2007. *The Mesoamerican Ritual Economy: Archaeological and Ethnological Perspectives*. Boulder: University Press of Colorado.
Wells, P. S. 1999. *The Barbarians Speak: How the Conquered Peoples Shaped Roman Europe*. Princeton: Princeton University Press.
West, S. A., C. El Mouden, and A. Gardner. 2011. "Sixteen Common Misconceptions About the Evolution of Cooperation in Humans." *Evolution and Human Behavior* 32(4):231–262. http://dx.doi.org/10.1016/j.evolhumbehav.2010.08.001.
Whalen, M. E. 1988. "Small Community Organization during the Late Formative Period in Oaxaca." *Journal of Field Archaeology* 15:291–306.
Wheeler, R. E. M. 1968. *The Indus Civilization*. Cambridge: Cambridge University Press.
Whittington, E. M., ed. 2001. *The Sport of Life and Death: The Mesoamerican Ballgame*. New York: Thames and Hudson.
Wilkinson, B. 1940. "The Chancery." In *The English Government at Work, 1327–1336, Volume I: Central and Prerogative Administration*, ed. W. A. Morris, 162–205. Cambridge: The Mediaeval Academy of America.
Wilks, I. 1975. *Asante in the Nineteenth Century: The Structure and Evolution of a Political Order*. Cambridge: Cambridge University Press.
Williams, E., ed. 2004. *Bienes Estratégicos del Antiguo Occidente de México: Producción e Intercambio*. Zamora: Colegio de Michoacán.
Winslow, D. 1980. "Rituals of First Menstruation in Sri Lanka." *Man N.S.* 15:603–625.
Winslow, D. 1996. "Pottery, Progress, and Structural Adjustments in a Sri Lankan Village." *Economic Development and Cultural Change* 44(4):701–733. http://dx.doi.org/10.1086/452242.
Winslow, D. 2003. "Potters' Progress: Hybridity and Accumulative Change in Rural Sri Lanka." *Journal of Asian Studies* 62(1):43–70. http://dx.doi.org/10.2307/3096135.

Winslow, D. 2009. "The Village Clay: Recursive Innovations and Community Self-Fashioning among Sinhalese Potters." *Journal of the Royal Anthropological Institute* 15(2):254–275. http://dx.doi.org/10.1111/j.1467-9655.2009.01552.x.

Winter, M. 1984. "Exchange in Formative Highland Oaxaca." In *Trade and Exchange in Early Mesoamerica*, ed. K. Hirth, 179–214. Albuquerque: University of New Mexico.

Winter, M. 1986. "Unidades Habitacionales Prehispánicas en Oaxaca." In *Unidades Habitacionales Mesoaméricanas y sus Areas de Actividad*, ed. L. Manzanilla, 325–374. México DF: INAH.

Winter, M., ed. 1995. *Entierros Humanos de Monte Albán*. Oaxaca: Centro INAH Oaxaca.

Winter, M. 2001. "Palacios, Templos y 1300 Años de Vida Urbana en Monte Albán." In *Reconstruyendo la Ciudad Maya: El Urbanismo en las Sociedades Antiguas*, ed. A. Ciudad, M. J. Iglesia, and M. del C. Martínez, 253–301. Madrid: Sociedad Española de Estudios Mayas.

Wittfogel, K. A. 1957. *Oriental Despotism: A Comparative Study in Total Power*. New Haven: Yale University Press.

Wolf, E. R. 1982. *Europe and the People without History*. Berkeley: University of California Press.

Wolfe, A. 1989. *Whose Keeper*. Berkeley: University of California Press.

Workinger, A. 2002. "Coastal/Highland Interaction in Prehispanic Oaxaca, Mexico: The Perspective from San Francisco de Arriba." PhD dissertation, Department of Anthropology, Vanderbilt University, Nashville.

Wray, L. R., ed. 2004. *Credit and State Theories of Money: The Contributions of A. Mitchel Innes*. Cheltenham: Edward Elgar Publishing. http://dx.doi.org/10.4337/9781843769842.

Wright, H. T. 1977. "Recent Research on the Origin of the State." *Annual Review of Anthropology* 6(1):379–397. http://dx.doi.org/10.1146/annurev.an.06.100177.002115.

Wright, H. T. 1977a. "Recent Research on the Origin of the State." *Annual Review of Anthropology* 6(1):379–397. http://dx.doi.org/10.1146/annurev.an.06.100177.002115.

Wright, H. T. 1977b. "Toward an Explanation of the Origin of the State." In *Explanation of Prehistoric Change*, ed. J. N. Hill, 215–230. Albuquerque: University of New Mexico Press.

Wright, H. T. 2009. "Past Action: Present and Future." *Native South* 2(1):121–125. http://dx.doi.org/10.1353/nso.0.0013.

Wright, R. P. 1989. "The Indus Valley and Mesopotamian Civilizations: A Comparative View of Ceramic Technology." In *Old Problems and New Perspectives in the Archaeology of South Asia*, ed. J. M. Kenoyer, 145–156. Wisconsin: Archaeological Reports.

Wright, R. P. 1991. "Women's Labor and Pottery Production in Prehistory." In *Engendering Archaeology: Women and Pre-History*, ed. J. Gero and M. Conkey, 194–223. Oxford: Basil Blackwell.
Wright, R. P. 2010. *Ancient Indus: Urbanism, Economy, and Society*. Cambridge: Cambridge University Press.
Wright, R. 2014. "Comparative Studies and Early Civilisations Revisited." In *Civilisation: Perspectives from the Engelsberg Seminar 2013*, ed. K. Almquist and A. Linklater, 41–50. Stockholm: Axel and Margaret Axeson Johnson Foundation.
Wrigley, C. 1996. *Kingship and State: The Buganda Dynasty*. Cambridge: Cambridge University Press. http://dx.doi.org/10.1017/CBO9780511584763.
Wyatt, A. R. 2008. "Gardens on Hills: Ancient Maya Terracing and Agricultural Production at Chan, Belize." Unpublished PhD dissertation, Department of Anthropology, University of Illinois, Chicago.
Yaeger, J. 2003. "Untangling the Ties that Bind: The City, the Countryside, and the Nature of Maya Urbanism at Xunantunich, Belize." In *The Social Construction of Ancient Cities*, ed. M. L. Smith, 121–155. Washington, DC: Smithsonian Institution Press.
Yaeger, J. 2010. "Shifting Political Dynamics as Seen from the Xunantunich Palace." In *Classic Maya Provincial Politics: Xunantunich and Its Hinterlands*, ed. L. J. LeCount and J. Yaeger, 145–160. Tucson: University of Arizona Press.
Yaeger, J., M. Peuramaki-Brown, and B. Cap. 2010. "Social and Economic Dynamics at Buenavista del Cayo during the Late and Terminal Classic Periods." *Research Reports in Belizean Archaeology* 7:161–170.
Yáñez Rosales, R. H. 1994. "Dinámica del Desplazamiento de las Lenguas Indígenas en el Occidente de México: Época Colonial." In *El Occidente de Mexico en el Tiempo*, ed. P. R. Avila, 59–75. Guadalajara: Universidad de Guadalajara.
Yáñez Rosales, R. H. 2004. "El Vocabulario de Gerónimo Cortés y Zedeño: Entre Antonio de Lebrija y Alonso de Molina." In *De Historiografía Linguística Historia de las Lenguas*, ed. I. Guzmán Betancourt, P. Máynez, and A. H. de León-Portilla, 81–95. México DF: Siglo XXI Editores/UNAM.
Yap, K. S. 1994. *The Community Construction Contract System in Sri Lanka*. Nairobi: United Nations Centre for Human Settlements (HABITAT).
Ylvisaker, M. 1979. *Lamu in the Nineteenth Century: Land, Trade, and Politics*. Boston: African Studies Center, Boston University.
Yoffee, N. 2005. *Myths of the Archaic State: Evolution of the Earliest Cities, States, and Civilizations*. Cambridge: Cambridge University Press. http://dx.doi.org/10.1017/CBO9780511489662.
Zagarell, A. 1989. "Pastoralism and the Early State in Greater Mesopotamia." In *Archaeological Thought in America*, ed. C. C. Lamberg-Karlovsky, 280–301.

Cambridge: Cambridge University Press. http://dx.doi.org/10.1017/CBO97805115 58221.018.
Zeitlin, R. N., and A. A. Joyce. 1999. "The Zapotec Imperialism Argument: Insights from the Oaxaca Coast." *Current Anthropology* 40:383–392.

Contributors

Lane F. Fargher, Departamento de Ecología Humana, Centro de Investigación y de Estudios Avanzados del Instituto Politécnico Nacional—Unidad Mérida, Merida, Yucatan 97310 (fargher@mda.cinvestav.mx)

Sarah B. Barber, Department of Anthropology, University of Central Florida, 4000 Central Florida Blvd., Phillips Hall Rm 309, Orlando, FL 32816 (Sarah.Barber@ucf.edu)

Cynthia L. Bedell, 221 E Navajo St., West Lafayette, IN 47906 (bbdell1@frontier.com)

Christopher S. Beekman, Department of Anthropology, University of Colorado-Denver, Campus Box 103, P.O. Box 173364, Denver, CO 80217 (Christopher.beekman@ucdenver.edu)

Frances F. Berdan, Department of Anthropology, California State University, San Bernardino, 5500 University Parkway, San Bernardino, CA 92407 (fberdan@csusb.edu)

Tim Earle, Department of Anthropology, Northwestern University, 1810 Hinman Avenue, Evanston, IL 60208 (tke299@northwestern.edu)

Carol R. Ember, Human Relations Area Files, Yale University, 755 Prospect St., New Haven, CT 06511 (Carol.Ember@Yale.edu)

Gary M. Feinman, Integrative Research Center, Field Museum of Natural History, 1400 South Lake Shore Drive, Chicago, IL 60605 (gfeinman@fieldmuseum.org)

Verenice Y. Heredia Espinoza, Centro de Estudios Arqueológicos, El Colegio de Michoacán, A.C., Cerro de Nahuatzen 85, Fracc. Jardines del Cerro Grande, 59379, La Piedad, Michoacán (herediav@colmich.edu.mx)

Arthur A. Joyce, Department of Anthropology, University of Colorado–Boulder, 1350 Pleasant St., Hale Science 350/233 UCB, Boulder, CO 80309 (Arthur.Joyce@Colorado.edu)

Stephen A. Kowalewski, Department of Anthropology, University of Georgia, Athens, GA 30602 (skowalew@uga.edu)

Lisa J. LeCount, Department of Anthropology, University of Alabama, 19 10 Hoor Hall, Tuscaloosa, AL 35487–0210 (llecount@as.ua.edu)

Linda M. Nicholas, Integrative Research Center, Field Museum of Natural History (lnicholas@fieldmuseum.org)

Peter N. Peregrine, Anthropology Department, Lawrence University, 711 E. Boldt Way, Appleton, WI 54911, and the Santa Fe Institute, 1399 Hyde Park Road, Santa Fe, NM 87501 (Peter.N.Peregrine@Lawrence.edu)

Peter Robertshaw, Department of Anthropology, California State University, San Bernardino, 5500 University Parkway, San Bernardino, CA 92407 (proberts@csusb.edu)

Barbara L. Stark, School of Human Evolution and Social Change, Arizona State University, PO Box 872402, Tempe, AZ 85287 (blstark@asu.edu)

T. L. Thurston, Department of Anthropology, 380 MFAC/Ellicott, State University of New York at Buffalo, Buffalo, NY 14261 (tt27@buffalo.edu)

Deborah Winslow, Cultural Anthropology Program, Directorate of Social, Behavioral, and Economic Sciences, National Science Foundation, 4201 Wilson Blvd., Arlington, VA 22230 (dwinslow@nsf.gov)

Rita Wright, Department of Anthropology, New York University, 25 Waverly Place, New York City 10003 (rita.wright@nyu.edu)

Index

Page numbers in italics indicate illustrations.